Y0-BXX-155

Volume 31

SOCIETY OF GENERAL PHYSIOLOGISTS SERIES

Biogenesis and Turnover of Membrane Macromolecules

Society of General Physiologists Series

Published by Raven Press

Vol. 31: Biogenesis and Turnover of Membrane Macromolecules
J. S. Cook, editor. 276 pp., 1976.

Vol. 30: Molecules and Cell Movement
S. Inoué and R. E. Stephens, editors. 460 pp., 1975.

Vol. 29: Cellular Selection and Regulation in the Immune Response
G. M. Edelman, editor. 299 pp., 1974.

Vol. 28: Synaptic Transmission and Neuronal Interaction
M. V. L. Bennett, editor. 401 pp., 1974.

Volume 31

SOCIETY OF GENERAL PHYSIOLOGISTS SERIES

Biogenesis and Turnover of Membrane Macromolecules

Editor:

John S. Cook, Ph.D.
Biology Division
Oak Ridge National Laboratory
and University of Tennessee–Oak Ridge
Graduate School of Biomedical Sciences
Oak Ridge, Tennessee

Raven Press ▪ New York

The publication of this volume was supported by grants from the National Institutes of Health and the National Science Foundation

Made in the United States of America

International Standard Book Number 0–89004–092–3
Library of Congress Catalog Card Number 75–25111

Peter F. Curran
(1931–1974)

Dedication

This volume is dedicated to Peter Curran in recognition of his contributions to the Society of General Physiologists as well as his stature as a scientist. Peter Curran died suddenly in the fall of 1974. Peter was President of our Society for the year 1972–1973, having served previously as a council member for the years 1968–1970. To put Peter's contributions in perspective, let me say a few words about our society. As you all must be aware, the Society of General Physiologists is a relatively small society and is rather uncomplicated in the tasks and objectives with which it is concerned. The primary activity of our society is to host an annual meeting and to sponsor a symposium (such as the one published here) around which the general meeting is held. The major activity of the officers and council of the society is to define the symposium topic and to see to its publication. As President, Peter Curran was pivotal both in selecting this year's symposium topic and in enjoining John Cook to organize it. It was also necessary in 1973 for the society to find a new publisher and it was under Peter's aegis that the current arrangement with Raven Press was negotiated. So Peter was very much involved with the workings of the society in furthering the cause of general physiology.

Peter Curran's principal scientific contributions were in the field of coupled transport phenomena in epithelial membrane systems; this aspect of Peter Curran's life has been notably summarized by Stanley Schultz in the February (1975) issue of *The Physiologist.* It can be said that Peter's contributions were of the highest quality and showed uncommon insight into the detailed workings of physiological processes at a quantitative level. His talents represented an unusual blend of precise thinking, biological intuition, and a rigorous experimental approach. And these qualities were exemplified in his teaching as well as in his research activities. He was also very much involved in leadership roles with various editorial boards, governmental committees, and other professional societies, roles that now serve to remind us of his emergence as one of the nation's leading scientists and educators.

Peter was born in 1931 in Waukesha, Wisconsin and died at the age of 42. He graduated from Harvard College in 1953 and received his Ph.D. from Harvard University in 1958, where he worked until he joined our faculty at Yale in 1967, becoming Professor of Physiology in 1969. His many students and his many colleagues cherish their association with him. He was a stimulating, devoted, and patient friend with selfless energies, as well as a paradigm of scholastic excellence. I think the Society of General Physiologists can be justly proud that we had among us Peter Curran.

Joseph F. Hoffman
Department of Physiology
Yale University School of Medicine
President, Society of General
Physiologists, 1975–1976

A MEMOIR ON HIS YEARS AT THE BIOPHYSICAL LABORATORY, HARVARD UNIVERSITY MEDICAL SCHOOL

I have been asked to write about Peter Curran as a scientist, although our bond was, in many senses, much more personal than the usual professional relationship between teacher and student. When Peter came to the Biophysical Laboratory in the Fall of 1952 in his senior year at Harvard College, I was forty years old—not quite as old as he was when he died—and he was twenty-one; our relationship deepened because this difference in our years was very close to that between father and son.

I would like to begin with a quotation from *Science and Human Values* by Bronowski.

> The society of scientists is simple because it has a directing purpose: to explore the truth. Nevertheless, it has to solve the problem of every society, which is to find a compromise between man and men. It must encourage the single scientist to be independent, and the body of scientists to be tolerant. From these basic conditions, which form the prime values, there follows step by step the spectrum of values: dissent, freedom of thought and speech, justice, honor, human dignity and self-respect.

Peter's first research in the Biophysical Laboratory led to his undergraduate honors thesis: *Experiments in search of a carrier molecule for the transport of Na and K across the red cell membrane.* He devoted an unusually large fraction of his energies to this problem, despite his other academic obligations and his duties as captain of the track team. As graduation from college approached, I tried my best to persuade Peter to enter graduate school instead of going to medical school as he intended, but I was unsuccessful and he entered Loyola Medical School.

Even though he was first in his class, and its president, Peter's experience in medical school did not satisfy his intellectual needs, and he returned to Harvard as a graduate student. This was before Harvard offered a degree in biophysics, so Peter embarked on a Ph.D. program in physiology. Although my own education had been in physical chemistry and physics, I had been asked to deliver the lectures on the gastrointestinal tract in the physiology course. Reading the literature made me increasingly dissatisfied with the explanations advanced for the mechanism of ion and solute transport in the intestine. The concepts put forward by Ussing in his studies on active transport in the frog skin seemed more relevant than those in the gastrointestinal literature. Consequently, as soon as Peter came back to Harvard we began to investigate the mechanism of ion and water movement in the intestine and we developed the perfusion method, which led to the demonstration that water flux was driven by the flux of NaCl on which it was linearly dependent.

It soon became apparent that the physical and physicochemical forces governing the movement of ions and water in the red cell and intestine were generally operative in a wide variety of epithelial and other tissues. Members of the laboratory began to study water flow in the stomach, and also to perfuse the proximal tubule of *Necturus* kidney, looking for responses to physical forces analogous to those Peter had found in the intestine. We studied ion and solvent transport in other tissues, primarily epithelia and heart muscle, but we also undertook an extensive investigation of cation metabolism in *Escherichia coli* when we found that bacterial mutations could affect ion transport. All of our work was related to membrane transport, and there was synergistic and contagious feedback as experience with one system suggested experiments in another.

Before the laboratory had extended its interests to these several systems, Peter had written his thesis *Sodium Chloride and Water Absorption from Rat Ileum* and received his doctorate. It was quite natural for him to choose Hans Ussing's laboratory in Copenhagen for his postdoctoral fellowship. My memories of these years are enlivened by recollections of a visit to Copenhagen—the high point, socially, if not intellectually, being a marvelous evening spent at Tivoli.

We had become aware of the importance of irreversible thermodynamics to membrane transport from the papers by Staverman on the reflection coefficient. Peter was always particularly interested in the basic physical principles that underlie the responses elicited from living tissue. The papers of Kedem and Katchalsky showed how the concepts of irreversible thermodynamics could be applied systematically to studies of coupled flows across biological membranes. Consequently, on his return to the Biophysical Laboratory at Harvard, Peter offered our first course on irreversible thermodynamics in 1960–1961. Two years later, the course was given again, and this time Aharon Katchalsky came to Harvard to deliver the lectures. From this grew the collaboration of Katchalsky and Curran in their book *Nonequilibrium Thermodynamics in Biophysics,* published by the Harvard University Press in 1965. This volume became a classic guide to the application of the concepts of irreversible thermodynamics to the solution of biological problems. One characteristic incident from these years has stuck in my memory. It was during a discussion on the derivation of the equations which govern the passage of permeants through an unstirred layer. The problem is mathematically complex in that it requires that the classic equation of diffusion be coupled to the irreversible thermodynamic treatment of solute and solvent fluxes. What stands out in my memory is that it was Aharon Katchalsky who wrote the initial steps of the derivation on the blackboard, but the final and correct solution had to wait until Peter solved it in detail, and finally got each of the terms correct not only in magnitude, but more importantly, in sign.

By this time, the importance of membrane transport had been recognized in the medical school physiology course and the Biophysical Laboratory was asked to take charge of the experiments on frog skin that illustrated the principles involved. Peter was very interested in teaching medical students and he was very good. Furthermore, his experiences in Ussing's laboratory had made him familiar with the properties of frog skin. The experiments for the medical students were carefully designed, not only to enlighten them about transport problems, but also, hopefully, to proselytize some of the most able to work in our laboratory, a process which Peter's personal magnetism promoted most effectively.

We often discussed one knotty problem that related to the coupling of solute and water flow in the intestine—the observation that salt transport could drive water movement up a water activity gradient, as the intestine could continue to pump water from the blood into very dilute solutions. We were convinced that water itself was transported passively and that there had to be a region of high Na content within the tissue into which water moved following its own osmotic pressure gradient. However, none of us had any idea of where this region could be, or how the flows were coupled, or what kinds of structures might serve to alter the direction of water flow. Peter first advanced a mechanism to explain how the problem could be solved and, subsequently, in an elegant paper with McIntosh, then himself an undergraduate at Harvard, demonstrated that two membranes in series, each with a different reflection coefficient, could be coupled to drive water passively in just one direction as a consequence of an active solute pump. This was an idea whose time had come, and others of the laboratory, themselves studying abroad, subsequently and independently, came to the same conclusion, but it was Peter who got there first.

Those years from 1960 to 1967, when Peter left to join the Physiology Department at Yale, were a seminal period in the Biophysical Laboratory. The problems we were studying were central to membrane biophysics, and those of us who were involved worked closely together in intense and easy collaboration. The laboratory was full of high spirits and our energies were spent in learning and teaching.

And now all this is over, and it is proper to ask: for what purpose? As I reflect on the purpose, it seems to me that it rests in the continuity of science. George Kistiakowsky was my teacher. He received his Ph.D. in 1925 in Berlin under Bodenstein and one of his examiners was Nernst. I received my degree as George's pupil in 1937—and Peter Curran received his degree under my direction in 1958. Peter has added other tributaries to this stream, which flowed out of his experiences with Aharon Katchalsky and Hans Ussing, and has passed his distillate on to his postdoctoral fellows. Our common devotion to science stretches in an unbroken line from Bodenstein and Nernst in Berlin to Peter's students in the United States. This represents the continuity of those prime values which Peter espoused and which I mentioned earlier, from which there "follows step by step the spectrum of values: dissent, freedom of thought and speech, justice, honor, human dignity and self-respect."

A. K. Solomon
Biophysical Laboratory
Harvard University Medical School
September 4, 1975

Preface

For a number of years most biologists have accepted the principle that membranes, both at the cell surface and bounding subcellular organelles, regulate the traffic of materials between the compartments of cells. Within the confines of the Second Law, it became clear that although many of these processes must require energy, some of them could be coupled to other reactions so that, with appropriate exogenous input (light, reduced metabolites), membranes might also be the site of energy production or, more accurately, energy transduction into forms utilizable by cells. If the properties of the cell's membrane systems then define a set of regulatory functions, it follows that these same properties are an important component in defining the differentiated and functional states of the cell. The distinction here between "differentiated" and "functional" lies in recognizing that although the membranes of a lymphocyte, a muscle cell, an erythrocyte, a chloroplast, or a liver mitochondrion may be radically dissimilar as differentiated structures, nevertheless, none is locked into an unvarying set of functions. They may also be differentially responsive to a variety of external stimuli, hormones being a prominent example.

Given this set of ideas as a premise, membrane research has developed simultaneously along several parallel lines. (1) There has been a rigorous and detailed definition of the kinetic parameters of the individual membrane transport systems. Outstanding achievements have been made in this area in the understanding of electrical activity in excitable cells, in the role of membranes in bioenergetics, in the descriptions of the many transport systems of the experimentally abundant red blood cells, and in the development of nonequilibrium thermodynamics as applied to transport systems. (2) Different functions among differentiated cells have been delineated, including the changing state of membranes during development, differentiation, transformation to malignancy, and blastogenesis. (3) Both chemical and physiological responses of cell membranes to hormones and neurotransmitters have been investigated, including the role of cyclic nucleotides. (4) Membrane composition and structure have been studied from both physical and chemical points of view. (5) Membrane synthesis and turnover, including problems of secretion and endocytosis, have been studied. (6) There have been investigations of reconstituted membranes from purified elements and, analogously, the construction of model systems from nonmembranous elements the properties of which may mimic those of membranous elements. Naturally, there has been a fair amount of overlap among these areas, but each is so large in itself that the overlap has been far from complete.

When the symposium represented by the present volume was being organized, it seemed an appropriate time to attempt a synthesis of these disparate approaches and to ask a question in the general form: Given that organized and oriented activities in cell membranes perform essential regulatory roles in the physiology of the cell, what, at the next higher level, controls these activities? Briefly, how are these regulators themselves regulated? In the broadest sense it appears that two mechanisms, acting together or individually, must be operative—either the number of molecules subserving a specific function in the membrane must be altered by net synthesis or net degradation, or the specific activity of existing molecules must be altered. Examples of both mechanisms are discussed in these pages. Nevertheless, neither of these two mechanisms can be treated with any generality at the present time. We cannot make a generally acceptable statement on membrane synthesis that explains how new molecules appear in the cell surface preferentially with respect to other molecules, e.g., how (assuming that it is true) new hexose transporters arise in the membranes of glucose-deprived cells, while Na^+-K^+ transporters are not increased simultaneously. The same question arises with the variable number of insulin-binding sites in states of altered insulin sensitivity, as well as in a host of other known examples.

The question of changing specific activities of existing membrane molecules is equally difficult. To measure specific activities one must be able to measure surface concentrations of the molecule under scrutiny. Despite the availability of specific ligands for many surface receptors, it is not always clear whether a change in binding reflects a change of state or a change in the number of binding molecules. Even when changes in specific activity seem the most sensible interpretation of the data—for example, when the stimulation of cells from quiescence to growth is accompanied by very rapid enhancement of transport without a requirement for protein synthesis—still there is no satisfactory accounting for the underlying mechanism. Changes in membrane fluidity seemed an attractive possibility at one time, but there are enough counter examples that this can no longer be accepted as a general hypothesis.

The symposium on which this volume is based brought together people with quite diverse inherent approaches to these problems but all of whom had a fundamental interest in the question of regulation of membrane function. We did not anticipate that the basic problems could be solved or that mechanisms could be definitively established. What we hoped to achieve was to bring the problems themselves more sharply into focus and thereby enhance our ability to ask clearer questions toward their resolution.

With respect to these published proceedings, we had initially planned two major discussions of the important roles of cyclic nucleotides. The extraordinarily unfortunate death of Gordon Tomkins led to the withdrawal of the contribution from his laboratory. Much less serious, a persistent illness of the second contributor made it impossible for him to submit his manuscript. For these regrettable reasons, the cyclic nucleotides are under-represented in this volume, although they were much discussed at the symposium.

Peter Curran, president of the Society of General Physiologists when the plan for this symposium was formulated, had an exceptionally broad and sophisticated view of the many aspects of membrane function in the life of the cell. We are proud to dedicate this volume to him.

Finally, I must acknowledge with great gratitude the assistance of a number of people whose calm and competent expertise brought everything to fruition at its appropriate moment. These include most particularly Mrs. Neva P. Hair of Oak Ridge, Mrs. Ruth Shephard of the Job Shop in Woods Hole, the staff of the Marine Biological Laboratory who provided the physical facilities for the symposium, and the many excellent people at Raven Press who expedited the publication. On behalf of all the participants, the Society of General Physiologists is especially appreciative of the supporting grants made by the National Science Foundation and the National Institute of General Medical Sciences.

John S. Cook
March 1976

Contents

Contributors

Raymond M. Baker
Department of Biology and Center for Cancer Research
Massachusetts Institute of Technology
Cambridge, Massachusetts 02139

Alan Bernstein
The Ontario Cancer Institute and The Department of Medical Biophysics
University of Toronto
500 Sherbourne Street
Toronto, Ontario M4X 1K9, Canada

Alastair S. Boyd
The Ontario Cancer Institute and The Department of Medical Biophysics
University of Toronto
500 Sherbourne Street
Toronto, Ontario M4X 1K9, Canada

Terrance M. Brady
Department of Microbiology
University of Illinois
Urbana, Illinois 61801

Emily Tate Brake
Cancer and Toxicology Program
Biology Division
Oak Ridge National Laboratory and University of Tennessee–Oak Ridge Graduate School of Biomedical Sciences
Oak Ridge, Tennessee 37830

Trent Buckman
Frederick Cancer Research Center
Frederick, Maryland 21701

Thomas A. Cebula
Department of Biology and The McCollum-Pratt Institute
The Johns Hopkins University
Baltimore, Maryland 21218

Zanvil A. Cohn
Laboratory of Cellular Physiology and Immunology
The Rockefeller University
New York, New York 10021

John S. Cook
Cancer and Toxicology Program
Biology Division
Oak Ridge National Laboratory and University of Tennessee–Oak Ridge Graduate School of Biomedical Sciences
Oak Ridge, Tennessee 37830

Valerie Crichley
The Ontario Cancer Institute and The Department of Medical Biophysics
University of Toronto
500 Sherbourne Street
Toronto, Ontario M4X 1K9, Canada

Peter N. Devreotes
Department of Biophysics
The Johns Hopkins University
Baltimore, Maryland 21210

Michael E. Dockter
Biocenter
University of Basel
Basel, Switzerland

Jacques Dornand
Department of Biology
University of Ottawa
Ottawa, Canada

I. S. Edelman
Cardiovascular Research Institute and The Departments of Medicine and Biochemistry and Biophysics
University of California
School of Medicine
San Francisco, California 94143

David Epel
Marine Biology Research Division
Scripps Institution of Oceanography
University of California, San Diego
La Jolla, California 92093

Douglas M. Fambrough
Department of Embryology
Carnegie Institution of Washington
Baltimore, Maryland 21210

Mary Catherine Glick
Department of Pediatrics
University of Pennsylvania School of Medicine
Children's Hospital of Philadelphia
Philadelphia, Pennsylvania 19104

Arthur H. Hale
Department of Microbiology
University of Illinois
Urbana, Illinois 61801

J. Kenneth Hoober
Department of Biochemistry
Temple University School of Medicine
Philadelphia, Pennsylvania 19140

James D. Johnson
Marine Biology Research Division
Scripps Institution of Oceanography
University of California, San Diego
La Jolla, California 92093

J. G. Kaplan
Department of Biology
University of Ottawa
Ottawa, Canada K1N 6N5

W. H. Knox
Departments of Medicine and Pharmacology
Faculty of Medicine
University of Toronto
Toronto, Ontario M5S 1A8, Canada

Valerie Lamb
The Ontario Cancer Institute and The Department of Medical Biophysics
University of Toronto
500 Sherbourne Street
Toronto, Ontario M4X 1K9, Canada

Denise D. LaRossa
Kodak Research Park
Eastman Kodak Company
Rochester, New York 14650

Abraham Novogrodsky
Department of Biophysics
The Weizmann Institute of Science
Rehovot, Israel

William R. Proctor
Department of Physiology
University of North Carolina
School of Medicine
Chapel Hill, North Carolina 27514

M. R. Quastel
Ben Gurion University and Soroka Medical Center
Beer Sheba, Israel

Elliott M. Ross
Department of Pharmacology
University of Virginia School of Medicine
Charlottesville, Virginia 22903

Stephen Roth
Department of Biology and The McCollum-Pratt Institute
The Johns Hopkins University
Baltimore, Maryland 21218

J. Sax
Departments of Medicine and Pharmacology
Faculty of Medicine
University of Toronto
Toronto, Ontario M5S 1A8, Canada

Gottfried Schatz
Biocenter
University of Basel
Basel, Switzerland

A. K. Sen
Departments of Medicine and Pharmacology
Faculty of Medicine
University of Toronto
Toronto, Ontario M5S 1A8, Canada

Andrew H. Soll
Department of Medicine
Division of Gastroenterology
Room 44–143
University of California at Los Angeles
Center for the Health Sciences
10833 LeConte Avenue
Los Angeles, California 90024

Ralph M. Steinman
Laboratory of Cellular Physiology and Immunology
The Rockefeller University
New York, New York 10021

Michael J. Weber
Department of Microbiology
University of Illinois
Urbana, Illinois 61801

Peter C. Will
Department of Anatomy
Case Western Reserve University
Cleveland, Ohio 44106

D. R. Wilson
Departments of Medicine and Pharmacology
Faculty of Medicine
University of Toronto
Toronto, Ontario M5S 1A8, Canada

Tom M. Yau
Department of Radiology
Case Western Reserve University
Cleveland, Ohio 44106

Introduction: A Quiet Revolution

The announcement of the symposium on Biogenesis and Turnover of Membrane Molecules stated as its objective the exploration of the regulation of synthesis and turnover of specific functional membrane components in eukaryotic cells. The format of this volume is a continuum that considers sequentially the biogenesis and turnover of membrane macromolecules; membrane genetics and the differentiation of functional components in specialized cell membranes; the molecular basis of membrane changes in response to hormones; and finally, the basis of changes related to cell differentiation and growth, both normal and abnormal. Our ultimate goal is a better understanding of the pericellular environment—which provides conditions under which the division of labor of cell specialization occurs—and its regulation.

Cell–cell interaction is a necessary condition for the formation of the cellular architecture of organs as well as for their organization and interconnection. The large questions remain unanswered: What is the role of cell–cell interactions in the regulation of morphogenetic movements? Do these interactions require the exchange of interactants via cytoplasmic bridges, specialized junctions or other mechanisms, or are signals generated at the membrane, transmitted via specific receptor sites and intracellular mediators like Ca^{2+} and cyclic AMP to specific genes? Ultimately we shall need to know how specific receptor sites are generated and how the cells that bear them are distributed over the early embryo.

In principle, cell–cell interaction involving the exchange of "information" at the point of impact by the exchange of vesicular material or by cytoplasmic bridges cannot be excluded. This volume, however, focuses primarily on surface interactions—on the nature of signals, of cell receptors, and of the known steps leading to new functional states of membranes. Its approach is not novel but it is timely. Our willingness, indeed our ability, to examine cell–cell interactions in this perspective has emerged slowly. In contrast to the almost cataclysmic effects on developmental biology of the "quest for the organizer," our current perception of cellular communication has evolved over several decades. It has been a "quiet revolution."

It is fitting that these questions were addressed at the Marine Biological Laboratory, for it was there that the early stages of the quiet revolution were fomented, and early findings were announced and discussed. It was at the Marine Biological Laboratory that Daniel Mazia (1) carried out the experiments that led to the publication of his dissertation in 1937, an article entitled "The Release of Calcium in *Arbacia* Eggs on Fertilization," a now classic statement that emphasized the possible importance of internal release and redistribution of inorganic ions to new binding sites.

It was there, too, in 1939 that Lester Barth (2) first presented his observations, subsequently published in 1941 as "Neural Differentiation without Organizer." But the revolution was quiet indeed, as students of developmental biology were swept up in the wave of extraordinary advances generated by the one gene–one enzyme hypothesis, leading to the emergence of molecular genetics and nucleic acid chemistry as focal fields of research. Barth's observations were largely forgotten or "explained away" by a variety of arguments, none of which has stood the test of time. Mazia's discovery was received somewhat more kindly—almost invariably it was included in that long list of changes in the egg upon activation.

Happily, revolutionary discoveries neither die nor fade away. Today the ideas of Barth and Mazia are fresh and meaningful. In a series of publications, Barth and Barth (3–10) showed that the course of differentiation of small aggregates of cells prepared from explants of ventral ectoderm of *Rana pipiens* gastrula depends on the ionic composition of the solution in which they are cultured. Aggregates treated for 3–4 hr in 71 mM lithium and then cultured in a standard solution containing 1.3 mM potassium differentiate into nerve and pigment cells. As the potassium concentration is raised gradually to 5.3 mM, the frequency of pigment cells declines and networks of vacuolated cells form which later form small notochordal masses. The Barths have attempted to formulate a general theory of ionic regulation of normal embryonic induction. They postulate a normal mobilization of cations from internal compartments; these cations would be trapped within the roof of the archenteron and the presumptive neural plate by the relatively impermeable outer surfaces of cells carried in during gastrulation. Concentration of cations could then result in local internal amounts and ratios of ions similar to those found to induce nerve, pigment cells, and notochord when applied to small aggregates of gastrula ectoderm cells in culture.

Today the Barths' general scheme is plausible, indeed attractive, although difficult to test. For those reared in the second messenger era the idea does not appear revolutionary, but to a generation brought up to believe that cell–cell interactions require the exchange of "information" contained in large molecules, putatively nucleic acids, it does smack of heresy.

We are indebted to the organizers of this volume for developing a format carefully constructed to permit us to examine in some detail the evidence supporting today's dogma. Perhaps I can render no greater service than to remind all of us that the quiet revolution goes on. Much of today's detail will be quickly forgotten. Yet it will hopefully inspire—or provoke—observations and questions leading to a new synthesis. As Phyllis McGinley put it:

Heretics choose for heresy
Whatever's the prevailing fashion.

Thus, it is currently attractive to speak of the specificity or selectivity of lectins. It is unlikely that in the immediate future we shall see a synthesis of the wealth of findings, frequently contradictory, that arises from the use of mitogens. The complexity of the problem is suggested by recent observations by Ozato et al. (11–14) that highly significant cell proliferation is observed even in *syngeneic* combinations, when either responding cortisone-resistant thymocytes or stimulating spleen cells (treated with mitomycin C) are treated with native concanavalin A before they are added to mixed lymphocyte cultures. Con A treatment alone was shown not to be mitogenic. The triggering of proliferation is dependent on both the presence of spleen cells as stimulating cells and on the binding of Con A to either one of the partners, the effect being removed by the specific inhibitor of Con A, α-mannopyranoside. Even more striking is the observation that cultures in which either one of the partners was pretreated with Con A in *allogeneic* combinations showed a strong suppression of the subsequent generation of cytotoxic lymphocytes. The Con A concentration required to trigger a proliferative response corresponded to that for suppressing the generation of cytotoxic lymphocytes. In what sense, then, can the membrane sites involved in lymphocyte activation be considered unique? In what sense can the lectins be considered selective in their action? Trypsinized Con A restores SV40-transformed 3T3 cells to a "normal" state of contact inhibition, at the same time restoring normal levels of sodium–potassium-dependent ATPase and adenylyl cyclase activity (15). How are these membrane receptors related to those in human lymphocytes the lectin-stimulated transformation of which is blocked by ouabain?

Nothing short of the isolation and characterization of a wide variety of receptor sites and a demonstration of their distribution on the cell surface will begin to answer the questions that now confront us. Considering the enormous range of specificities ascribed to putative receptor sites, there may yet be surprises in store for us.

REFERENCES

1. Mazia, D. (1937): The release of calcium in *Arbacia* eggs on fertilization. *J. Cell. Comp. Physiol.,* 10:291–304.
2. Barth, L. G. (1941): Neural differentiation without organizer. *J. Exp. Zool.,* 87:371–384.
3. Barth, L. G., and Barth, L. J. (1959): Differentiation of cells of the *Rana pipiens* gastrula in unconditioned medium. *J. Embryol. Exp. Morphol.,* 7:210–222.
4. Barth, L. G., and Barth, L. J. (1962): Further investigations of the differentiation *in vitro* of presumptive epidermis cells of the *Rana pipiens* gastrula. *J. Morphol.,* 110:347–373.
5. Barth, L. G., and Barth, L. J. (1963): The relation between intensity of inductor and type of cellular differentiation of *Rana pipiens* presumptive epidermis. *Biol. Bull.,* 124:125–140.
6. Barth, L. G. (1966): The role of sodium chloride in sequential induction of the presumptive epidermis of *Rana pipiens* gastrulae. *Biol. Bull.,* 131:415–426.

7. Barth, L. G., and Barth, L. J. (1969): The sodium dependence of embryonic induction. *Dev. Biol.,* 20:236–262.
8. Barth, L. G., and Barth, L. J. (1972): ^{22}Na and ^{45}Ca uptake during embryonic induction in *Rana pipiens. Dev. Biol.,* 28:18–34.
9. Barth, L. G., and Barth, L. J. (1974): Ionic regulation of embryonic induction and cell differentiation in *Rana pipiens. Dev. Biol.,* 39:1–22.
10. Barth, L. J., and Barth, L. G. (1974): Effect of the potassium ion on induction of notochord from gastrula ectoderm of *Rana pipiens. Biol. Bull.,* 146:313–325.
11. Ozato, K., Adler, W. H., and Ebert, J. D. (1975): Synergism of bacterial lipopolysaccharides and concanavalin A in the activation of thymic lymphocytes. *Cell. Immunol.,* 17:532–541.
12. Ozato, K., Ebert, J. D., and Adler, W. H. (1975): Pretreatment of murine thymocytes by phytohemagglutinin inhibits the binding of ^{3}H-concanavalin A. *J. Immunol.,* 115:339–344.
13. Ozato, K., Ebert, J. D., and Adler, W. H. (1975): The differentiation of suppressor cell populations as revealed by studies of the effects of mitogens on the mixed lymphocyte reaction and on the generation of cytotoxic lymphocytes. (*Manuscript under review.*)
14. Ozato, K., and Ebert, J. D. (1975): Concanavalin A potentiates syngeneic response in murine lymphocytes. (*Manuscript under review.*)
15. Yoshikawa-Fukada, M., and Nojima, T. (1972): Biochemical characteristics of normal and virally transformed mouse cell lines. *J. Cell. Physiol.,* 80:421–430.

James D. Ebert
President, Marine Biological Laboratory
Woods Hole, Massachusetts

and

Carnegie Institution of Washington
Department of Embryology
115 West University Parkway
Baltimore, Maryland 21210

Biogenesis and Turnover of Membrane Macromolecules,
edited by John S. Cook. Raven Press, New York 1976.

Endocytosis and the Vacuolar Apparatus

Ralph M. Steinman and Zanvil A. Cohn

Laboratory of Cellular Physiology and Immunology, The Rockefeller University, New York, New York 10021

In this chapter we will discuss the interactions that take place between the components of the vacuolar system, with particular emphasis on the plasma membrane and its derived organelles. For the sake of brevity, we will focus on mammalian cells, and, in particular, on the characteristics of cells *in vitro,* i.e., conditions in which environmental stimuli can be better controlled. We will compare the properties of two cells derived from murine sources, which will illustrate some of the more general questions in the field. The first is a highly endocytic, nonreplicating mouse macrophage obtained from the resident peritoneal population and freshly explanted into a tissue culture environment. The second is a continuous cell line of fibroblastic origin, the "L" cell, maintained either in suspension or on plastic substrates. There have been a number of more comprehensive reviews concerning this and related areas (1–3).

THE VACUOLAR APPARATUS—GENERAL CONSIDERATIONS

A simplified, schematic representation of the vacuolar apparatus can be found in Fig. 1. Exogenous macromolecules of both a soluble and particulate nature enter the cell within endocytic vacuoles derived from the plasma membrane. These vacuoles then flow centripetally through the cytoplasm in a saltatory fashion, apparently directed by means of microtubules which radiate from the centrioles. Their final destination is the perinuclear region, where they congregate in the vicinity of the Golgi apparatus (4). Such vesicles contain, in addition to plasma membrane components, a variety of adsorbed and fluid-phase constituents of the medium. At both the light and electron microscopic level, the vesicles of pinocytic origin are lucent and often fuse with one another during their transit to form larger structures with similar optical properties. In the case of phagocytic vesicles, the size, shape, and density of these structures is governed by the properties of the particulates contained within them.

Once within the peri-Golgi region, endocytic vesicles demonstrate alterations in both composition and structure. Initially devoid of demonstrable lysosomal acid hydrolases, they rapidly begin to express activity during their conversion to secondary lysosomes. The transfer of lysosomal hydrolases is

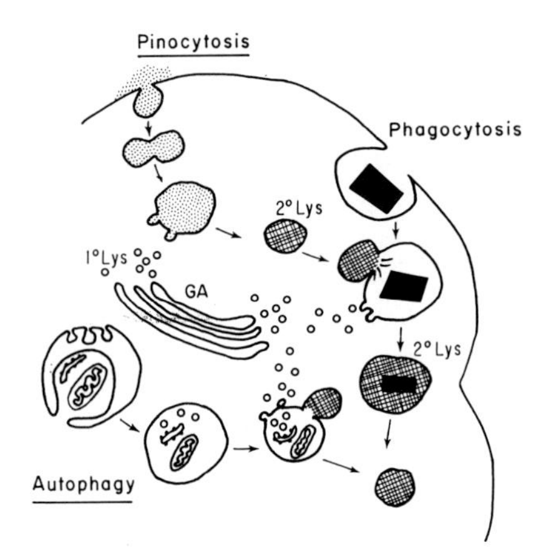

FIG. 1. The vacuolar apparatus, its components and interrelationships. Endocytic vacuoles, containing extracellular fluid droplets (pinocytosis) or particles (phagocytosis), fuse with primary and secondary lysosomes. 1° lysosomes, by definition, have not previously fused with endocytic vacuoles, and probably correspond to tiny vesicles formed in the Golgi apparatus (GA). 2° lysosomes (cross-hatching) contain both acid hydrolases and endocytosed substrates, and can fuse with additional vacuoles. 1° or 2° lysosomes also acquire endogenous substrates, e.g., rough endoplasmic reticulum and mitochondria, by autophagy.

accomplished by a series of membrane fusions in which both primary and secondary lysosomes participate. In the macrophage, primary lysosomes appear to be derived from the Golgi saccules and are present in the form of tiny vesicles which bud from the lamellae. Acid hydrolases of endogenous origin are presumably synthesized in the rough endoplasmic reticulum, flow into the Golgi saccules, and are then packaged within Golgi-derived membrane in a manner akin to secretory granules or the primary, storage lysosomes of polymorphonuclear leukocytes. In some cells, a more highly integrated and structured GERL (Golgi–endoplasmic reticulum–lysosome) is evident on thin sections as described by Novikoff and his colleagues (5). Because the Golgi is also involved in the packaging of proteins destined for secretion, it is unclear how this organelle maintains product compartmentalization. Evidence for this general pathway comes from a variety of techniques, including electron-microscopic autoradiography, cytochemistry, and ultrastructural analysis (6,7). Once enzymes are packaged within Golgi vesicles, they may then fuse with endocytic vacuoles, discharging their contents into this composite structure and initiating the process of intracellular digestion. The secondary lysosome then represents a digestive organelle surrounded by a semipermeable membrane derived from both the Golgi and plasmalemma, and containing within it endogenous hydrolases and exogenous substrates. Once formed, these organelles may themselves fuse with endocytic vacuoles, thereby mixing and redistributing both new substrates and preformed enzymes. The flow of substrate and membrane into these structures, as well as the size of this cytoplasmic compartment, will be examined in more detail in a later section.

Figure 1 also describes an additional organelle, the autophagic vacuole, which is a secondary lysosome or digestive body in nature, but contains within it substrates derived from the cell's own cytoplasm. Organelles such as mitochondria, lipid droplets, endoplasmic reticulum, or even nonstructured elements of the cytosol may be incorporated within a membrane system and subsequently digested. The mechanism by which these "self" constituents are enveloped is not fully understood, but may encompass the invagination of cytoplasm into a preformed vacuole or lysosome, or the encirclement of structures by endoplasmic reticulum with subsequent fusion to form a vacuole. Autophagic vacuoles may, in addition, fuse with endocytic vacuoles, mixing both endogenous and exogenous substrates of the lysosomal hydrolases.

It is generally considered that the fusions taking place between the various cytomembranes of the vacuolar apparatus, i.e., plasma membrane and Golgi membrane, demonstrate relative specificity and directionality. That is, the organelles of the vacuolar apparatus do not readily fuse with other cytoplasmic structures, nor do endocytized macromolecules enter other cytoplasmic compartments, i.e., cisternae of the Golgi and endoplasmic reticulum. Similarly, there is little evidence for the fusion of either primary or secondary lysosomes with the unmodified plasmic membrane, although this occurs commonly in *Protozoa* (exocytosis). The release of hydrolytic enzymes into the environment of the cells is often associated with phagosome–lysosome fusion, prior to complete closure of the endocytic vacuole, and the release of previously endocytized macromolecules has not, in our experience, been unequivocally documented.

Once substrates and hydrolases come together within the confines of a secondary lysosome, the process of intracytoplasmic digestion is initiated. This apparently takes place in an acidic milieu, consistent with the pH optima of many of the lysosomal hydrolases, which, acting in concert, can degrade most naturally occurring and biologically important macromolecules. A variety of evidence indicates that intralysosomal hydrolysis is extensive and gives rise to small molecular weight degradation products. Most information has come from studies with labeled proteins, and in this instance intralysosomal hydrolysis proceeds to the amino acid and dipeptide level prior to the diffusion or transport of digestion products into the cytosol (8–10). The permeability properties of the secondary lysosomal membrane are imperfectly understood, but depend on factors such as charge, molecular weight, and lipophilicity. Whether or not active transport systems function in this milieu remains uncertain.

ENDOCYTOSIS

Endocytosis is a generic term that describes the quantal uptake by cells of both fluid droplets and solid particles. Classically, pinocytosis has been associated with fluid droplet interiorization, whereas phagocytosis implies the uptake of particulates. In both instances, vesicles are generated from the plasma membrane and pinch off and flow into the cytoplasm. In fact, it is difficult to differentiate the two processes in any meaningful way, and adsorptive forms of pinocytosis or the uptake of tiny virus particles have been referred to interchangeably.

Pinocytosis, a property of most, if not all, mammalian cells, was described by Warren Lewis some 44 years ago (11). Although considerable descriptive information has appeared over the past years, we are still woefully ignorant of its basic control mechanisms, as well as its importance in cellular physiology. This is, in part, related to our general lack of information concerning the properties of the plasma membrane and its interaction with structural elements of the cytosol and energy-generating systems. Two forms of pinocytosis can be clearly distinguished. The first is termed fluid-phase or nonadsorptive pinocytosis, in which all constituents of the environment are interiorized in a direct relationship to their extracellular concentration. In the idealized situation little or no interaction between solutes and the plasmalemma is thought to take place. Recent studies have suggested that the glycoprotein enzyme horseradish peroxidase (HRP) is an excellent and sensitive marker for fluid-phase pinocytosis, and its interiorization, localization, and degradation can be followed with both biochemical and cytochemical techniques (12,13). In the macrophage and L cell fibroblast, 10^6 cells/hr interiorize solute at a rate of 0.01 and 0.0035% of the administered load of HRP/ml culture fluid over a wide range of concentrations. On a cell protein basis, the pinocytic activity of macrophages was approximately 10-fold greater than cultured fibroblasts. Modification of the environmental temperature from 2° to 38°C gave a Q_{10} of ~3 and an activation energy of 18 kcal/mole. The uptake of peroxidase was, in general, related to the intracellular levels of ATP. In the fibroblast, the level of this high-energy intermediate is controlled by both glycolytic and respiratory pathways. A combination of glycolytic and respiratory inhibition was therefore necessary to depress ATP levels and pinocytic activity to 10–20% of control values. It was of interest that, as L cells reached the confluent state, their pinocytic rate increased by a factor of 2- to 4-fold. This was associated with an increased complexity of cell-surface projections.

Adsorptive pinocytosis occurs when solutes bind to the plasma membrane, probably via specific recognition sites, and are then interiorized bound to the pinocytic vesicle membrane (and likely in the fluid phase of the vesicle as well). It has been evident in both amoeba (14) and mammalian cells (15) that binding of solutes to the cell surface is associated with more rapid rates of interiorization. Jon Silver and ourselves have recently studied an example of this process in cultivated macrophages and L cells. The marker was succinylated concanavalin A (succ-Con A) produced by treatment of native tetrameric Con A with succinic anhydride (16). The resultant dimers retain the affinity of native lectin for cell surface glycoproteins, i.e., for mannopyranosyl and glucopyranosyl residues in α-glycosidic linkages. However, they do not appear to alter either the formation or fate of pinocytic vesicles (P. J. Edelson, *personal communication*). We then studied the rate of interiorization of [^{125}I]succ-Con A, following equilibration with the cell surface, and compared it to fluid-phase uptake of this marker. The latter can be measured by administering the succ-Con A in the presence of 0.01 M α-Me-mannoside, which inhibits binding to surface glycoproteins. With the dose of succ-Con A employed, we found that adsorptive uptake proceeded 100 times faster than fluid-phase uptake.

It is to be pointed out that the increased uptake of solutes which follows surface binding need not represent stimulated or selective pinocytosis. Rather the enhancement of uptake may simply be a "passenger" or "piggyback" effect, in which bound solute accompanies a large influx of cell surface during normal ongoing pinocytic activity (*vide infra*).

It should be clear that many questions concerning the pinocytic mechanism remain to be answered. These include (*i*) the factors underlying the fusion of plasma membrane to form vesicles; (*ii*) the forces leading to membrane invagination; (*iii*) the mechanisms whereby cytoplasmic vesicle flow takes place and the directionality of the flow; and (*iv*) the role of pinocytosis in cell nutrition and the interiorization of regulatory macromolecules.

PHAGOCYTOSIS

Because of its more apparent association with the area of host resistance to infection, more detailed information has accrued concerning the determinants and consequences of particle uptake. These have been summarized in a number of reviews (3,17–19). For the purposes of this discussion a few more recent observations will be mentioned. Although many cultured cells demonstrate the ability to interiorize solid particles, their rate and efficiency are strikingly different. In general, the so-called "professional" phagocytic cells of the white cell series differ both qualitatively and quantitatively. Polymorphonuclear leukocytes and macrophages demonstrate receptors on their membranes, whereby they bind and interiorize particles coated with immunologically important molecules. In this regard, one can characterize an Fc receptor which recognizes the Fc portion of IgG and a complement receptor which interacts with activated components of C′3 (20). Fibroblasts lack these receptors but interiorize relatively inert particles such as polystyrene latex and aldehyde-fixed erythrocytes (21). Recently, Griffin and Silverstein (22) examined the question of whether the trigger initiating the phagocytic process is localized to that area of the membrane underlying the attached particle, or whether the stimulus is more generalized. They employed two particles, both of which attached to the macrophage membrane, but only one of which could be ingested. They then covered the surface with the noningestible particle and subsequently added the ingestible particle. Only the ingestible particle was interiorized, with the other particles remaining on the cell surface. This suggests that the phagocytic stimulus is highly segmental in nature and is presumably the consequence of very localized events. The nature of these localized events is still unclear, but is under study. One factor which appears to be of importance is a requirement for the circumferential binding of membrane receptors to ligands on the particle (23). After attaching a particle to the membrane, the removal or blocking of receptors adjacent to the attachment zone, or the removal of ligands on the particle surface, blocks the phagocytic process. This implies that the plasma membrane must flow around the particle, continuing to make receptor–ligand complexes until, finally reaching the apex, membrane fusion and phagocytic vacuole formation occurs. It also suggests that particle uptake is accompanied by little or no fluid uptake. Possibly the processes of phagocytosis and pinocytosis can then be differentiated in a very literal sense, i.e., phagocytosis involves particle uptake alone, while pinocytosis involves the influx of a fluid droplet plus or minus solutes, or small particles, or both, adsorbed to the cell surface.

The role of subplasmalemmal contractile filaments in the phagocytic process remain poorly understood. It is likely that some of these filaments represent actin, and Korn et al. (24) have presented micrographs of phagocytizing *Acanthamoeba,* illustrating a thick cortical accumulation of actin-like filaments immediately beneath the invaginating plasma membrane. This is in keeping with the segmental nature of the process, but the factors leading to this morphological alteration and the possible role of other components of the contractile system remain unknown.

MEMBRANE FLOW DURING ENDOCYTOSIS

The fact that plasma membrane surrounds the invaginating pinocytic and phagocytic vacuole automatically raises two sorts of questions: (*i*) Which components of the cell surface and what proportion of the total are included in the forming vacuole? (*ii*) What is their fate following formation of the secondary lysosome?

We have just started to obtain information on both of these issues. Many people suspect that the endocytic vacuole membrane may be a highly selected vs. representative sample of the cell surface. Tsan and Berlin (25) first established the possibility. They induced phagocytic cells to interiorize latex spheres, and then asked if any amino acid and purine transport sites had been ingested. Their kinetic data indicated that the number of surface transport sites in normal vs. latex-laden cells were similar, i.e., these transport sites must have been excluded from the phagocytic vacuole. Charalampous and Gonatas (26) reported contrasting data when they found that KB cells had a marked decrease in aminoisobutyric acid transport sites following phagocytosis of latex beads. Taylor et al. (27) looked at a type of pinocytosis in lymphocytes. Binding of multivalent ligands to specific surface components, e.g., antiimmunoglobulin (Ig) antisera binding to surface Ig, seemed to induce selective pinocytosis of the surface Ig. Following treatment with anti-Ig they found that surface Ig was no longer detectable, whereas other markers, like histocompatibility antigens, were still expressed.

Studies such as these are intriguing, but obviously are not easily interpreted. In neither case do we have either quantitative or direct data, or both, on defined membrane components, either on or in the cells; nor do we have information on the surface area of the cell and endocytic vacuoles. In our laboratory we are looking at both phagocytosis and pinocytosis with these problems in mind. We are measuring the actual amount of defined cell surface components, and then determining the proportion interiorized during endocytosis. We then compare these data with data on the surface area of newly formed endocytic vacuoles vs. the surface area of the whole cell. Finally, we are looking at the fate of the interiorized surface membrane. We will first consider methods for determining the overall area of cell vs. endocytic vacuoles, and then look at some specific cell-surface components.

SURFACE AREAS OF WHOLE CELLS

In most previous studies, it has been assumed that cells are smooth spheres with a surface area of $4\pi R^2$, where R is the cell radius. Clearly, cell surfaces are highly irregular with microvilli, ruffles, blebs, microspikes, etc. Stereologic approaches are available for obtaining a precise cell surface area (28). For cells in culture, one requires (*i*) enough thin sections through randomly sampled cells; (*ii*) a value for the average cell volume; and (*iii*) data on the number of intersections a grid of test lines makes with cell surface, as well as the number of test points falling on the cell. Our data indicate that mouse macrophages (monolayer cultures maintained in 20% fetal calf serum for 24 hr) have a volume of 395 μm^3 and surface area of 825 μm^2 (Table 1). In contrast, L cells, a continuous fibroblast line maintained either in suspension or on plastic surfaces, have a volume of 1,765 μm^3 and area of 2,100 μm^2 (Table 1). These surface areas are both threefold greater than the area of smooth spheres having the same volume.

SURFACE AREAS OF ENDOCYTIC VACUOLES

For phagocytosis, latex spheres are a most useful marker. Both macrophages and L cells will interiorize large numbers (100–200) of these particles under appropriate conditions, and in a short time (1 to 2 hr). The spheres are uniform in diameter (1.1 μm), and are taken up as single particles. In thin sections, the vacuole membrane is closely applied to the sphere; the studies of Silverstein, Griffin, and associates (23) suggest that the plasma membrane is closely apposed to ingested particles throughout the interiorization process itself. The number of spheres interiorized per cell can be determined either by direct counts or spectrophotometric assays, so that this number

TABLE 1. *Dimensions of the vacuolar system*

	Macrophages	L cells
Cell measurements		
Volume (μm^3)	395	1,765
Surface area (μm^2)	825	2,100
Compartment measurements		
Pinocytic vesicles		
Area (μm^2)	103	89
Volume (μm^3)	10	5.7
Secondary lysosomes		
Area (μm^2)	148	136
Volume (μm^3)	10	15.7
Fractional influx during pinocytosis		
Cell surface area/min (%)	3.1	0.9
Cell volume/min (%)	0.43	0.05

This table summarizes data from a recent study of the vacuolar system using HRP as a marker (28). Cell volume was determined from the diameter of spherical cells in suspension. Surface area values are from electron microscopic specimens analyzed by stereology. The volume and area of the total pinocytic vesicle and secondary lysosome space were measured by stereologic techniques (point and intersection counting) on cells labeled with HRP to equilibrium. Fractional influx data are derived from point and intersection counting of cells pulsed for brief periods ($^1/_2$ to 5 min) with HRP.

times the surface area per sphere (3.8 μm^2) gives the amount of cell surface interiorized during phagocytosis. Finally, the spheres have a low density, so that clean phagolysosome preparations can be obtained with flotation techniques (e.g., refs. 29 and 30).

For pinocytosis, HRP appears to be a suitable marker for estimating the surface area of incoming pinocytic vesicle membrane. Again we have studied both macrophages and L cells *in vitro*. As mentioned previously, the enzyme is taken up in the fluid phase of pinocytic vesicles (12,13). It does not alter ongoing pinocytic activity, and the incoming pinocytic vesicles containing enzyme can be identified cytochemically. We have recently completed a study in which we use stereologic techniques to determine the rates at which pinocytic vesicles were formed per unit time, as well as their number, surface area, and volume (28). For example, we use a grid of test points to obtain volume measurements. We count the number of points falling on HRP reactive pinocytic vesicles vs. the cell itself. In kinetic studies, the cytochemically reactive space expands at a rate of 0.43 and 0.05% of the cell volume per minute in macrophages and L cells (Table 1). This gives a volume of fluid uptake which compares favorably with data on fluid uptake, using an enzymic assay (since HRP is interiorized in the fluid phase, measurements of solute uptake by an enzymic assay can be converted to fluid uptake). The cytochemical technique must therefore be fully sensitive. One can then use a grid of test lines to obtain data on the relative surface areas of reactive pinocytic vesicles vs. plasma membrane (by counting intersections). The result is that macrophages and L cells take in the equivalent of 3.1 and 0.9% of their cell surface area each minute (Table 1).

PHAGOCYTOSIS OF SURFACE MARKERS IN L CELLS

Hubbard and Cohn (30) fed L cells latex spheres for 1 hr. By direct counts of sectioned material, they found that an average of 170 particles were taken in per cell, i.e., 170×3.8 μm^2 per particle or 646 μm^2 total, which is the equivalent of 31% (646/2,100) of the cell surface. The cell surface had been iodinated with lactoperoxidase prior to feeding. Latex phagolysosome fractions, in near-perfect yield, were obtained, and in fact contained 28.9% of the acid-insoluble label and 30.6% of a membrane marker enzyme, phosphodiesterase. These data state very clearly that the phagocytic vesicle membrane in L cells is representative of the cell surface, for the same proportion of the cell surface is interiorized by all three assays. Confirmation involved an SDS gel electrophoresis analysis in which all cell-surface iodinated polypeptides were interiorized to the same relative extent.

PHAGOCYTOSIS OF SURFACE MARKERS IN MACROPHAGES

The studies of Werb and Cohn (27,29) relate to the issue of membrane flow during latex phagocytosis by macrophages. Precise data of the surface area of cells grown in 20% newborn calf serum are not yet available, so that we will use the value of 829 μm^2 obtained from cells grown in 20% fetal calf serum, which may be an underestimate. Good data on latex uptake are also not available, and we will use a value of 150 to 200 latex particles per cell, which is probably reasonable. Roughly then, the cells interiorize some 570–760 μm^2 (3.8 μm^2 per latex bead) of surface in latex phagolysosomes, which is very close to the cell surface area itself.

The lactoperoxidase iodination system has not yet been applied to macrophages, but other interesting markers are available. A fascinating one has to do with exchangeable cholesterol pools. Werb and Cohn (31) first established that macrophages depend on exogenous sources (e.g., serum lipoprotein) for their cholesterol content. The exchange of radiolabeled sterol indicated that cells have a rapidly and slowly exchanging compartment. Good evidence exists that the former represents the plasma membrane and the latter vacuolar membranes (32).

The size of the cell surface vs. phagolysosome cholesterol pools was then investigated before and after latex ingestion (31). Prior to phagocytosis, 60.9% of the cell's cholesterol was in plasma membrane and 39.1% in the lysosomal apparatus, corresponding to 0.73 and 0.47 μg cholesterol, respectively, per flask of cells. Within 24 hr after phagocytosis, the total cell cholesterol had increased by 0.65 μg. This increase involved new protein synthesis, which may be related to the fact that the surface receptor for cholesterol exchange is trypsin-sensitive. The increase in cholesterol was entirely in the slow or phagolysosome compartment. The data can be interpreted to state that the cell monolayers interiorize a large proportion (0.65 μg) of the plasma membrane cholesterol pool (0.73 μg) during latex phagocytosis, and then replace the plasma membrane cholesterol by new membrane assembly. Werb and Cohn (29) also noted that a plasma membrane marker enzyme, 5′-nucleotidase, was diminished 50–60% in activity several hours after phagocytosis, and then returned to normal over the next day. Initially, much of the nucleotidase was present in a latex phagolysosome fraction. It is difficult to extrapolate data on enzyme activity to actual amounts of surface vs. intracellular components; but like the cholesterol, it seems quite likely that a representative amount of this plasma membrane marker is interiorized during a huge latex meal.

PINOCYTOSIS OF SURFACE MARKERS IN MACROPHAGES AND L CELLS

We have begun to obtain data on the rate of interiorization of cell membrane markers during pinocytosis, so that we can compare them to the stereologic data on bulk membrane flow. So far, the marker we have used is the binding site for [^{125}I]succ-Con A. The amount of succ-Con A on the cell surface, which reflects the number of succ-Con A binding sites, is determined by the amount releasable by 0.1 M α-methylmannoside (α-MM), a specific competing sugar for its binding; non-α-MM releasable counts are presumably intracellular, and, in fact, we have convincing cell fractionation data which show that the nonreleasable pool is in lysosomes. The experiment is to expose macrophage monolayers continuously to 3 μg/ml of radiolabeled succ-Con A. Within 15 to 30 min, the surface pool, or α-MM releasable compartment, reaches equilibrium. It then remains constant for the duration of the experiment. The non-α-MM releasable counts, however, increase linearly following equilibration of the cell surface. The rate corresponds to 0.74 ± 0.20% (10 experiments) and 0.59 ± 0.07% (four experiments) of the cell surface pool per minute in macrophages and L cells, respectively. Under the conditions we used, the amount of succ-Con A taken in per cell per unit time in the fluid phase is 50–100 times less than that entering via adsorptive pinocytosis. The presumptive rates of interiorization of succ-Con A binding sites are somewhat smaller than that of bulk membrane flow predicted by stereologic techniques. Unfortunately, both approaches include assumptions and data that may not be precise, and it is entirely possible that succ-Con A binding sites and bulk surface membrane are being interiorized at similar rates.

THE FATE OF PLASMA MEMBRANE SURROUNDING INGESTED LATEX PARTICLES

A puzzling unknown in the physiology of the vacuolar system has to do with the fate of the incoming vesicle membrane. After all, the newly formed secondary lysosome contains a spectrum of hydrolases capable of attacking most biologic substrates, and yet all the evidence indicates that the membranous confines of the vacuolar system remain impermeable to most small particles, macromolecules, and even oligopeptides (9,10). Data are now available suggesting that acid hydrolases do attack the incoming membrane. L cell phosphodiesterase (30) and macrophage 5′-nucleotidase (29) are both inactivated following interiorization into latex phagolysosomes, and both have a $t_{1/2}$ of about 2 hr. In addition, L cell plasma membrane polypeptides iodinated by Hubbard and Cohn (30), underwent extensive (down to the level of monoiodotyrosine) and rapid (70% of the interiorized label was degraded with a half-life of ~2 hr) degradation following latex uptake. We obviously have a lot to learn in this latex system. We do not know yet that the inactivation of surface enzymes relates to degradation of the enzyme; in addition, the degradation of exteriorly disposed polypeptides accessible to lactoperoxidase iodination may occur with relatively little effect on the other plasma membrane components.

THE FATE OF PLASMA MEMBRANE FOLLOWING PINOCYTOSIS

A more intriguing possibility concerning the fate of endocytosed membrane has emerged from a recent stereologic analysis of pinocytosis in cultivated macrophages and L cells, i.e., that incoming membrane may be recycled largely intact back to the cell surface (28). The approach was as follows. In cells exposed continuously to HRP, it is possible to distinguish incoming pinocytic vesicles from those which have fused to form secondary lysosomes. In the former, cytochemical reaction product is distributed peripherally in the vacuole, while it is found throughout the vacuole following secondary lysosome formation (Fig. 2). We then measured the rates at which these two types of vacuoles fill with HRP, as well as the total dimensions of the vesicle and lysosome spaces at equilibrium (Table 1). In macrophages we established that both compartments are saturated after a 1-hr exposure to HRP.

The findings were that macrophages have a pinocytic vesicle space that is 10 μm^3 in volume and 103 μm^2 in total area, whereas the corresponding secondary lysosome measurements are 10 μm^3 and 148 μm^2. However, each hour we found that pinocytic vesicle fluid and membrane moves into the cell at a rate of 100 μm^3 and 1,560 μm^2—values that are obviously much larger than the combined size of the entire pinocytic vesicle and secondary

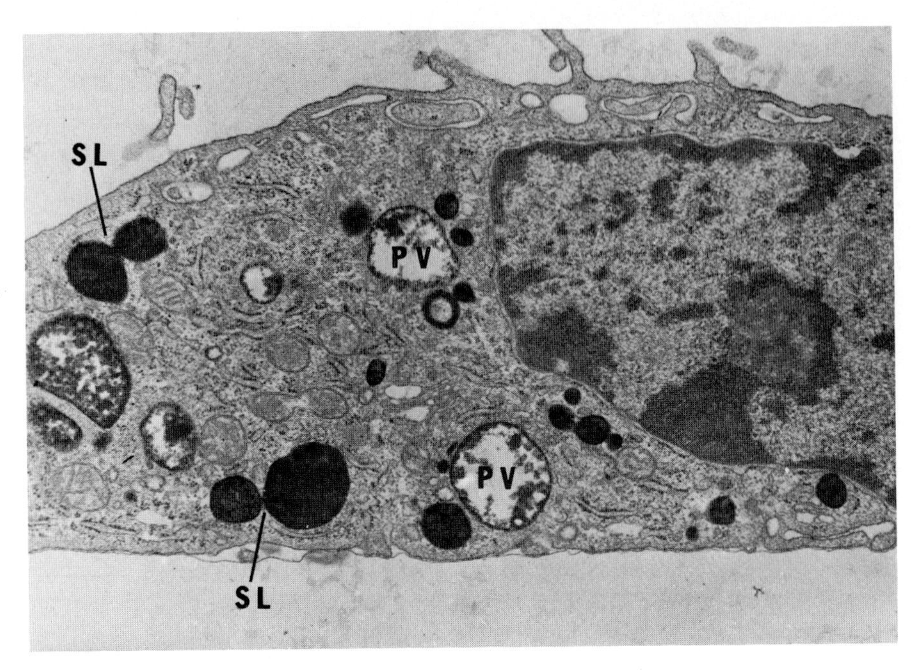

FIG. 2. A macrophage exposed for 1 hr to 1 mg/ml of soluble HRP, followed by cytochemical localization of cell bound enzyme. Reaction product is seen in two types of vacuole. Presumptive incoming pinocytic vesicles (PV) have a sparse content of product, primarily found around the periphery of the vacuole. Following fusion with preexisting lysosomes, secondary lysosomes with a dense content of reactivity are noted (SL). The cell surface does not exhibit here. ×17,100

lysosome spaces. In fact, more than 90% of the interiorized fluid and membrane cannot be accounted for in the secondary lysosome space. A very similar picture was obtained in L cells. We previously had obtained evidence that all incoming HRP enters the secondary lysosome space (12,13), so that the incoming vesicle must rapidly shrink following lysosome formation. We think that the incoming fluid egresses across the vacuole membrane and then out of the cell proper, whereas the pinocytosed membrane is recycled back to the cell surface. It is our present bias that the incoming membrane is not degraded significantly, although we have little direct evidence as yet. This interpretation also makes the assumption that membrane components are not resynthesized at a rate rapid enough to account for the ability of the macrophage to interiorize twice its surface area per hour.

ACKNOWLEDGMENT

This work was supported by grants A101831 and A107012 from the United States Public Health Service. R.M.S. is a scholar of the Leukemia Society of America.

REFERENCES

1. Cohn, Z. A., and Fedorko, M. E. (1969): The formation of lysosomes. In: *Lysosomes in Biology and Pathology,* Vol. 1, p. 43, edited by J. Dingle and H. Fell. North-Holland, London.
2. Gordon, S., and Cohn, Z. A. (1973): The macrophage. *Int. Rev. Cytol.,* 36:171–214.
3. Steinman, R. M., and Cohn, Z. A. (1974): The metabolism and physiology of the mononuclear phagocytes. In: *The Inflammatory Process,* Vol. 1, pp. 449–510, edited by B. W. Zweifach, L. Grant, and R. T. McCluskey. Academic Press, New York.
4. Cohn, Z. A. (1975): Macrophage physiology. *Fed. Proc.,* 34:1725–1729.
5. Novikoff, A. B., Essner, E., and Quintana, N. (1964): Golgi apparatus and lysosomes. *Fed. Proc.,* 23:1010–1022.
6. Cohn, Z. A., Fedorko, M. E., and Hirsch, J. G. (1966): The *in vitro* differentiation of mononuclear phagocytes. V. The formation of macrophage lysosomes. *J. Exp. Med.,* 123:757–766.
7. Nichols, B. A., Bainton, D. B., and Farquhar, M. G. (1971): Differentiation of monocytes. Origin, nature and fate of their azurophil granules. *J. Cell Biol.,* 50:498–515.
8. Ehrenreich, B. A., and Cohn, Z. A. (1967): The uptake and digestion of iodinated human serum albumin by macrophages. *J. Exp. Med.,* 126:941–958.
9. Cohn, Z. A., and Ehrenreich, B. A. (1969): The uptake, storage and intracellular hydrolysis of carbohydrates by macrophages. *J. Exp. Med.,* 129:201–225.
10. Ehrenreich, B. A., and Cohn, Z. A. (1969): The fate of peptides pinocytosed by macrophages *in vitro. J. Exp. Med.,* 129:227–245.
11. Lewis, W. H. (1931): Pinocytosis. *Johns Hopkins Med. J.,* 49:17–36.
12. Steinman, R. M., and Cohn, Z. A. (1972): The interaction of soluble horseradish peroxidase with mouse peritoneal macrophages *in vitro. J. Cell Biol.,* 55:186–204.
13. Steinman, R. M., Silver, J. M., and Cohn, Z. A. (1974): Pinocytosis in fibroblasts. Quantitative studies *in vitro. J. Cell Biol.,* 63:949–969.
14. Schumaker, V. N. (1958): Uptake of protein from solution by *Amoeba proteus. Exp. Cell Res.,* 15:314–331.
15. Von Figura, K., and Kresse, H. (1974): Quantitative aspects of pinocytosis and the intracellular fate of *N*-acetyl-α-D-glucosaminidase in Sanfilipo B fibroblasts. *J. Clin. Invest.,* 53:85–90.
16. Gunther, G. R., Wang, J. L., Yahara, I., Cunningham, B. A., and Edelman, G. M. (1973): Concanavalin A derivatives with altered biological activities. *Proc. Natl. Acad. Sci. USA,* 70:1012–1016.
17. Rabinovitch, M. (1970): Phagocytic recognition. In: *Mononuclear Phagocytes,* p. 299, edited by R. van Furth. Davis, Philadelphia, Pa.
18. Elsbach, P. (1974): Phagocytosis. In: *The Inflammatory Process,* Vol. 1, pp. 363–410, edited by B. W. Zweifach, L. Grant, and R. T. McCluskey. Academic Press, New York.
19. Stossel, T. P. (1974): Phagocytosis. *N. Engl. J. Med.,* 290:717–723; 774–780; 833–839.
20. Griffin, F. M., Jr., Bianco, C., and Silverstein, S. C. (1975): Characterization of the macrophage receptor for complement and demonstration of its functional independence from the receptor for the Fc portion of immunoglobulin G. *J. Exp. Med.,* 141:1269–1277.
21. Rabinovitch, M. (1969): Uptake of aldehyde-treated erythrocytes by L 2 cells. *Exp. Cell Res.,* 54:210–216.
22. Griffin, F. M. Jr., and Silverstein, S. C. (1974): Segmental response of the macrophage plasma membrane to a phagocytic stimulus. *J. Exp. Med.,* 139:323–336.
23. Griffin, F. M., Jr., Griffin, J. A., Leider, J. E., and Silverstein, S. C. (1975): Studies on the mechanism of phagocytosis. I. Requirements for circumferential attachment of particle-bound ligands to specific receptors on the macrophage plasma membrane. *J. Exp. Med.,* 142:1263–1282.
24. Korn, E. D., Bowers, B., Batzri, S., Simmons, S. R., and Victoria, E. J. (1974): Endocytosis and exocytosis: role of microfilaments and involvement of phospholipids in membrane fusion. *J. Supramol. Struct.,* 2:517–528.
25. Tsan, M. F., and Berlin, R. D. (1971): Effect of phagocytosis on membrane transport of nonelectrolytes. *J. Exp. Med.,* 134:1016–1035.

26. Charalampous, F. C., and Gonatas, N. K. (1975): The plasma membrane of KB cells; isolation and properties. *Methods Cell Biol.*, 9:260–280.
27. Taylor, R. B., Duffus, W. P. H., Raff, M. C., and de Petris, S. (1971): Redistribution and pinocytosis of lymphocyte surface immunoglobulin molecules induced by anti-immunoglobulin antibody. *Nature* [*New Biol.*], 233:225–229.
28. Steinman, R. M., Brodie, S. E., and Cohn, Z. A. (1976): Membrane flow during pinocytosis. A stereologic analysis. *J. Cell Biol.*, 68:665–687.
29. Werb, Z., and Cohn, Z. A. (1972): Plasma membrane synthesis in the macrophage following phagocytosis of polystyrene latex particles. *J. Biol. Chem.*, 247:2439–2446.
30. Hubbard, A. L., and Cohn, Z. A. (1975): Externally disposed plasma membrane proteins. II. Metabolic fate of iodinated polypeptides of mouse L cells. *J. Cell Biol.*, 64:461–479.
31. Werb, Z., and Cohn, Z. A. (1971): Cholesterol metabolism in the macrophage. II. Alteration of subcellular exchangeable cholesterol compartments and exchange in other cell types. *J. Exp. Med.*, 134:1570–1590.
32. Werb, Z., and Cohn, Z. A. (1971): Cholesterol metabolism in the macrophage. I. The regulation of cholesterol exchange. *J. Exp. Med.*, 134:1545–1569.

Biogenesis and Turnover of Membrane Macromolecules,
edited by John S. Cook. Raven Press, New York 1976.

Turnover of Ouabain-Binding Sites and Plasma Membrane Proteins in HeLa Cells

John S. Cook, Peter C. Will, William R. Proctor,* and Emily Tate Brake

Cancer and Toxicology Program, Biology Division, Oak Ridge National Laboratory, and University of Tennessee–Oak Ridge Graduate School of Biomedical Sciences, Oak Ridge, Tennessee 37830

Macromolecules at the cell surface are, to a great extent, responsible for the specificity of interactions of cells with their environment, and maintenance of the integrity of these macromolecules is an important aspect of physiological regulation. In the absence of turnover or repair, some surface molecules may be subject to thermal inactivation at significant rates with respect to the lifetime of the cell and its progeny, even at body or ambient temperatures. In addition, the surface is subject to the action of chemical or physical agents in the immediate environment, many of which may inactivate or perturb surface functions. Particularly interesting agents in this regard are drugs that react with specific membrane molecules. Apart from the direct reversal of the perturbing event, e.g., the dissociation of a drug from its receptor or the reduction of an oxidized sulfhydryl group, the maintenance of integrity appears principally to depend on molecular turnover. The perturbed molecule, either alone or together with a larger portion of the membrane, is in some manner removed from the surface and replaced with a functional molecule. The study of such repair-effecting turnover processes may be greatly complicated by the simultaneous occurrence of non-steady-state processes such as (*i*) cell growth; (*ii*) changes in the physiological state of the cell (differentiation, transformation); (*iii*) specific activation or induction or repression of the surface function as a consequence of the perturbation; (*iv*) preferential, rather than random, turnover of the inactivated molecules; and (*v*) in some cases the inhibition of the turnover processes as a consequence of the inhibition of the surface function.

The study described here focuses on the turnover of a specific molecule, inactivated by a specific drug, in the plasma membrane of a cloned cell line. The molecule is the Na^+–K^+ transport enzyme, which we label with the inactivating ligand [^{3}H]ouabain. The cell line is HeLa S_3, which, as a human

* *Present address:* Department of Physiology, School of Medicine, University of North Carolina, Chapel Hill, N.C., 27514.

cell, binds ouabain tightly (as opposed to cultured rodent cells; see, e.g., the chapter by Baker, *this volume*), but not covalently. The fact that the line has been cloned (1) eliminates the heterogeneity of responses that can occur among the mixed cell types of a more complex tissue. Although we use ouabain primarily as a marker to locate the transport enzyme, ouabain is also representative of an important class of drugs, the cardiac glycosides, and the manner in which such drugs are handled by cells is a matter of pharmacological interest in its own right.

OUABAIN BINDING TO HeLa CELLS

Ouabain binds to the catalytic subunit of the Na^+–K^+-ATPase, a polypeptide that, in a variety of purified preparations, has a molecular weight of about 90,000–125,000 (2–5). When ouabain is bound to the enzyme, the Na^+–K^+-dependent ATPase activity is inhibited, and when it is bound to HeLa cells, $^{86}Rb^+$ (a faithful K^+ analog) transport at the binding site is blocked (6–10). Under the conditions of our experiments, in which we most frequently use ouabain at concentrations between 10^{-8} and 10^{-6} M, the binding appears specific for the transport site. Extrapolation from the so-called nonspecific binding observed at higher concentrations suggests that less than 2% of that bound at 2×10^{-7} M is associated with sites other than those concerned with Na^+–K^+ transport.

Typical binding and drug-release curves for whole HeLa cells are given in Fig. 1.* As has been noted in a number of systems, ouabain binding is sensitive to and inhibited by K^+ in the medium. However, provided that binding is not so slow that internalization of the binding sites becomes a complicating factor (see below), there is no difference in the final extent of binding at K^+ concentrations between 0 and 20 mM or ouabain concentrations between 2×10^{-7} and 10^{-6} M. There is likewise no noticeable effect of the presence or absence of serum or dialyzed serum, glucose, or HCO_3^- levels in the binding medium when fresh cells are assayed. Binding as measured in a standard, complete growth medium is indistinguishable from binding measured in a Krebs ringer solution at the same temperature.

In normal growth medium at 37°C with a K^+ concentration of 5.5 mM, binding occurs over a time scale of minutes and release from the cells occurs over a time scale of hours or even days. As may be seen in Fig. 1b,

* Essential experimental details are given in the legends to the figures. We have used both attached cells and suspension cells; there is no important difference between them in rates or extent of binding. For simplicity, the medium in which the cells were grown is usually referred to here as "normal growth medium." This has been basically Eagle's minimal essential medium with Earle's salts or spinner salts plus 10% calf serum or 5–10% fetal calf serum, equilibrated with an environment of 95% air–5% CO_2. For many of the later experiments the cells were grown in stoppered bottles with 20 mM Hepes buffer. None of these variations in growth media had any noticeable effect on ouabain binding or other aspects of the results.

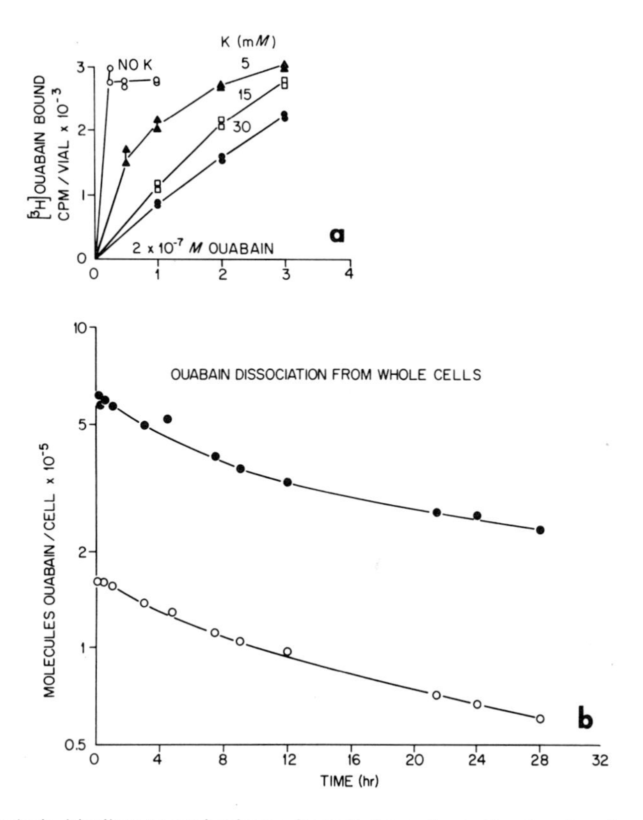

FIG. 1. $[^3H]$ouabain binding to and release from HeLa cells. **a:** Binding to attached cells. Cells were seeded in a number of replicate scintillation vials and allowed to grow 24 hr. At the time of the experiment the medium in each vial was aspirated and fresh medium with various K^+ concentrations plus $[^3H]$ouabain at 2×10^{-7} M was added. At the indicated times, paired vials were rinsed in ice-cold buffered saline and counted for radioactivity. **b:** Release from suspension cells. Ouabain was allowed to bind to the extent indicated on the ordinate at 0 time (these cells have a total of about 8×10^5 binding sites), then washed and returned to growth medium. Ouabain per cell was determined at the indicated times thereafter. Note that release is nonlinear on these semilogarithmic coordinates.

dissociation from cells is not a single exponential, a fact that we shall return to below. Nevertheless, if we estimate a dissociation rate constant from the early time points and compare it to association data at 5.5 mM K^+ in Fig. 1a, we may calculate an equilibrium constant for binding, K_D, to be $\sim 2\text{–}4 \times 10^{-8}$ M.

Some of the important physiological consequences of ouabain binding to these cells are shown in Fig. 2, where several parameters are plotted as a function of the logarithm of the ouabain concentration. In this experiment the cells have been continuously exposed to the ouabain up to and including the time of assay, and since the binding is slow at low concentrations, the

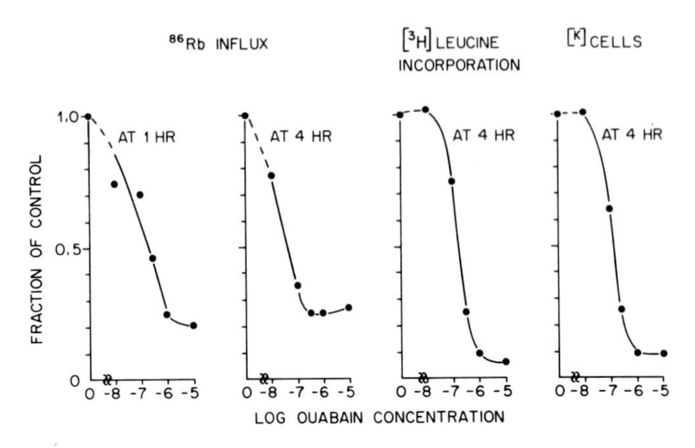

FIG. 2. Concentration-dependent effects of ouabain on several HeLa cell functions. On the abscissa are plotted on a logarithmic scale the ouabain concentrations from 10^{-8} to 10^{-5} M. Attached cells were in growth medium with 5.5 mM K^+, 37°C. $^{86}Rb^+$ influx was measured as 10-min uptake. [^{3}H]leucine incorporation into trichloroacetic acid-insoluble material was measured over 15 min. Cell K^+ concentration was measured by flame photometry, and is expressed as a fraction of the control value which is 160 meq/liter-cell water. [From Cook et al. (10).]

most meaningful measurements are those made at 4 hr (compare first and second panels of Fig. 2). The uptake of $^{86}Rb^+$ is inhibited, with the maximum inhibition reducing the uptake to about 25% of the control level. The K_I, or concentration of ouabain at which its half-maximum effect is observed, is about 2×10^{-8} M. The correspondence between this value and the binding constant K_D is an important part of the evidence that the binding we observe is specifically related to the inhibition of the Na^+–K^+ transport system. Uninhibited cells take up K^+ at rates that, if there were no efflux, would double the K^+ content of the cells in 1 hr (10). The membrane is K^+-leaky, however, and the passive efflux is nearly as great (96–97%) as the total influx, so that the net accumulation is only 3–4% per hour. Under the conditions of our experiments, the cells grow and double their K^+ content in a generation time of about 28 hr. Because of this K^+ leakiness, after 4 hr in ouabain at concentrations greater than 10^{-7} M the intracellular K^+ falls from its normal value of 160 meq/liter-H_2O to 10% or less of that value (Fig. 2, fourth panel). In addition to being an important component in osmotic regulation, intracellular K^+ over a very broad concentration range is an essential factor in protein synthesis (11,12). The loss of K^+ from cells in ouabain-containing medium leads to a parallel decrease (Fig. 2, third panel) in the rate of [^{3}H]leucine incorporation into acid-insoluble material, a decrease that we could not account for by a change in leucine uptake by the cells. It is important to recognize these secondary effects of ouabain, since they can have an important influence on turnover as measured in our subsequent experiments.

RECOVERY FROM OUABAIN BINDING AND TURNOVER OF THE BINDING SITE

Our first evidence that the cells recover from subtotal ouabain blockade by a more complex mechanism than simple dissociation of the drug from its surface receptor is shown in Fig. 3. From the control binding curve in this experiment we measure 7×10^5 binding sites per cell, a number that varies somewhat with the growth state of the cells (see also Baker, *this volume*). If we stop the ouabain binding midway by washing the cells and returning them to normal medium, in which case there remains a substantial fraction of normal transport and protein synthesis, we observe the usual very slow loss of ouabain from the cells. After 5 or 8 hr we challenge them again so as to measure the total number of binding sites on the surface, and we find that the increment of binding is approximately equal to the number of sites initially present. In other words, although very little ouabain has been lost, the normal number of titratable binding sites has somehow been restored to the cell surface. The functional correlates are shown in Fig. 3b. With this subtotal ouabain blockade, $^{86}Rb^+$ transport capacity has been halved at the time that further binding is stopped by replacing the medium,

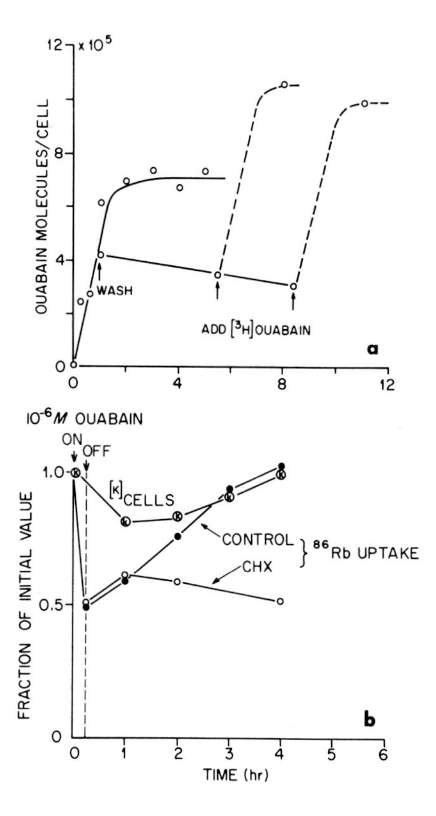

FIG. 3. Recovery of HeLa cells from partial ouabain blockade. Attached cells were in normal growth medium, 5.5 mM K^+, 37°C. **a:** Curve on left shows typical ouabain binding with ouabain in medium at 10^{-6} M. When cells are washed (*first arrow*) and returned to ouabain-free medium, the bound ouabain is released slowly. Second challenges at 5 or 8 hr after washing show that, despite the cells' retention of most of the initially bound ouabain, there is now a number of additional binding sites approximately equal to the number initially present. *Dashed lines,* drawn parallel to the first binding curve. [From Vaughan and Cook (9).] **b:** Same experimental conditions as in (**a**). Following 15 min exposure to 10^{-6} M ouabain, the cells are returned to fresh ouabain-free medium with or without 100 μg/ml cycloheximide (Chx). $^{86}Rb^+$ uptake is measured over 10 min and expressed here as a fraction of the initial control value. Encircled Ks refer to the K^+ concentration in the cells in the cycloheximide-free medium, which at 0 time in the controls is 160 meq/liter-water. [From Cook et al. (10).]

and the K^+ concentration of the cells drops slightly. Over the next several hours, the transport capacity and K^+ content return to normal, despite the fact that very little of the inhibitor is lost from the cells. This functional recovery is cycloheximide-sensitive, although cycloheximide alone appears to have no direct effect on the transport system or the dissociation of ouabain from these cells.

There are a number of *a priori* hypotheses that might account for these observations. For clarity, we present in Fig. 4 the model that we currently favor and that best accounts for the available data; we shall then describe the principal data on which the model is based.

The model depicts in cross section a unit of membrane area. With respect to the Na^+–K^+-ATPase, there is a continual synthesis and insertion of the transport enzyme into this unit of area from the cytoplasmic side. In the membrane, the enzyme is susceptible to attack by the glycoside. The enzyme molecules are removed from the membrane by a first order-degradative process that does not distinguish between native and ouabain-bound molecules. The removal is accomplished by an internalization process that carries the Na^+–K^+-ATPase subunit together with any ouabain that may be bound to it into the cell, where the ouabain is released, possibly as a consequence of the degradation of the enzyme. The turnover of the enzyme in the unit area of the membrane is continuous, and its density per unit area is constant. In addition to this turnover there is a continuous net synthesis of new membrane (i.e., growth), each unit area of which contains

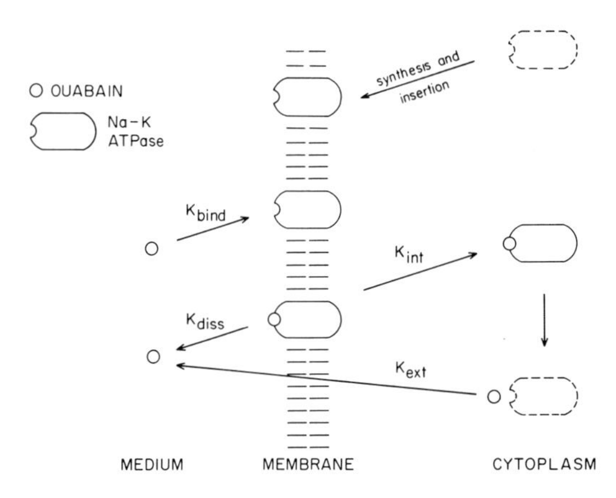

FIG. 4. Model for turnover of ouabain and ouabain-binding sites in HeLa cells. From the cytoplasmic side on the right the binding site (catalytic subunit of the Na^+–K^+-ATPase) is inserted into the membrane where it is susceptible to ouabain in the medium. As part of turnover, the site is eventually internalized with or without bound ouabain. If the former, the ouabain is eventually released, possibly as the binding site is degraded, and escapes back into the medium. From the medium side, bound ouabain may either be released by dissociation back into the medium or be cycled through the site-internalization and intracellular-release process. See text for fuller discussion.

the same density of enzyme; these new enzyme molecules are equally subject to turnover.

In this same process with respect to the drug, ouabain in the medium binds to its specific site on the ATPase subunit. The site that is thus inactivated may be restored to activity by simple dissociation of the drug, or an equivalent activity may be regenerated by the turnover process in which bound ouabain is internalized with the catalytic subunit while new enzyme is inserted into the membrane. Inside the cell, the drug is released without further metabolism, and exists in the cytosol as free ouabain until it escapes back into the medium. Very little drug enters the cell by passive diffusion, and the probability is vanishingly small that a ouabain molecule in the process of escaping from the cell will rebind at the surface.

OUABAIN BINDING AND TRANSPORT DURING THE CELL CYCLE

Consider first the question of growth and constancy of binding sites per unit area. There are reports that in some cells the activity of the Na^+–K^+-ATPase in isolated membranes may fluctuate over as much as a 10-fold range between G1 and G2 (13). Because of the contribution of the alkali cations to the osmotic properties of the cell, the growth of the cells is directly influenced by this transport system (14). Such very large fluctuations, if they are reflected by parallel fluctuations in transport rates *in vivo,* should be readily detectable either as bursts of growth, or relatively high cellular Na^+ and low cellular K^+ when transport is at a minimum. An alternative possibility is that, together with the increased transport capacity, there is an increased passive leak flux, so that the system accomplishes, for whatever functional reason, a high turnover of cations without change in the other parameters. This latter possibility should be easily discernible as a markedly increased passive K^+ influx at the same time as the transport increases. In HeLa cells synchronized by a double-thymidine block technique (15) and followed from late G1 at the time of release from the block through S, G2, mitosis, and well into the subsequent G1, we see no such reproducible discontinuities. The cell volume, as determined by an electronic analyzer (Coulter), closely follows the curve calculated for a population growing logarithmically with respect to mass without change in number, except during the division period, and the K^+ : protein ratio for the cells is virtually constant throughout (data not shown). Some of our cell-cycle data relevant to the present discussion are shown in Fig. 5. The upper curve is K^+ transport capacity, measured as $^{86}Rb^+$ influx from a medium 10 mM in K^+. In parallel experiments, the ouabain (10^{-4} M)-insensitive $^{86}Rb^+$ influx was $20 \pm 5\%$ of the uninhibited flux throughout. The lower curve in Fig. 5 gives the number of ouabain-binding sites per cell. The lines are fitted to the data only near the peak point at the beginning of cell division. The remainder of the lines are calculated on the basis of (*i*) the population having a generation

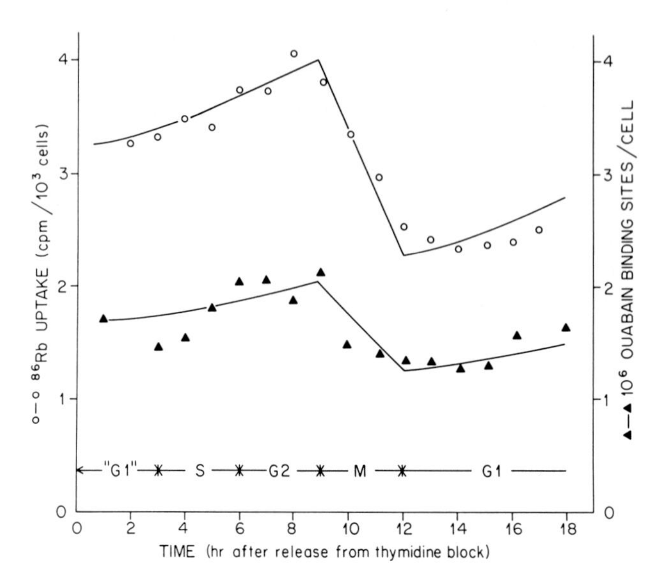

FIG. 5. $^{86}Rb^+$ transport capacity (*open circles*) and ouabain binding sites per cell (*closed triangles*) through the cell cycle following synchronization by double-thymidine block. $^{86}Rb^+$ uptake was measured over 10 min from a medium containing 10 mM K^+ to ensure saturation of transport from the K^+ side. Ouabain was measured as 45-min binding in 10^{-6} M glycoside. The 9-hr points were fitted by eye, and the drawn lines calculated from them as described in the text.

time of 27 hr (independently determined for this experiment); (*ii*) the cells growing logarithmically with respect to mass, the mass halving at division, and the surface growing as the two-thirds power of the mass; and (*iii*) division occurring over 3 hr and involving 80% of the population (independently determined in this experiment). Despite some scatter, particularly in the ouabain-binding data, the data are in reasonable agreement with the calculations, together with the assumption that the density of binding sites (and hence transport capacity) is constant per unit area of surface. An additional implicit assumption in the above is that the cells are isomorphous throughout the cell cycle. In contrast to the case for attached, nontransformed cells, we find from scanning electron microscopy that this is a reasonable assumption for suspended HeLa cells. They remain rounded and covered with microvilli throughout, although there may be systematic fluctuations in microvillar and other surface structures. We conclude that the net growth of HeLa cell membrane involves an increase in surface area adequate to accommodate the increased cell mass without an important change in cell geometry, and that the increases in transport, in passive leak, and in ouabain-binding sites are those to be expected on the basis of the increased total area with no change in mean density of transport enzyme per unit area.

MEMBRANE-BOUND AND INTRACELLULAR OUABAIN

In order to test our model for turnover of the binding site, we needed to distinguish membrane-bound ouabain from internal ouabain. For this purpose a modification of the membrane preparation technique of Atkinson and Summers (16,17) has proven very useful. Cells are swollen in hypotonic medium and lysed by a few strokes in a Dounce homogenizer; the membranes come off in large sheets. The homogenate is then made up to 5% in sucrose together with restoration of salts to approximately "normal" concentrations (150 mM NaCl; 5 mM KCl; 3 mM $MgCl_2$; 10 mM Tris, pH 7.3). Following a very brief low-speed spin to remove whole cells and nuclei, the homogenate is layered on 30% sucrose, which, in turn, has already been layered on a 45% sucrose cushion. The two-step gradient is centrifuged at $8{,}000 \times g$ for 20 min. Membranes are recovered at the 30/45% interface (Fig. 6), where we find 40–70% of the recovered 5′-nucleotidase (a surface enzyme not affected by ouabain), 6–12% of the total protein, less than 3% of the acid phosphatase (lysosomal marker), and no detectable glucose 6-phosphatase (microsomal marker). The entire preparation can be done in 40 min. Since the preparation is made at 0–4°C, and since ouabain dissociates from whole cells and isolated membranes at less than 1% per hour at these temperatures, this is a convenient technique for localizing bound ouabain. It may be seen in Fig. 6a that if the cells are treated briefly with [^{3}H]ouabain, washed to remove extracellular drug, and immediately subjected to the gradient preparation, the ^{3}H is generally codistributed with

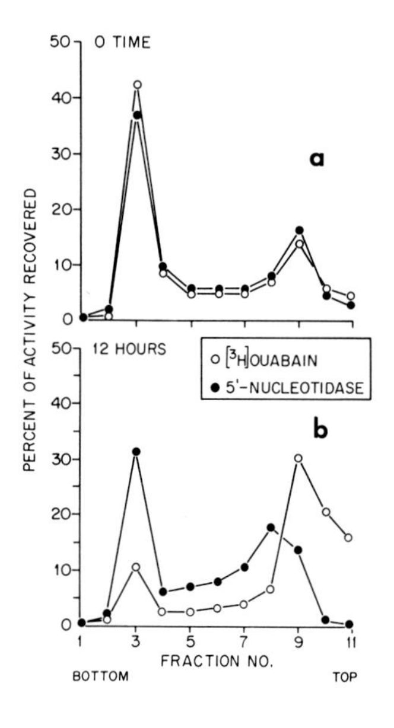

FIG. 6. Distributions in discontinuous sucrose gradients [Atkinson and Summers (16)] of cell-associated [^{3}H]ouabain (*open circles*) and 5′-nucleotidase (*closed circles;* measured as 5′-AMPase) activity following gentle Dounce homogenization as described in the text. The interface between 45 and 30% sucrose is in tube 3, and between 30 and 5% sucrose is in tube 9. *Upper,* Cells had been exposed to subtotal binding by [^{3}H]ouabain, washed and homogenized immediately. *Lower,* After washing, the cells were returned to growth medium for 12 hr before homogenization.

the 5′-nucleotidase. The change in this pattern with time as the cells are incubated, shown in Fig. 6b, is discussed below. Note that at the top of the gradient in Fig. 6a, even with this relatively low-force centrifugation, the ^3H activity has pulled away from the meniscus, i.e., there is little soluble or free ouabain in the gradient. If the homogenization is overvigorous, the two markers remain codistributed, but a much greater fraction of both remains at the upper step. We have concluded that most, if not all, of the 5′-nucleotidase at the upper step arises from small pieces or vesicles of plasma membrane generated by the homogenization, and too small to be spun down to the lower step. The activity of this enzyme at the lower step is our quantitative marker for membranes from log-phase cells.

The analysis of our turnover model relies heavily on comparisons between the membrane-rich fraction and the whole cell. Some of our experiments in this regard simultaneously test a "derepression" hypothesis that could, in principle, account for the reappearance of ouabain-binding sites depicted in Fig. 3a. Briefly described, a number of cell types, when starved for certain essential metabolites, develop an enhanced transport capacity for these metabolites. This is true for the uptake of amino acids (18,19) and sugars (20–26) and has been proposed for Na^+–K^+ transport as well (27–30). Such a derepression-like phenomenon is thought to be induced either by growing the cells in low-K^+ medium, or preventing normal K^+ uptake by growing the cells in sublethal concentrations of a transport inhibitor. Derepression could, if it occurred rapidly enough, account for the extra bound ouabain seen in Fig. 3a. At one time we suggested this might be the mechanism (9), although it is now reasonably certain that, in these short-term experiments, induction of transport does not play a significant role (10). Since other authors (28–30) had suggested that the derepression might take significantly longer time, one to two cell generations, we have made a direct search for enhanced ouabain-binding sites on the surfaces of glycoside-treated cells and have not found them. An experiment of this kind is shown in Fig. 7. Here control cells are first acutely challenged with 2×10^{-7} M [^3H]ouabain in low-K^+ medium to count the number of surface binding sites. Other cells from the same population are grown several days in low ouabain, 10^{-8} M, of the same specific radioactivity. Although the experiment in Fig. 7 lasted only 40 hr, which was a little more than one generation for these slightly inhibited cells, longer experiments give similar results. At the end of the 40 hr of chronic treatment, we find that the total ouabain per cell is about 80% of what we had found in the acute challenge. When these chronically treated cells are now acutely challenged at the higher drug concentration, the total drug per cell is found to be 160% of the initial control. If all the ouabain were on the surface, this is the expected result of an induction or derepression-type response. With each of these determinations, however, we also prepared membranes and looked at the [^3H]ouabain : 5′-nucleotidase ratio in the membrane fraction. Now we find that after the 40-hr

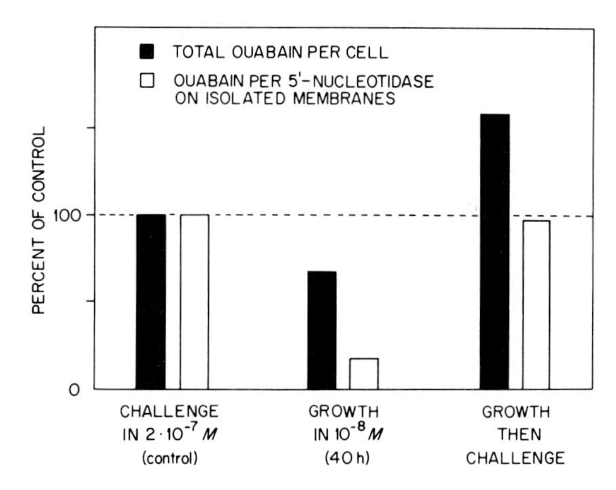

FIG. 7. Distinction between amount of whole-cell-associated (*filled bars*) and membrane-associated (*open bars*) [^{3}H]ouabain after acute, saturating challenge (2×10^{-7} M, 90 min, 0.5 mM K^+); chronic, subsaturating challenge (10^{-8} M, 40 hr, 5.5 mM K^+); and chronic + acute treatments. Membrane-associated drug was measured in terms of the [^{3}H]ouabain : 5′-nucleotidase ratio on the lower shelf of gradients like those shown in Fig. 6. See text.

growth period this ratio is only 20% of what we had observed in the initial challenge, a result showing that most of the ouabain is not on the membrane, but is elsewhere in the cell. When challenged with the higher drug concentration, the [^{3}H]ouabain : 5′-nucleotidase ratio in the membrane fraction comes back to nearly the same level observed in the initial controls. Our conclusions are that there has been no induction in these cells, but that, on the contrary, the number of ouabain-binding sites per unit of membrane (i.e., the density of the transport enzyme) has remained constant. The "excess" ouabain is inside the cell and is not associated with the transport enzymes on the surface.

COMPUTER MODELING OF OUABAIN AND OUABAIN-BINDING SITE TURNOVER IN INTACT CELLS

Accepting that neither acute nor chronic ouabain blockade alters the site density, we test the turnover model in two complementary ways. The first is to subject the cells to a brief subtotal binding and to observe for prolonged periods in intact cells the quantity of cell-associated drug; these drug-release data are analyzed in terms of the model with an appropriate computer program. The advantage of this approach is that we deal with intact cells. The second method utilizes fractionation and other disruptive techniques to analyze the system, from which analysis we synthesize a picture of the events in the intact cell. The extent to which the two methods yield corresponding results reinforces the interpretation.

We noted in Fig. 1b that, after a brief subtotal challenge, the loss of ouabain from the intact cells is not a single exponential, a result that suggests heterogeneity of the processes responsible for glycoside release. We have written a formal quantitative description of this dissociation in terms of the three compartments in our model—the membrane, the cell interior, and the medium. Following the short challenge, all of the drug is to be found on the membrane. After the cells are returned to drug-free medium, the drug leaves the membrane by either simple dissociation (diss) into the medium or by internalization (int) in the turnover process. Formally

$$\frac{dX_{\mathrm{Mb}}}{dt} = -k_{\mathrm{diss}}X_{\mathrm{Mb}} - k_{\mathrm{int}}X_{\mathrm{Mb}} \tag{1}$$

where X_{Mb} is the quantity of the drug in the membrane compartment and the ks are first-order rate coefficients. This formalization assumes no heterogeneity in the rate-limiting steps in either process. From our point of view, the estimate of k_{int} is the result of major interest, since this gives information on the rate of turnover of this surface enzyme. In the cytoplasmic compartment, the amount of drug increases as internalization from the membrane compartment progresses, and falls again as the drug is externalized (ext), i.e., is released from the bound form and ultimately is allowed to escape into the medium

$$\frac{dX_{\mathrm{Cy}}}{dt} = +k_{\mathrm{int}}X_{\mathrm{Mb}} - k_{\mathrm{ext}}X_{\mathrm{Cy}} \tag{2}$$

where X_{Cy} is the quantity of drug in the nonmembrane cytoplasmic compartment. The intracellular processing of the binding sites, the release of the ouabain into the cytosol and its subsequent escape is obviously a complex sequence of events, and its description by a single first-order rate coefficient is at best an approximation for whichever step is rate-limiting. The two equations may be summed:

$$\frac{dX_{(\mathrm{Mb+Cy})}}{dt} = -k_{\mathrm{diss}}X_{\mathrm{Mb}} - k_{\mathrm{ext}}X_{\mathrm{Cy}} \tag{3}$$

This is a formal statement of the processes whereby the drug is released from the cell, and differs from Eq. (1) which, given steady-state turnover, is a formal statement of physiological recovery of membrane function.

Using the FORTRAN program CRICF devised by Chandler, Hill, and Spivey (31) for coupled enzymic reactions, we have taken whole-cell drug-release data, like those in Fig. 1b, and fit them by least squares to an integral of Eq. (3), a calculation that requires estimations of k_{int} as well as k_{diss} and k_{ext} and the time-dependent magnitudes of X_{Mb} and X_{Cy}. The results of such a fit are shown in Fig. 8, where the points are experimental data. From these whole-cell data alone, the computed analysis states that (*i*) the half-time for simple dissociation is about 10 hr; (*ii*) the half-time for internalization is about 5 hr; (*iii*) as a consequence of these two parallel processes, the half-

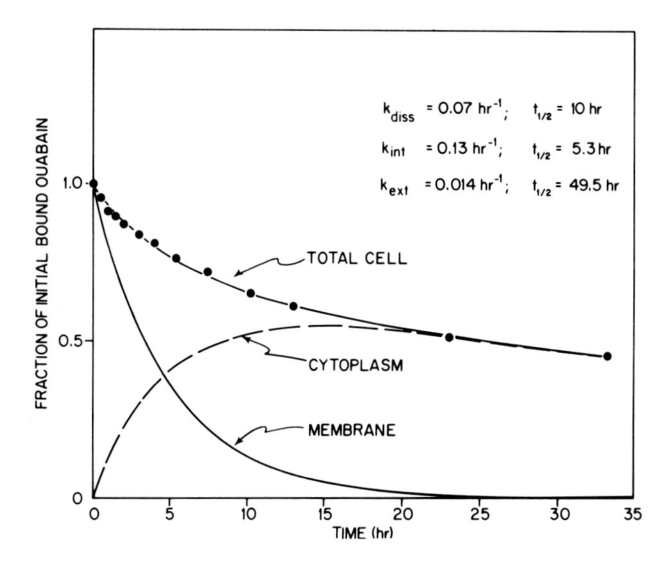

FIG. 8. CRICF (31) computations for the fate of ouabain bound to HeLa cells, based on the model discussed in the text and in Fig. 4. Unconstrained fit to three-compartment model. The best fit is given to the experimental points on the curve for total cell-associated ouabain following brief subtotal binding and return of the cells to normal ouabain-free growth medium. Also given for this fit are the computed rate constants for dissociation of ouabain from the surface (k_{diss}), for internalization of ouabain-binding sites (k_{int}), and for externalization of internalized ouabain (k_{ext}) as listed in the figure. As computed from these rate constants, the fractions of initially bound ouabain remaining on the membrane or taken into the cytoplasm are plotted as a function of time.

time for ouabain on the surface membrane is about 3 hr; (*iv*) between 12 and 30 hr, about half of the initially bound ouabain is inside the cell; and (*v*) after 15 hr, when very little of the cell-associated ouabain is on the surface, ouabain is lost from the cytoplasmic compartment very slowly, with a half-time of about 50 hr.

EXPERIMENTAL TESTS OF THE MODEL

In general, we have confirmed these statements by direct measurement on cells subjected to various manipulations, although the various rate coefficients from experiment to experiment are not precisely those we have computed above. In Fig. 9, we compare the total ouabain per cell, following a pulse binding, with the ouabain on the plasma membrane as defined by the [^{3}H]ouabain : 5′-nucleotidase criterion on the lower shelf of Atkinson-Summers gradients (compare Fig. 6a,b). Membrane-ouabain is linear on this semilogarithmic plot, although the complexity of the manipulations restricts the number of points that can be taken in a single experiment. The half-time for ouabain on the membrane in this experiment is 3.5 hr. Allow-

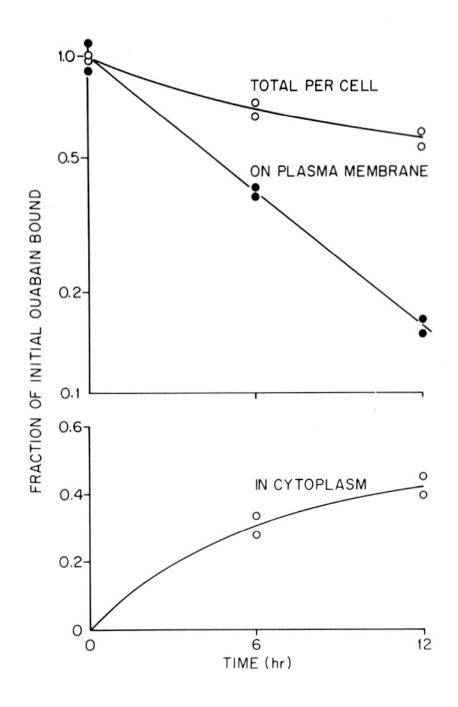

FIG. 9. Fate of [^{3}H]ouabain bound to HeLa cells, cellular distribution of [^{3}H]-ouabain with time after subtotal binding. Suspension cells washed to remove free extracellular glycoside and returned to growth medium. *Upper curves,* Total ouabain per cell (*open circles*) is curvilinear on these semilogarithmic coordinates; ouabain on the membrane (*closed circles*), calculated from the [^{3}H]ouabain : 5′-nucleotidase ratio on the lower shelf of gradients like those of Fig. 6, is linear with time. *Lower curve,* The difference between the upper curves, cell-associated, but not plasma membrane-bound ouabain, plotted on arithmetic coordinates.

ance for growth (see below) makes only a small difference. It does not affect the single-exponential prediction, but suggests that in this experiment the half-time in terms of the sum of only the two processes described in Eq. (1) should be corrected to about 3.8 hr. After 12 hr, only about 15% of the initially bound ouabain is still on the membrane. In the lower part of Fig. 9, where the difference between the two upper curves is plotted, it appears that after 12 hr more than 40% of the initially bound ouabain is now inside the cell.

A further prediction of the model is that if turnover, i.e., internalization, could be stopped, the loss of ouabain from the cells would be by dissociation from the membrane alone. This should be (*i*) a faster loss from the cells than when the internalization and slow externalization parallel pathway is operative, but (*ii*) a slower loss of ouabain from the membrane compartment alone than when both processes are functioning, and finally (*iii*) a single exponential if there is no heterogeneity of the surface binding sites. To examine these points experimentally, it was necessary to find a very fast-acting metabolic inhibitor to stop turnover—puromycin, cycloheximide, and NaN_3, by themselves, were too slow. We finally chose a combination of NaN_3 and 2-deoxyglucose. The former inhibits ATP synthesis, while the latter rapidly enters the cells and is phosphorylated without undergoing any further metabolism of quantitative significance. The phosphorylation is an ATP sink, and the two compounds together can effectively deplete the cell of its ATP within a few minutes. Such cells do not, of course, survive very long, and experiments done with them are limited to a few hours at most. One such experiment is shown in Fig. 10 in which HeLa cells were treated with [^{3}H]ouabain and subsequently resuspended in ouabain-free medium, but with the other inhibitors present. Control cells were treated with ouabain only, and the release of the glycoside from these control cells followed the usual pattern. In the deoxyglucose-azide-treated cells the drug release was first order, faster than from the controls, and slower than from the membrane compartment or normal whole cells as shown in Fig. 9. The predictions of the model were thus substantiated. There are two important controls on this experiment, for which the data are not given here. First, ouabain dissociates from isolated membranes with the same first-order rate coefficient observed for deoxyglucose-azide-treated cells as shown in Fig. 10, and is not influenced by deoxyglucose-azide. Second, there is no increase of ouabain in the intracellular soluble fraction of the deoxyglucose-azide-treated cells, although such an increase is observed in controls. This latter experiment confirms that the internalization of ouabain is an ATP-requiring event.

Strictly speaking, the equations and curves in Fig. 8 deal with distributions of quantities of drug in the total population of cells in a closed system irrespective of growth. The fall in specific activity ([^{3}H]ouabain : 5′-nucleotidase ratio) plotted in Fig. 9 can, in fact, be affected by growth; even with no dissociation or internalization this specific activity would decrease as 5′-nucleotidase is added in the growing surface. Allowance for the population's growing with a generation time τ leads to the expression:

$$\ln\left(\frac{X/N}{X_0/N_0}\right) = -\left(k_{\text{diss}} + k_{\text{int}} + \frac{\ln 2}{\tau}\right)t \tag{4}$$

where X/N is the [^{3}H]ouabain : 5′-nucleotidase ratio. This is what is plotted in Fig. 9, its slope being the sum of the terms in the parentheses. From data

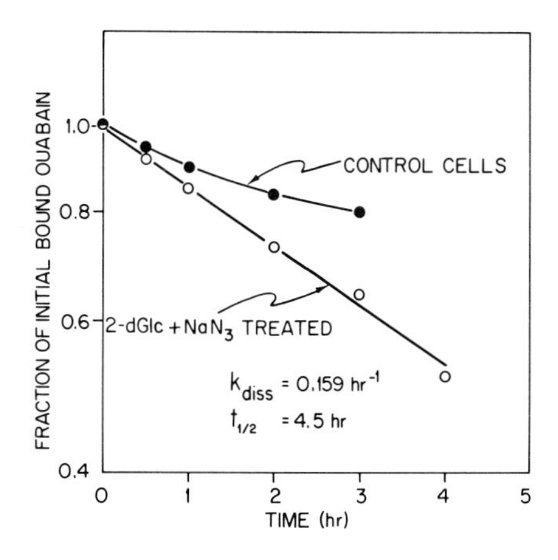

FIG. 10. Ouabain release from cells treated with 10 mM NaN_3 plus 5 mM 2-deoxyglucose to inhibit the internalization process. Note the linearity of this release (*open circles*) on these semilogarithmic coordinates, compared to the curvilinearity in the controls (*closed circles*).

in nongrowing cells like those in Fig. 10 we have estimates of k_{diss}, and since we know τ we may calculate k_{int}. For cells growing with a generation time of 28 hr, ln $2/\tau$ is about 0.02 hr^{-1}. Values from a number of experiments yield estimates of k_{int} between 0.06 and 0.10 hr^{-1}, with a mean of about 0.08 hr^{-1}. This result is not greatly different from the computed estimate of 0.13 hr^{-1} in the intact-cell experiment (Fig. 8), and indicates that, in addition to net synthesis associated with growth, the binding site turns over approximately twice per 28-hr generation.

In what form and by what mechanism is the binding site internalized? Our answer to this question is not complete. During the postbinding recovery period we find cell-associated ouabain principally in two forms. The first may be sedimented from whole-cell homogenates in 10% sucrose under relatively low G forces—17,000 × g for 12 min. This is sufficient to bring down the large plasma membrane fragments, as well as mitochondria and lysosomes, but not microsomes. The [^{3}H]ouabain that remains in the supernatant is less than 5% bound, according to the criterion of exclusion from Sephadex G-50. The other 95% is not at all excluded from Sephadex and appears to be native ouabain in solution. With slightly different experimental conditions, if we define intracellular ouabain as cell-associated glycoside that is not bound to the plasma membrane, we observe that, at any given time, about one-third of the intracellular ouabain is bound to the large sedimentable particulates and the remainder is free in the cytosol. The glycoside that is eventually released from the cells and recoverable from the medium is chromatographically indistinguishable from authentic ouabain. Recognizing that these cells are growing and turning over their membrane molecules at normal or near-normal rates, and allowing for dissociation of drug from the surface, we estimate that these cells are capable of holding intracellularly in bound form 250,000–400,000 ouabain molecules per cell in addition to those on the cell surface (Table 1). However, we find few or no intracellular binding sites in homogenates if we challenge with [^{3}H]ouabain after the cells have been dounced. In these latter measurements the surface binding sites are blocked with unlabeled drug and the cells washed before homogenization. In parallel controls whole cells are added back to homogenates to show that the conditions are favorable for binding

TABLE 1. *Steady-state numbers of ouabain-binding sites per cell (1000s)*

Cell surface	800–1,000
Incorporated by internalization into the cytoplasm (sedimentable at 17,000 × g × 12 min)	250–400
Sites other than cell surface accessible to ouabain in whole-cell homogenates	Less than 15

if binding sites are there and available. In other words, bound intracellular glycoside can be found only after internalization by the intact cell. The results are consistent with a model in which the surface sites are internalized by endocytosis, yielding a vesicle with the binding site facing inward. We assume, as a working hypothesis, that the binding site is subsequently degraded (after fusion of the vesicle with a lysosome?) and the ouabain is released free into the cytosol. The plasma membrane appears to have a very low passive permeability to ouabain. Its final release into the medium is a very slow process, and we have observed that free intracellular ouabain builds up to concentrations as high as 2×10^{-7} M with no apparent effect on cell growth or metabolism.

TURNOVER VS. SPECIFIC REPAIR

The extent to which our experiments measure the actual turnover of the native enzyme remains a problem. It is known, for example, that cellular proteins containing a few "incorrect" amino acid substitutions are subject to faster turnover than their normal counterparts (32). The catalytic subunit of Na^+–K^+-ATPase, when bound with ouabain, may be regarded by the cell as a partially denatured or nonnative polypeptide and so be degraded at a faster than normal rate. We have two correlative observations that suggest, but by no means prove, that the turnover of the bound site as we measure it is not accelerated, but is the same as the turnover of the unbound site. In the first place, if [^{3}H]ouabain is a general label of the site, but an innocuous one from the turnover point of view, then the rate at which the cells process [^{3}H]ouabain should appear as a random sampling of the processing of all the catalytic subunits and the rate constants for the loss of ouabain from the cells should be independent of the extent of binding. The curves for ouabain loss from the cells should then be parallel on a semilog plot. This is what is observed (Fig. 1b). In contrast, this result may mean no more than that the degradation rate of the ouabain-bound site is not saturated even at high binding. The result is only consistent with the model but not a proof of it.

Our second correlative approach has been to look in the membrane-rich fractions for polypeptides that turn over at the rate observed for the ouabain-binding site, and more specifically to look for such polypeptides in the vicinity of MW 100,000. When the highly membrane-enriched material from the lower shelf of the Atkinson–Summers gradient is electrophoresed on SDS-acrylamide gels (Fig. 11), we reproducibly find 26 peaks (see also ref. 33), many of which are clearly compound. We measure turnover in this material by a double-label method that minimizes reutilization of the label as a potential artifact. Cells are grown several days in [^{14}C]leucine, after which time all cellular proteins are uniformly labeled. While the cells are still in log growth, they are pulsed briefly with [^{3}H]leucine and harvested. The

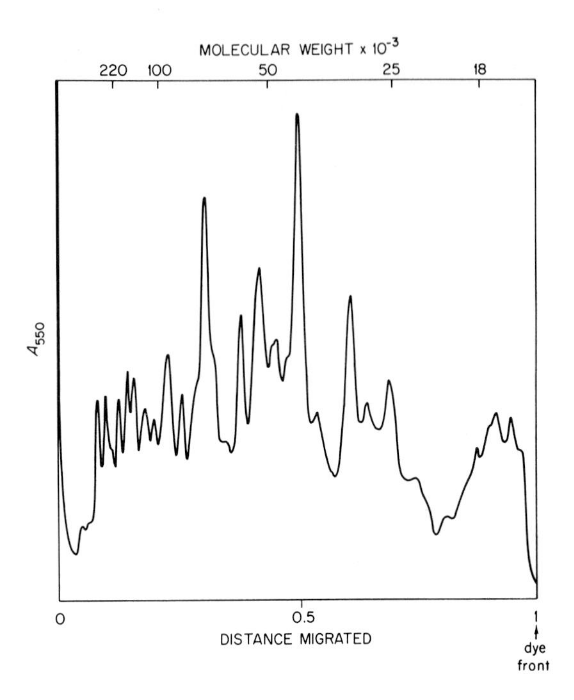

FIG. 11. SDS-polyacrylamide-gel electrophoresis of HeLa membrane proteins. Membrane-enriched fraction, like that in tubes 3 of Fig. 6, was boiled 2 min in 1% sodium dodecyl sulfate (SDS) and electrophoresed in 6.5 cm gel of 7.5% acrylamide and 0.1% SDS at 3 mA/gel. Polypeptides in the gel were stained with Coomassie blue and the gel scanned at 550 nm.

membrane-rich fraction is prepared and analyzed electrophoretically. Since the cells were in steady growth throughout the experiment, the $^3H:^{14}C$ ratio at each point on the gel is a measure of the rate of synthesis of the corresponding polypeptides. This ratio is highest for polypeptides that turn over most rapidly and lower for those that turn over slowly. More quantitatively, it is possible to calculate a second ratio (total synthesis per generation) : (net synthesis per generation). By definition, the net synthesis is the doubling of any protein or polypeptide in one generation. If there is no turnover, (i.e., no degradation), total synthesis is the same as net synthesis, and the ratio is 1.0. A ratio of 3.0 indicates that the polypeptide is synthesized three times and degraded twice in one generation. Figure 12 shows the measured $^3H:^{14}C$ values (left ordinate) and calculated turnover values (right ordinate). We attach little significance to the specific ups and downs between adjacent points, much of which reflects scatter in the counting data. What is clear, and independent of the order in which the isotopes are used, is that (*i*) all polypeptides are turning over with a mean synthetic rate twice that required to double all the proteins in one generation, and (*ii*) the larger

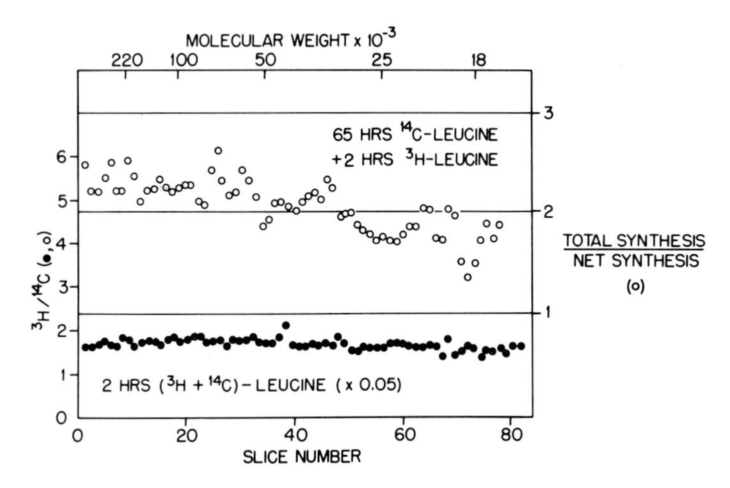

FIG. 12. Turnover, as estimated by a double-label technique, of polypeptides in the membrane-enriched fraction from HeLa cells. *Open circles,* Cells were grown 65 hr in 0.1 μCi/ml [^{14}C]leucine (when the radioactivity of free leucine in the medium had decreased by less than 25%), at which time [^{3}H]leucine was added to an activity of 5 μCi/ml for 2 hr. A membrane-enriched fraction was prepared and electrophoresed as in Fig. 11, and the ^{3}H/^{14}C ratio was determined for each slice of the gel. The right-hand ordinate expresses a turnover-related parameter—the ratio of total to net synthesis—and applies only to the open circles. *Closed circles,* As above, except that 2 μCi/ml [^{14}C]leucine was used, and both the ^{3}H and ^{14}C activities were added simultaneously for 2 hr. The constancy of ^{3}H/^{14}C ratio in this control demonstrates that the experimental results are not caused by isotope effects. Cells were in log growth throughout the experiment.

polypeptides in general turn over more rapidly than the smaller. The latter result is in agreement with the earlier results of Arias et al. (34), Dehlinger and Schimke (35), and Schmidt-Ullrich et al. (36), but not with those of Roberts and Yuan (37) nor Hubbard and Cohn (38). The latter pairs of authors found that in a variety of cells all membrane polypeptides turn over at the same rate. The reason for the differences between laboratories is not resolved. Nevertheless, in the membrane-rich fraction of HeLa cells, there are polypeptides at molecular weights near 100,000 that turn over at or near the same rate we estimate from ouabain-binding experiments. This second correlative observation is again consistent with our determination that ouabain-binding sites, in addition to increasing with growth of the membrane, turn over about twice per generation.

CONCLUSION

We have presented an argument, not yet complete in all its details, that at least one set of surface molecules of functional importance is not statically emplaced in the cell membrane and left there, susceptible to adverse environmental effects, as well as probable thermal inactivation over an extended period of time. On the contrary, these transport molecules appear to be continuously turning over (or possibly even specifically replaced when inactivated) in a fashion that maintains their functional capacity at nearly constant levels, always provided that they have not been subjected to "overkill" by the drug we use as a marker.

One final point in conclusion: We have proposed that the polypeptides in the membrane-rich fraction are turning over at different rates, but that they appear to be internalized as particulates or membrane-bound vesicles. Although in our case this point is not firmly established, there is a clear precedent for the idea that endocytic internalization of membrane need not involve a cross-sectional assortment of all membrane proteins. Tsan and Berlin (39) in experiments confirmed in our laboratory (Krogsrud and Cook, *unpublished*) have shown that phagocytosis of latex beads by mouse macrophages involves internalization of considerable surface membrane without loss of adenosine transport capacity. In similar experiments Schwartz et al. (40) showed that phagocytosis in human monocytes also resulted in no loss of insulin receptors on the cell surface. In experiments closer to ours, Schmidt-Ullrich et al. (36) have shown that following concanavalin A treatment, the proteins of lymphocyte membranes turn over at different rates, again the larger ones generally faster, with the concanavalin A binding site turning over fastest. The latter high rate is clearly related to the capping induced by the lectin (41–43). Whether the relatively high rate of turnover of ouabain-bound sites in HeLa cells may be related to anything like a capping phenomenon has not yet been determined.

ACKNOWLEDGMENTS

P.C.W. is a Postdoctoral Investigator, supported by subcontract No. 3322 from the Biology Division of the Oak Ridge National Laboratory to the University of Tennessee. W.R.P. is a Predoctoral Fellow supported by Grant GM 1974 from the National Institute of General Medical Sciences, National Institutes of Health. Research was sponsored jointly by the National Cancer Institute and the U.S. Energy Research and Development Administration under contract with the Union Carbide Corporation.

We are indebted to James W. Longworth for bringing to our attention the CRICF Program and consulting with us on its use. Margarita K. Churchich was also exceptionally helpful with computations. Donald W. Salter, who has been working in our laboratory on a separate problem, was a constant source of good suggestions and could always be counted on to be generous and helpful.

REFERENCES

1. Puck, T. T., Marcus, P. I., and Cieiura, S. (1956): Clonal growth of mammalian cells *in vitro*. *J. Exp. Med.*, 103:273–284.

2. Kyte, J. (1971): Purification of the sodium- and potassium-dependent adenosine triphosphatase from canine renal medulla. *J. Biol. Chem.*, 246:4157–4165.
3. Hokin, L. E., Dahl, J. L., Deupree, J. D., Dixon, J. F., Hackney, J. F., and Perdue, J. F. (1973): Studies on the characterization of the sodium–potassium transport adenosine triphosphatase. X. Purification of the enzyme from the rectal gland of *Squalus acanthias*. *J. Biol. Chem.*, 248:2593–2605.
4. Jørgensen, P. L. (1974): Purification of (Na^{+}–K^{+})-ATPase: Active site determinations and criteria of purity. *Ann. NY Acad. Sci.*, 242:36–52.
5. Nakao, M., Nakao, T., Hara, Y., Nagai, F., Yagasaki, S., Koi, M., Nakagawa, A., and Kawai, K. (1974): Purification and properties of Na, K-ATPase from pig brain. *Ann. NY Acad. Sci.*, 242:24–35.
6. Baker, P. F., and Willis, J. S. (1970): Potassium ions and the binding of cardiac glycosides to mammalian cells. *Nature*, 226:521–523.
7. Baker, P. F., and Willis, J. S. (1972): Binding of the cardiac glycoside ouabain to intact cells. *J. Physiol. (Lond.)*, 224:441–462.
8. Lamb, J. F., and McCall, D. (1972): Effect of prolonged ouabain treatment on Na, K, Cl, and Ca concentration and fluxes in cultured human cells. *J. Physiol. (Lond.)*, 225:599–617.
9. Vaughan, G. L., and Cook, J. S. (1972): Regeneration of cation-transport capacity in HeLa cell membranes after specific blockade by ouabain. *Proc. Natl. Acad. Sci., USA*, 69:2627–2631.
10. Cook, J. S., Vaughan, G. L., Proctor, W. R., and Brake, E. T. (1975): Interaction of two mechanisms regulating alkali cations in HeLa cells. *J. Cell. Physiol.*, 86:59–70.
11. Lubin, M. (1967): Intracellular potassium and macromolecular synthesis in mammalian cells. *Nature*, 213:451–453.
12. Pestka, S. (1970): Protein biosynthesis: mechanisms, requirements, and potassium-dependency. In: *Membranes and Ion Transport*, Vol. 3, pp. 279–296, edited by E. E. Bittar. Wiley-Interscience, New York.
13. Graham, J. M., Sumner, M. C. B., Curtis, D. H., and Pasternak, C. A. (1973): Sequence of events in plasma membrane assembly during the cell cycle. *Nature*, 246:291–295.
14. Jung, C., and Rothstein, A. (1967): Cation metabolism in relation to cell size in synchronously grown tissue culture cells. *J. Gen. Physiol.*, 50:917–932.
15. Petersen, D. F., and Anderson, E. C. (1964): Quantity production of synchronized mammalian cells in suspension cultures. *Nature*, 203:642–645.
16. Atkinson, P. H., and Summers, D. F. (1971): Purification and properties of HeLa cell plasma membranes, *J. Biol. Chem.*, 246:5162–5175.
17. Atkinson, P. H. (1973): HeLa cell plasma membranes. In: *Methods Cell Biol.*, 7:157–188.
18. Frengley, P. A., Peck, W. A., and Lichtman, M. A. (1974): Accelerated active transport of alpha-aminoisobutyric acid by human leukemic leukocytes in an amino acid deficient environment. *Exp. Cell Res.*, 88:442–444.
19. Hume, S., and Lamb, J. F. (1974): Effect of growth in various concentrations of amino acids on the properties of the A-mediated amino acid uptake system in cultured cells. *J. Physiol. (Lond.)*, 239:46P–47P.
20. Martineau, R., Kohlbacher, M., Shaw, S. N., and Amos, H. (1972): Enhancement of hexose entry into chick fibroblasts by starvation: differential effect on galactose and glucose. *Proc. Natl. Acad. Sci. USA*, 69:3407–3411.
21. Shaw, S. N., and Amos, H. (1973): Insulin stimulation of glucose entry into chick fibroblasts and HeLa cells. *Biochem. Biophys. Res. Commun.*, 53:357–365.
22. Hatanaka, M. (1973): Sugar effects on murine sarcoma virus transformation. *Proc. Natl. Acad. Sci. USA*, 70:1364–1367.
23. Plagemann, P. G., and Richey, D. P. (1974): Transport of nucleosides, nucleic acid bases, choline and glucose by animal cells in culture. *Biochim. Biophys. Acta*, 344:263–305.
24. Kletzien, R. F., and Perdue, J. F. (1975): Induction of sugar transport in chick embryo fibroblasts by hexose starvation. *J. Biol. Chem.*, 250:593–600.
25. Salter, D. W., and Cook, J. S. (1975): Altered glucose transport and metabolism in cultured human cells deprived of glucose. *Biophys. J.*, 15:14a.
26. Salter, D. W., and Cook, J. S. (1976): Reversible independent alterations in glucose transport and metabolism in cultured human cells deprived of glucose. *J. Cell. Physiol.* (*in press*).
27. Chan, P. C., and Sanslone, W. R. (1969): The influence of a low-potassium diet on rat-

erythrocyte-membrane adenosine triphosphatase. *Arch. Biochem. Biophys.*, 134:48–52.
28. Boardman, L., Huett, M., Lamb, J. F., Newton, J. P., and Polson, J. M. (1974): Evidence for the genetic control of the sodium pump density in HeLa cells. *J. Physiol.* (*Lond.*), 241:771–794.
29. Boardman, L., Hume, S. P., Lamb, J. F., McCall, D., Newton, J. P., and Polson, J. M. (1976): Genetic control of sodium pump density. In: *Developmental and Physiological Correlates of Cardiac Muscle*, pp. 127–138, edited by M. Lieberman and T. Sano. Raven Press, New York.
30. Cuff, J. M., and Lichtman, M. A. (1975): Adaptation of potassium metabolism and restoration of mitosis during prolonged treatment of mouse lymphoblasts with ouabain. *J. Cell. Physiol.*, 85:217–226.
31. Chandler, J. P., Hill, D. E., and Spivey, H. O. (1972): A program for efficient integration of rate equations and least-squares fitting of chemical reaction data. *Computers Biomed. Res.*, 5:515–534.
32. Goldberg, A. L. (1972): Degradation of abnormal proteins in *Escherichia coli. Proc. Natl. Acad. Sci. USA*, 69:422–426.
33. Johnson, S., Stokke, T., and Prydz, H. (1975): HeLa cell plasma membranes—changes in membrane protein composition during the cell cycle. *Exp. Cell Res.*, 93:245–251.
34. Arias, I. M., Doyle, D., and Schimke, R. T. (1969): Studies on the synthesis and degradation of proteins of the endoplasmic reticulum of rat liver. *J. Biol. Chem.*, 244:3303–3315.
35. Dehlinger, P. J., and Shimke, R. T. (1971): Size distribution of membrane proteins of rat liver and their relative rates of degradation. *J. Biol. Chem.*, 246:2574–2583.
36. Schmidt-Ullrich, R., Wallach, D. F. H., and Ferber, E. (1974): Concanavalin A augments the turnover of electrophoretically defined thymocyte plasma membrane proteins. *Biochim. Biophys. Acta*, 356:288–299.
37. Roberts, R. M., and Yuan, B. O.-C. (1974): Chemical modification of the plasma membrane polypeptides of cultured mammalian cells as an aid to studying protein turnover. *Biochemistry*, 13:4846–4855.
38. Hubbard, A. L., and Cohn, Z. A. (1975): Externally disposed plasma membrane proteins. II. Metabolic fate of iodinated polypeptides of mouse L cells. *J. Cell Biol.*, 64:461–479.
39. Tsan, M.-F., and Berlin, R. D. (1971): Effect of phagocytosis on membrane transport of nonelectrolytes. *J. Exp. Med.*, 134:1016–1035.
40. Schwartz, R. H., Bianco, A. R., Handwerger, B. S., and Kahn, C. R. (1975): Demonstration that monocytes rather than lymphocytes are the insulin-binding cells in preparations of human peripheral blood mononuclear leucocytes: Implications for studies of insulin-resistant states in man. *Proc. Natl. Acad. Sci. USA*, 72:474–478.
41. Taylor, R. B., Duffus, W. P. H., Raff, M. C., and dePetris, S. (1971): Redistribution and pinocytosis of lymphocyte surface immunoglobulin molecules induced by anti-immunoglobulin antibody. *Nature* [*New Biol.*], 233:225–229.
42. Unanue, E. R., Perkins, W. D., and Karnovsky, M. J. (1972): Ligand-induced movement of lymphocyte membrane macromolecules. I. Analysis by immunofluorescence and ultrastructural radioautography. *J. Exp. Med.*, 136:885–906.
43. Yahara, I., and Edelman, G. M. (1973): The effect of concanavalin A on the mobility of lymphocyte surface receptors. *Exp. Cell Res.*, 81:143–155.

Biogenesis and Turnover of Membrane Macromolecules,
edited by John S. Cook. Raven Press, New York 1976.

Synthesis and Degradation of Mitochondrial Cytochromes

Elliott M. Ross, Michael E. Dockter,* and Gottfried Schatz*

*Department of Pharmacology, University of Virginia School of Medicine, Charlottesville, Virginia 22903 and *Biocenter, University of Basel, Basel, Switzerland*

Mitochondria are synthesized by a close cooperation between two distinct genetic systems—the mitochondrial genetic system and the nucleocytoplasmic genetic system (1–3). The nucleocytoplasmic system manufactures all mitochondrial lipids, the proteins of the mitochondrial outer membrane and of the matrix, most proteins of the mitochondrial inner membrane, and most, if not all, proteins of the mitochondrial genetic system itself. Taken together, these components represent more than 90% of the mitochondrial mass. The mitochondrial system synthesizes probably all of its own nucleic acids and about one dozen hydrophobic polypeptides which are tightly associated with the mitochondrial inner membrane. In lower eukaryotes, three of these polypeptides are associated with cytochrome *c* oxidase, one or two with cytochrome *b,* and two to four with the oligomycin-sensitive ATPase complex (1,2,4,5). The remaining polypeptides have not yet been identified. If the synthesis of mitochondrially made polypeptides is prevented by antibiotics or by mutations, mitochondrial structures are still made, but these structures have lost a functional respiratory chain, as well as an energy-coupling system (1–3). Experiments on mitochondrial biogenesis have thus provided some interesting insights into how two genetic systems may interact at the level of individual polypeptide chains.

In contrast, very little is known about how mitochondria are degraded *in vivo.* Whereas the early studies of Fletcher and Sanadi (6) had suggested that mitochondria turn over as a unit, later experiments revealed that the individual mitochondrial components turn over at widely different rates (7). Virtually nothing is known about the mechanisms governing mitochondrial turnover, particularly the turnover of the tightly membrane-bound mitochondrial proteins. Do the individual subunits of oligomeric membrane-bound complexes turn over at identical rates? How can one explain the degradation of membrane polypeptides which are either buried in the interior of the membrane or facing the matrix side of the mitochondrial inner membrane?

We have decided to approach these questions by studying the synthesis and the degradation of two chemically and functionally defined enzyme complexes of the mitochondrial inner membrane: cytochrome *c* oxidase and cytochrome c_1. On the one hand, these components can be easily monitored by absorption spectroscopy in mitochondria or intact cells and are simple enough to allow detailed chemical and physical investigation. On the other hand, they are important structural components of the mitochondrial inner membrane and any information obtained on these components should be of immediate relevance to the mitochondrial inner membrane as a whole.

In this contribution, we shall first discuss the structure and the synthesis of cytochrome *c* oxidase and cytochrome c_1 in the yeast *Saccharomyces cerevisiae*. Following that, we shall describe the loss of these cytochromes *in vivo* if their synthesis is arrested by lack of oxygen.

STRUCTURE AND SYNTHESIS OF CYTOCHROME *c* OXIDASE

Purified cytochrome *c* oxidase from baker's yeast contains seven dissimilar polypeptide subunits (Fig. 1 and Table 1). The three large subunits are

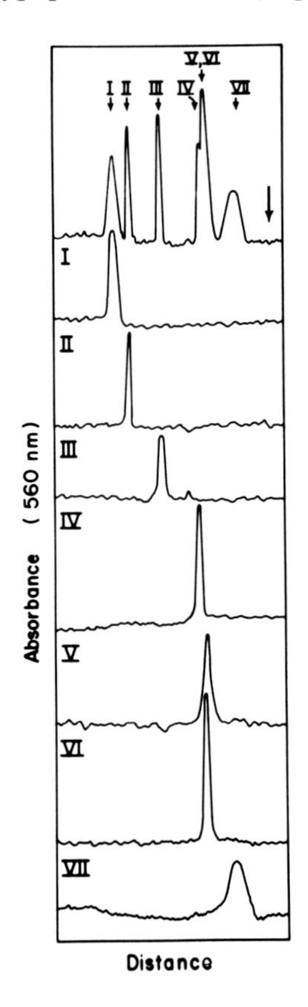

FIG. 1. SDS-acrylamide gel electrophoresis of yeast cytochrome *c* oxidase (*top*) and of the individual, purified subunits (I–VII). (From ref. 10.)

TABLE 1. *Properties of the subunits of yeast cytochrome* c *oxidase*

Properties	Subunit no.						
	I	II	III	IV	V	VI	VII
Molecular weight							
By SDS-acrylamide gel electrophoresis	40,000	33,000	23,500	14,500	13,500	12,500	4,700
By gel filtration in guanidine HCl	41,600	28,300	21,700	14,700	13,100	13,100	4,500
By amino acid analysis	36,930	33,800	—[a]	14,000	12,100	12,100	4,600
N-Terminal amino acid	—[a]	—[a]	—[a]	Glx	Ala	Ser	Ala
C-Terminal amino acid	—[a]	—[a]	—[a]	Leu	Ala	Ala	Lys
Polarity index[b]	35	42	—[a]	48	50	51	41
Site of translation	Mitochondria			Cytoplasm			

[a] Not determined.
[b] Defined as mole percent of hydrophilic amino acids Lys, His, Arg, Asp, Thr, Ser, Glu.

quite hydrophobic and are synthesized on mitochondrial ribosomes, whereas the four small subunits are more hydrophilic and are synthesized on cytoplasmic ribosomes (8–10). The individual subunits have been purified and studied by chemical and immunochemical methods (Table 1). Work on the amino acid sequence of subunit VI is nearly completed.

The arrangement of these subunits in the membrane-bound enzyme was investigated as follows (11,12). In the first step, exposed subunits were coupled to membrane-impermeable radioactive "surface probes." In the second step, the membranes were lysed with cholate and cytochrome *c* oxidase was isolated from the lysate by immunoprecipitation. In the third step, the cytochrome *c* oxidase subunits were separated from each other by SDS-polyacrylamide gel electrophoresis and scanned for radioactivity. These studies indicated that in yeast mitochondria, the two largest mitochondrially made subunits (I and II) are localized in the interior of the membrane, whereas all other subunits are at least partially accessible to externally added reagents.

In order to study the asymmetric arrangement of cytochrome *c* oxidase across the mitochondrial inner membrane, analogous labeling experiments were carried out with bovine heart mitochondria (12). In contrast to yeast mitochondria, bovine heart mitochondria can be prepared intact and can then be inverted by sonication. (The subunit composition of bovine cytochrome *c* oxidase is closely similiar to that of the yeast enzyme, except that the bovine enzyme lacks the second largest subunit.) These experiments showed clearly that the arrangement of cytochrome *c* oxidase in the mitochondrial inner membrane is transmembranous and asymmetric; using the terminology of Table 1, subunits III, VI, and VII are situated on the outer face of the mitochondrial inner membrane, subunit IV is situated on the matrix side, and subunits I and V are buried in the interior of the membrane.

This information is important for understanding the function of cytochrome *c* oxidase and its assembly and degradation in living cells.

EFFECT OF OXYGEN ON THE SYNTHESIS AND THE STABILITY OF CYTOCHROME *c* OXIDASE

The synthesis of cytochrome *c* oxidase requires oxygen (7). If yeast cells are grown under anaerobic conditions, they lose respiring mitochondria and instead accumulate cytochrome- and respiration-deficient promitochondria (13). Promitochondria still contain a functional mitochondrial genetic system and an energy-coupling system; upon aeration of the anaerobic cells, they differentiate into respiring mitochondria (13). We have chosen this system to study the effect of oxygen on the synthesis and the degradation of mitochondrial cytochromes.

Early experiments (9) showed that the accumulation of the two largest cytochrome *c* oxidase subunits is dependent on oxygen (Fig. 2). If aerobically grown, respiration-competent yeast cells are pulse-labeled under non-growing conditions with radioactive leucine in the absence of oxygen, the two largest mitochondrially made subunits are no longer labeled. This "shift-down" experiment shows clearly that oxygen is involved in the assembly of cytochrome *c* oxidase; however, the experiment is complicated

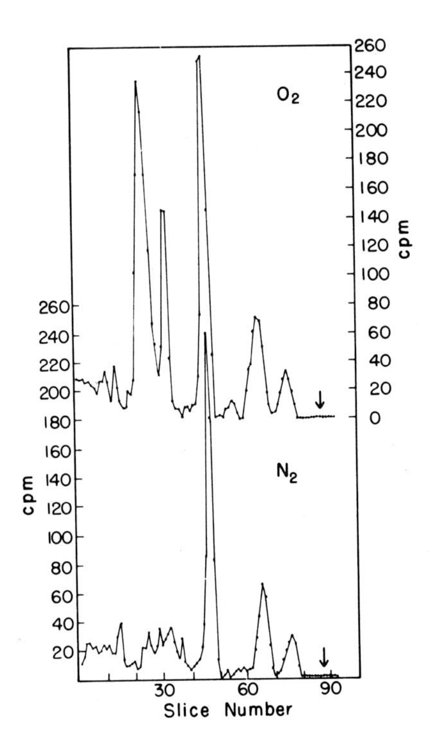

FIG. 2. Effect of anaerobiosis on the synthesis of cytochrome *c* oxidase subunits in the absence of antibiotics. Aerobically grown yeast cells were suspended in glucose-containing phosphate buffer and labeled with [^{3}H]leucine aerobically (*top*) or anaerobically (*bottom*). The mitochondria were then isolated and subjected to immunoprecipitation with an antiserum against the four small cytochrome *c* oxidase subunits. Aliquots of the immunoprecipitates were analyzed by SDS-acrylamide gel electrophoresis. (From ref. 9.)

by the possibility that the cells may have contained excess pools of aerobically made cytochrome *c* oxidase subunits. If analogous experiments are carried out with cells grown anaerobically for at least 20 generations, none of the three mitochondrially made subunits can be detected after pulse-labeling under nitrogen. In contrast, the accumulation of mitochondrially made subunits associated with oligomycin-sensitive ATPase complex is not appreciably affected by lack of oxygen. We are currently trying to establish whether oxygen is necessary for the *synthesis,* or merely for the integration of mitochondrially made cytochrome *c* oxidase subunits.

What is the fate of *preformed* cytochrome *c* oxidase upon shift from aerobic to anaerobic conditions? Earlier studies by Luzikov and colleagues (14) have suggested that the respiratory capacity of yeast mitochondria decreases if the intact cells are incubated anaerobically or in the presence of metabolic inhibitors. More recently, Stone and Wilkie (15) have reported that yeast cells growing on a nonfermentable substrate may lose their preformed cytochrome *c* oxidase upon exposure to inhibitors of mitochondrial protein synthesis. Prompted by these observations, we decided to search for a system which would allow us to study the degradation of preformed cytochrome c oxidase in the absence of any metabolic inhibitors. Such a system is outlined in Fig. 3. Yeast cells growing aerobically on a nonfermentable car-

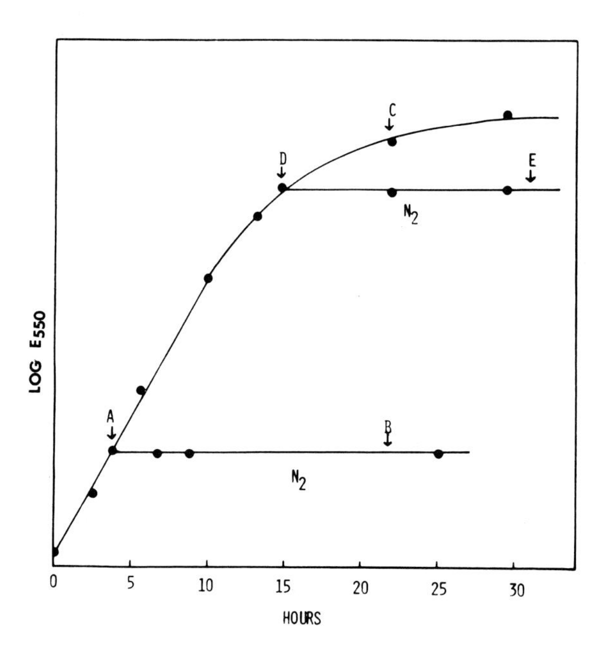

FIG. 3. Effect of anaerobiosis on the degradation of mitochondrial cytochromes. The experimental system. Yeast cells were grown aerobically at 28°C on semisynthetic medium containing 4% glycerol. At A and D, aliquots of the culture were shifted to anaerobic conditions. The final cell density at stationary phase was 3×10^8/ml.

bon source such as glycerol are shifted to anaerobic conditions either in early log phase or in early stationary phase and kept under anaerobiosis for up to 20 hr. At the points indicated in Fig. 3A–E, aliquots of the cells are monitored for mitochondrial cytochromes by low-temperature absorption spectroscopy (Fig. 4). All the aerobically grown cells exhibit a normal cytochrome spectrum consisting of cytochrome *c*, c_1, *b*, and aa_3, regardless of the point in the growth cycle (Fig. 4A,D,C). In contrast, the samples shifted to anaerobiosis exhibit altered spectra, and the alteration is dependent on the point in the growth phase at which the shift has occurred. Cells shifted to anaerobiosis in early log phase lose cytochrome *c* oxidase (and to a lesser extent cytochrome *b*) (Fig. 4B), whereas cells shifted in early stationary phase lose cytochrome *c* (Fig. 4E). Similar results are obtained if anaerobiosis is replaced by the addition of 1 mM cyanide. If growth on the nonfermentable carbon source is arrested by cycloheximide, or by oligomycin, the cytochrome spectra are not noticeably altered. This simple system, which does not require addition of any inhibitors, opens many experimental possibilities for studying the turnover of the individual cytochrome c oxidase subunits and for identifying the enzyme(s) mediating the degradation.

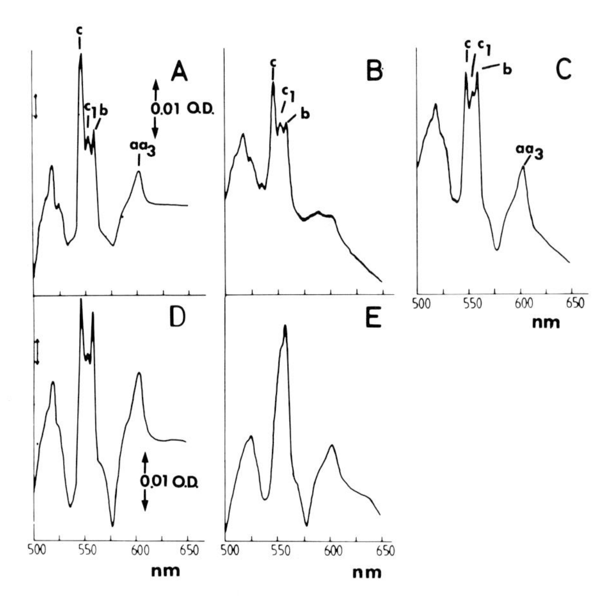

FIG. 4. Effect of anaerobiosis on the degradation of mitochondrial cytochromes. Absorption spectra. At points A–E indicated in Fig. 3, cells were isolated from 2 liters of culture by centrifugation and were washed twice with 0.1 M NaP_i buffer pH 7.4 at 4°C. Packed cells (0.2 g wet wt) were suspended in 1.5 ml of 50% glycerol – 50 mM NaP_i buffer pH 7.4 and dithionite-reduced minus ferricyanide-oxidized difference spectra were recorded at liquid nitrogen temperature with an Aminco DW-2 split beam spectrophotometer at a bandpass of 0.5 nm.

STRUCTURE AND BIOSYNTHESIS OF CYTOCHROME c_1

Cytochrome c_1 has been purified from *Saccharomyces cerevisiae* mitochondria (16,17) by solubilization with cholate, fractionation with ammonium sulfate, disruption of the cytochrome bc_1 complex with mercaptoethanol plus detergents and chromatography on DEAE-cellulose. The final product is spectrally pure (Fig. 5), contains up to 32 nmoles of covalently bound heme per milligram of protein, and does not measurably react with carbon monoxide. SDS disaggregates the purified cytochrome into a single 31,000-dalton subunit which carries the covalently attached heme. Many cytochrome c_1 preparations contain in addition an 18,500-dalton polypeptide which is free of covalently bound heme. However, this polypeptide can be removed from the heme-carrying subunit by ion exchange chromatography or by immunoprecipitating cytochrome c_1 from a mitochondrial extract with an antiserum raised against the heme-polypeptide (Fig. 6). It is therefore still open whether the 18,500-dalton protein is an essential subunit of cytochrome c_1.

Cytochrome c_1 is extremely sensitive to proteolysis. If it is purified in the absence of protease inhibitors, a family of heme polypeptides with molecular weights of 29,000, 27,000, and 25,000 daltons is obtained. In order to show

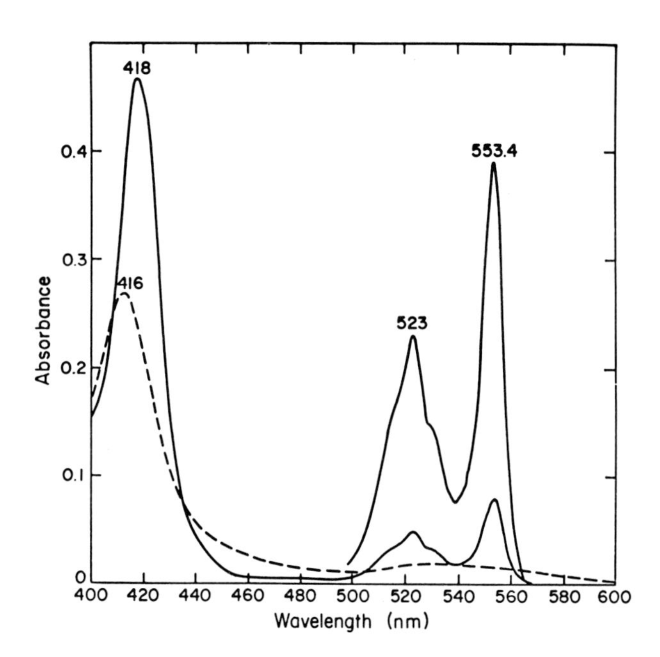

FIG. 5. Absolute room temperature spectra of yeast cytochrome c_1. *Solid lines,* reduced with $Na_2S_2O_4$. *Dashed line,* oxidized with $Na_2S_2O_8$. Protein concentration was 0.11 mg/ml with the lower curves and 0.54 mg/ml with the upper curve. *Lightpath,* 10 mm; 0.1 M NaP_i–0.5 mM EDTA pH 7.5.

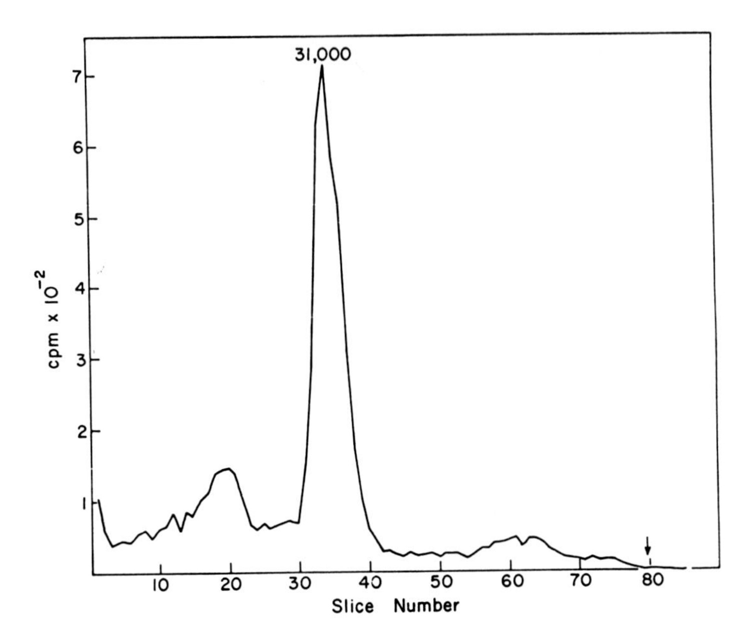

FIG. 6. Gel electrophoresis of immunoprecipitated yeast cytochrome c_1. Yeast cells were grown aerobically in the presence of [^{3}H]leucine and subjected to immunoprecipitation with cytochrome c_1 antiserum. The washed immunoprecipitate was solubilized with SDS and analyzed by SDS-polyacrylamide electrophoresis (17).

that only the 31,000-dalton heme polypeptide is the native species, yeast cells were labeled with the heme precursor [^{3}H]δ-aminolevulinic acid, converted to protoplasts and directly lysed with SDS in the presence of protease inhibitors. Subsequent electrophoresis of the lysate in the presence of SDS shows the covalently bound heme of cytochrome c_1 as a single symmetrical peak at 31,000 daltons.

To determine the site of translation of the heme-carrying polypeptide, yeast cells were labeled with [^{3}H]leucine under the following conditions: (*i*) in the absence of inhibitors; (*ii*) in the presence of acriflavin (an inhibitor of mitochondrial translation); (*iii*) in the presence of cycloheximide (an inhibitor of cytoplasmic translation). The incorporation of radioactivity into the heme protein was measured by immunoprecipitating it from mitochondrial extracts and analyzing it by SDS-polyacrylamide gel electrophoresis. Label was incorporated into the cytochrome c_1 apoprotein only in the presence of acriflavin or in the absence of inhibitor, but not in the presence of cycloheximide. Cytochrome c_1 is thus a cytoplasmic translation product (18). This conclusion was also supported by the observation that a cytoplasmic petite mutant lacking mitochondrial protein synthesis still contained holocytochrome c_1 that was indistinguishable from cytochrome c_1 of wild-type yeast with respect to absorption spectrum, molecular weight, antigenic properties, and the presence of a covalently bound heme group. The concentration of cytochrome c_1 in the mutant mitochondria approached that typical of wild-type mitochondria. However, the lability of the mitochondrially bound cytochrome c_1 to proteolysis was increased, suggesting that a mitochondrial translation product may be necessary for the correct conformation or orientation of cytochrome c_1 in the mitochondrial inner membrane (18).

EFFECT OF OXYGEN ON THE SYNTHESIS OF CYTOCHROME c_1

Promitochondria of anaerobically grown yeast appear to lack holocytochrome c_1, since no distinct absorption peaks attributable to cytochrome c_1 can be detected in these organelles. However, spectral measurements do not give any information on the possible presence of the protein moiety of cytochrome c_1. In an attempt to study this question, promitochondria were isolated from yeast cells grown anaerobically in the presence of [^{3}H]leucine and mixed with mitochondria from cells that had been grown aerobically in the presence of [^{14}C]leucine. The mixed particles were extracted, and the extract was immunoprecipitated with cytochrome c_1 antiserum. An electrophoretic analysis of the dissociated radioactive immunoprecipitate is shown in Fig. 7. No cytochrome c_1 protein is found in the promitochondria under conditions which would have detected less than 10% of the amount present in aerobic mitochondria (18). It is therefore unlikely that cytochrome c_1 or apocytochrome c_1 accumulates in the promitochondria of anaerobically grown yeast.

It remains possible, however, that apocytochrome c_1 is made in the absence of oxygen, but not properly integrated into the promitochondrial membranes. To test this possibility, we assayed for apocytochrome c_1 in whole-cell extracts of anaerobically grown yeast. Wild-type yeast cells were

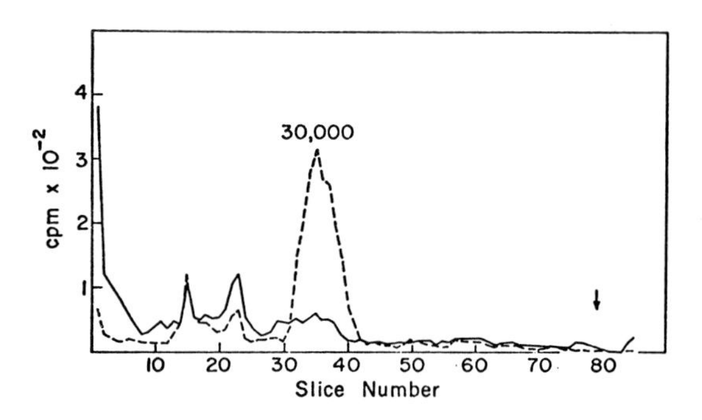

FIG. 7. Promitochondria lack any detectable apocytochrome c_1. Yeast cells were grown anaerobically in the presence of [^{3}H]leucine and broken by shaking with glass beads (8). The promitochondria (specific activity = 2.9×10^7 cpm mg^{-1}) were isolated and mixed with mitochondria (spec. act. = 6.3×10^5 cpm mg^{-1}) from cells that had been grown aerobically in the presence of [^{14}C]leucine. The mitochondrial mixture was subjected to immunoprecipitation with cytochrome c_1 antiserum and the immunoprecipitate was analyzed by SDS-polyacrylamide gel electrophoresis. *Solid line,* ^{3}H counts (promitochondria); *dashed line,* ^{14}C counts (mitochondria).

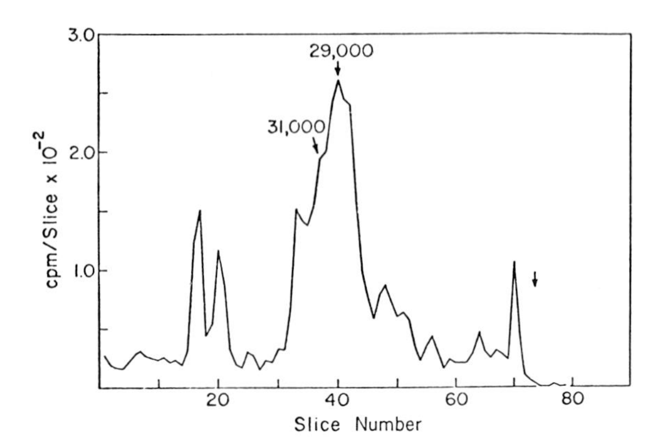

FIG. 8. Tentative evidence for the presence of cytochrome c_1 protein in a whole-cell extract of anaerobically grown yeast. Wild-type cells were grown anaerobically in the presence of [^{3}H]leucine. Spheroplasts were prepared, lysed, and subjected to immunoprecipitation with cytochrome c_1 antiserum. The immunoprecipitate was analyzed by SDS-polyacrylamide gel electrophoresis (18).

grown anaerobically in the presence of [^{3}H]leucine, converted to spheroplasts, and the lysed spheroplasts were immunoprecipitated with cytochrome c_1 antiserum. The analysis of the immunoprecipitate by SDS-acrylamide gel electrophoresis (Fig. 8) suggests that cytochrome c_1 protein may be present in anaerobically grown yeast cells, even though it cannot be detected in promitochondria. In view of the considerable noise level of the experiment depicted in Fig. 8, identification of the material immunoprecipitated from whole-cell extracts is still tentative. Nevertheless, these preliminary studies suggest that the loss of cytochrome c_1 from promitochondria may reflect a defect in assembly rather than a transcriptional or translational control.

Interestingly, closely similar results were obtained by analyzing the cytochrome-free mitochondria from a heme-requiring yeast mutant (19), which was kindly donated to us by Drs. Sprinson and Golub. Again, the defective mitochondria of this mutant did not contain any detectable cytochrome c_1 protein, whereas a whole-cell extract did. Since the synthesis of heme appears to require molecular oxygen, it is possible that the inhibitory effect of anaerobiosis on the incorporation of cytochrome c_1 into mitochondria is mediated via a deficiency of heme. We are presently attempting to use similar immunoprecipitation experiments to ascertain the fate of preformed cytochrome c_1 during an aerobic-to-anaerobic shift as outlined in Fig. 3.

CONCLUSION

These studies serve to emphasize the immense complexity of events which culminate in the synthesis or the degradation of a mitochondrial organelle. Still, we are confident that at least some of these events can be analyzed if they are studied with defined membrane components such as cytochrome *c* oxidase and cytochrome c_1.

ACKNOWLEDGMENTS

This study was supported by Grants GM16320 from the U.S. Public Health Service, GB40541X from the U.S. National Science Foundation, NY(C) 181406 from the U.S. Department of Agriculture, 3.2350.74 from the Swiss National Science Foundation, a predoctoral fellowship by the U.S. National Science Foundation to E.M.R. and a postdoctoral fellowship (ALTF 194–1974) by the European Molecular Biology Organization to M.E.D.

REFERENCES

1. Schatz, G., and Mason, T. L. (1974): The biosynthesis of mitochondrial proteins. *Annu. Rev. Biochem.*, 43:51–87.
2. Tzagoloff, A., Rubin, M. S., and Sierra, M. F. (1973): Biosynthesis of mitochondrial enzymes. *Biochim. Biophys. Acta*, 301:71–104.
3. Borst, P. (1972): Mitochondrial nucleic acids. *Annu. Rev. Biochem.*, 41:333–369.
4. Weiss, H. (1974): Mitochondrial translation of cytochrome b in *Neurospora crassa* and *Locusta migratoria*. *Z. Physiol. Chem.*, 355:300–304.
5. Jackl, G., and Sebald, W. (1975): Identification of two products of mitochondrial protein synthesis associated with mitochondrial adenosine triphosphatase from *Neurospora crassa*. *Eur. J. Biochem.*, 54:97–106.
6. Fletcher, M. J., and Sanadi, D. R. (1961): *Biochim. Biophys. Acta*, 51:356.
7. Schatz, G. (1970): Biogenesis of mitochondria. In: *Membranes of Mitochondria and Chloroplasts*, pp. 251–314, edited by E. Racker. Van Nostrand-Reinhold, New York.
8. Mason, T. L., Poyton, R. O., Wharton, D. C., and Schatz, G. (1973): Cytochrome c oxidase of baker's yeast. I. Isolation and properties. *J. Biol. Chem.*, 248:1346–1354.
9. Mason, T. L., and Schatz, G. (1973): Cytochrome *c* oxidase of baker's yeast. II. Site of translation of the protein components. *J. Biol. Chem.*, 248:1355–1360.
10. Poyton, R. O., and Schatz, G. (1975): Cytochrome *c* oxidase of baker's yeast. III. Physical characterization of the isolated subunits and chemical evidence for two different classes of polypeptides. *J. Biol. Chem.*, 250:752–761.
11. Eytan, G., and Schatz, G. (1975): Cytochrome *c* oxidase of baker's yeast. V. Arrangement of the subunits in the isolated and membrane-bound enzyme. *J. Biol. Chem.*, 250:767–774.
12. Eytan, G. D., Carroll, R. C., Schatz, G., and Racker, E. (1975): Arrangement of the subunits in solubilized and membrane-bound cytochrome *c* oxidase from bovine heart. *J. Biol. Chem.*, 250:8598–8603.
13. Plattner, H., Salpeter, M. M., Saltzgaber, J., and Schatz, G. (1970): Promitochondria of anaerobically grown yeast IV. Conversion into respiring mitochondria. *Proc. Natl. Acad. Sci. USA*, 66:1252–1259.
14. Luzikov, V. N., Zubatov, A. S., Rainina, E., and Bakeyeva, L. E. (1971): Degradation and restoration of mitochondria upon deaeration and subsequent aeration of aerobically-grown *Saccharomyces cerevisiae* cells. *Biochim. Biophys. Acta*, 245:321–334.
15. Stone, A. B., and Wilkie, D. (1976): Loss of cytochrome oxidase in *Saccharomyces cerevisiae* during inhibition of mitochondrial protein synthesis. *J. Gen. Microbiol., in press.*
16. Ross, E., Ebner, E., Poyton, R. O., Mason, T. L., Ono, B., and Schatz, G. (1973): The biosynthesis of mitochondrial cytochromes. In: *The Biogenesis of Mitochondria*, pp. 477–490, edited by A. M. Kroon and C. Saccone. Academic Press, New York.

17. Ross, E., and Schatz, G. (1975): Cytochrome c_1 of baker's yeast. I. Isolation and properties. *J. Biol. Chem.,* 251:1991–1996.
18. Ross, E., and Schatz, G. (1975): Cytochrome c_1 of baker's yeast. II. Synthesis on cytoplasmic ribosomes and influence of oxygen and heme on accumulation of the apoprotein. *J. Biol. Chem.,* 251:1997–2004.
19. Golub, E. G., Trocha, P., Liu, P. K., and Sprinson, D. B. (1974): Yeast mutants requiring ergosterol as the only lipid supplement. *Biochem. Biophys. Res. Commun.,* 56:471–477.

Biogenesis and Turnover of Membrane Macromolecules,
edited by John S. Cook. Raven Press, New York 1976.

Synthesis of the Major Polypeptides of Thylakoid Membranes in *Chlamydomonas reinhardtii* y-1 in Response to Light

J. Kenneth Hoober

Department of Biochemistry, Temple University School of Medicine, Philadelphia, Pennsylvania 19140

The chlorophyll-containing thylakoid membranes within plant cells carry out perhaps the most important functions in nature, namely, the production of oxygen from H_2O and the conversion of light energy into chemical energy. The production of oxygen permits oxidative metabolism, from which is obtained the bulk of the energy used for anabolic reactions in aerobic, nonphotosynthetic organisms. However, oxidative metabolism can proceed only at the cost of consuming available organic substrates, yielding CO_2 and H_2O. Although anaplerotic CO_2 fixation can occur with energy derived from oxidative phosphorylation, the inability to synthesize net amounts of carbohydrate and lipid determines this process as one with diminishing returns. In contrast, the ability of thylakoid membranes to use light energy for the production of ATP and NADPH provides a source of energy independent of oxidative metabolism. These substances permit the reductive incorporation of net amounts of carbon from CO_2 into carbohydrate. Moreover, the CO_2-fixing reaction is catalyzed by an enzyme not found in nonphotosynthetic organisms. The compartmentation of the enzymes within the chloroplast, in juxtaposition to the energy-supplying membranes, leads to an efficient anaplerotic system in eukaryotic plant cells. Therefore, as a result of photosynthesis, the organic pool of carbon is replenished.

Because of the importance of these photosynthetic functions carried out by thylakoid membranes, it is in our vital interest to understand the properties of these membranes. Investigations are proceeding actively toward an understanding of the photosynthetic functions of the membrane (1–5) and of the structural organization of its components (6–11). However, for the functions to be realized, the structures must be assembled from the component parts. Little information of significance has been obtained concerning formation of the structure beyond descriptions that the overall process occurs (12–21). An understanding of the mechanisms by which the components interact during assembly, as well as of the mechanisms by which synthesis of these components is controlled, remains at an embryonic stage.

A study of the assembly of thylakoid membranes is dependent upon the availability of a suitable system. Several types of plant cells have been used for this purpose. The seeds of most higher plants yield etiolated seedlings when germinated in the dark, and chloroplast development has been studied extensively upon exposure of these seedlings to light (20–30). Moreover, conditions which support a linear rate of greening have been developed for higher plants (30). But in species commonly used for these studies, the rate at which this process proceeds is relatively slow (see Table 1). Etiolated strains of a number of single-celled green algae have been isolated, of which most notably those of *Euglena, Ochromonas,* and *Chlorella* also show a relatively slow rate of greening (Table 1). In contrast, an etiolated strain of *Chlamydomonas* exhibits rapid kinetics of greening (Table 1). Intermediate in rate is a mutant strain of *Scenedesmus.* These latter two organisms have been the source of a large number of mutant strains defective in the formation and photosynthetic functions of thylakoid membranes (42–44). But the ability to analyze mutant strains of *Chlamydomonas* genetically, as well as biochemically, has permitted extensive studies on the genetics of photosynthesis in that organism (44). The ease of handling *Chlamydomonas* cells and their rapid rate of growth and division allow simple experimental designs. Finally, under appropriate conditions, etiolated cells of *Chlamydomonas*

TABLE 1. *Approximate rates of greening of various etiolated plant cells*

Organism	Temperature (°C)	Chlorophyll accumulation[a] (μg chlorophyll/mg protein-hr)	Ref.
Higher plants			
Canavalia ensiformis (jack bean)	28	1[b]	30
Phaseolus vulgaris (bean)	25	1[b]	22
Hordeum vulgare (barley)	25	1[b]	20
Algae			
Euglena gracilis	25	0.5–1[c]	19,31,32
Chlorella mutants	25–28	1–2[d]	12,16
Ochromonas danica	29	1[e]	18
Scenedesmus obliquus C-2A′	30	8[f]	17,33
Chlamydomonas reinhardtii y-1	26	12	14,15
Chlamydomonas reinhardtii y-1	38	18	34

[a] The rates were calculated from published values on chlorophyll and protein measurements. Unfortunately, both measurements generally were not made on the same material. The values must be considered only as approximations.

[b] In the absence of values for mg protein/g fresh weight in the greening studies, a value of 50 was used (see ref. 35).

[c] An average value of 3 mg protein/10^7 cells was used (36–39).

[d] A value of 70 μg protein/10^7 cells was used (40,41).

[e] An approximate value for mg protein/10^7 cells was obtained by a comparison of the cell volume of *O. danica* (18) with that of *C. reinhardtii* (13), using a value of 0.32 mg protein/10^7 cells of *C. reinhardtii* (15).

[f] The published value of 2.5 mg dry weight/10 μl packed cell volume was corrected for total carbohydrate content (17). It was assumed that protein comprised about 70% of the remainder.

reinhardtii y-1 also exhibit linear kinetics of membrane formation (34), a factor which, in conjunction with the rapid rate, promotes this organism as a favorable choice for studying the greening process.

Cells of *C. reinhardtii* contain a single, large chloroplast, which in cells grown in continuous light is nearly filled with thylakoid membranes (13). In the y-1 strain, formation of these membranes, but not cell growth, requires light. Following transfer of cultures to the dark, cell growth and division eventually yield cells whose chloroplast is nearly depleted of membranous structures (14). These etiolated cells, which contain less than 5% of the amount of chlorophyll in green cells, respond rapidly when returned to the light. The full complement of chlorophyll in green cells (20–25 μg chlorophyll/10^7 cells) is achieved generally after about 8 hr of exposure to light at 25°C or after about 3 hr at 37°–40°C.

Assembly of functional thylakoid membranes in *C. reinhardtii* y-1 requires the production of chlorophyll, a process dependent upon light (14, 24,27), the synthesis of other membrane lipids and, in particular, glycolipids in the chloroplast (45,46), and the synthesis of proteins on both chloroplastic and cytoplasmic ribosomes (15,47,48). This system can be manipulated easily by a variety of means, including (i) the use of selective inhibitors, (ii) the intensity and quality of light, and (iii) the incubation temperature. The following discussion will review certain aspects of the greening of *C. reinhardtii* y-1, with particular attention to regulation by light of the synthesis of the major polypeptides of thylakoid membranes.

THE THYLAKOID MEMBRANES IN *CHLAMYDOMONAS REINHARDTII*

Thylakoid membranes are composed of unique lipids as well as of unique proteins. The gross composition is about 45% lipid and 55% protein by weight. About 25% by weight of the lipid fraction is chlorophyll, with the remainder comprised largely of mono- and digalactosyldiglycerides (50% of membrane lipids). Sulfolipids, phospholipids, and carotenoids each comprise 3–5% of membrane lipids (49–51). The protein : chlorophyll ratio varies according to the method of preparation; in membranes washed with 8 M urea to remove loosely associated proteins, the ratio of protein to chlorophyll is 4.7 mg/mg (52). An analysis of those proteins, remaining with urea-washed membranes, by polyacrylamide gel electrophoresis in the presence of sodium dodecyl sulfate (53), revealed the pattern shown in Fig. 1. Two polypeptide fractions, designated for reference purposes as *b* and *c*, are predominant in the membranes purified from *C. reinhardtii*. The polypeptides in fractions *b* and *c* have masses of 28,000 ± 1,000 and 24,000 ± 1,000 daltons, respectively (34). Although apparently homogeneous in size, homogeneity in type has not been established.

Some evidence has recently been obtained on the functions of the polypeptides in fractions *b* and *c*. In strain ac-5 of *C. reinhardtii*, a change in

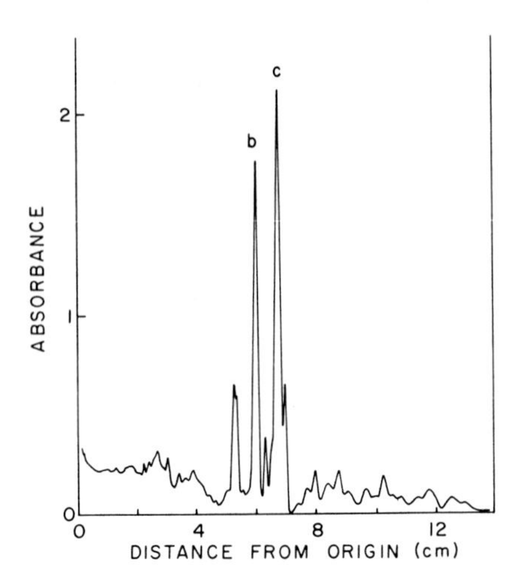

FIG 1. Pattern of absorbance obtained after electrophoresis of thylakoid membrane polypeptides from *C. reinhardtii* y-1. Membranes were purified (53), washed with 8 M urea, and extracted with 90% acetone. The protein was dissolved in 0.1 M Tris-acetate (pH 9.0), containing 2% SDS, 0.5 M urea, and 0.01% sodium ethylenediaminetetraacetate (EDTA), and treated with mercaptoethanol (final concentration, 3%, v/v). The sample was heated at 70°C for 1 min and then incubated for 30 min at 40°C. A 10-μl sample (about 40 μg of protein) was applied to a 10% polyacrylamide gel and subjected to electrophoresis (53) at 7.5 V/cm for 6 hr. The gel was stained with coomassie blue and, after destaining in 7% acetic acid, was scanned at 560 nm.

nuclear DNA causes the coordinate, but conditional, loss of both major polypeptides in thylakoid membranes (54). When grown mixotrophically, these cells are deficient in polypeptides *b* and *c* and in chlorophyll *b*, but nevertheless retain photosynthetic activity. However, membranes lacking these polypeptides show no propensity to form grana (7). Polypeptides *b* and *c* are probably analogous to the single major polypeptide of higher plant thylakoid membranes, which has a mass of 23,000 to 25,000 daltons (55–58). By the judicious use of detergents, this polypeptide was purified as a complex with chlorophyll (59). In higher plants, most if not all the chlorophyll *b* is associated, along with an equal amount of chlorophyll *a*, with this major polypeptide (59,60). Thornber et al. (60,61) have suggested that this complex has a role in harvesting light energy. After fractionation of membranes with detergents, the major polypeptides are found associated with particles rich in photosystem II activity (10,62–65).

On the basis of studies on thylakoid membranes of *C. reinhardtii* by electron microscopy, Kretzer (9) suggested that these membranes have a lipoprotein structure devoid of extensive lipid bilayer regions. Although less extreme, a similar interpretation was made of the images produced by freeze-fracture (6,7). The latter process revealed a dense array of large particles (160–175 Å in diameter) within the interior of the membrane (6,7,66). Such particles apparently result from the presence of membrane proteins; they can be produced experimentally by the introduction of proteins into lipid bilayers (67,68). Because of the marked deficiency of these particles in membranes of mixotrophically grown cells of *C. reinhardtii* ac-5 (7), Levine and Duram (54) suggested that the large particles in thylakoid membranes are an expression of the presence of polypeptides *b* and *c*. From the results shown in Fig. 1, these polypeptides are appropriate candidates for the components responsible for formation of the intramembranous particles. But the dimensions of a single polypeptide the size of *b* or *c* are insufficient to account for the size of the particles (8). Evidence for an association of these potential subunits to form larger proteins has not been obtained, but Menke and Ruppel (56) reported on the preparation from *Antirrhinum* chloroplasts of a protein, molecular weight 470,000 at pH 3, which dissociated into subunits of 25,000 daltons in the presence of sodium dodecyl sulfate. Nevertheless, the situation remains unclear. Henriques and Park (69) reported that no deficiency in intramembranous particles was observed in a chlorophyll *b*-less mutant of barley, which, like the ac-5 strain of *C. reinhardtii,* lacks the major polypeptide of thylakoid membranes (69,70).

ASSAY OF SYNTHESIS OF THE MAJOR POLYPEPTIDES OF THYLAKOID MEMBRANES

In green cells of *C. reinhardtii* y-1 the integral proteins of thylakoid membranes (see Fig. 1) account for about 25% of the cellular protein (34). An electrophoretic analysis of *total* protein of green cells provided the absorbance pattern shown in Fig. 2. Two prominent peaks were present in the pattern corresponding to the major membrane polypeptides *b* and *c*. As expected, these peaks were absent in scans of gels after electrophoresis of total protein from etiolated cells, which contain only small amounts of thylakoid membranes (13,14).

Synthesis of the major polypeptides of thylakoid membranes during greening has been studied by the incorporation of radioactivity from labeled acetate or arginine (47,48,53,71–73). Recently, it was found advantageous to use labeled leucine as a tracer (34). Although there is apparently no facili-

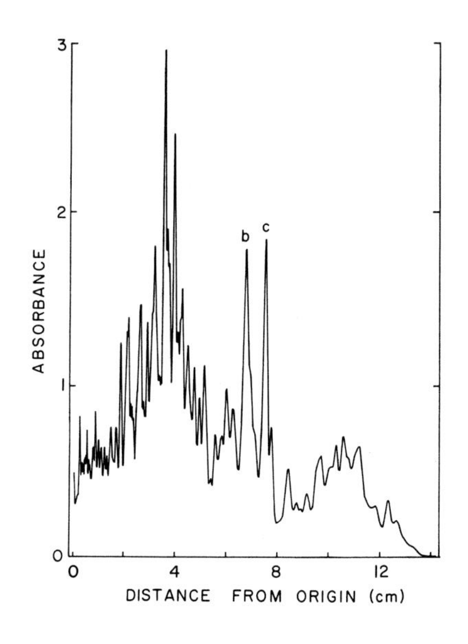

FIG. 2. Pattern of absorbance obtained after electrophoresis of total protein from green cells of *C. reinhardtii* y-1. 2×10^7 cells (about 0.8 mg protein) were dissolved in 0.1 ml of the Tris buffer-SDS solution described under Fig. 1. A 10-μl sample was applied to a 10% polyacrylamide gel and subjected to electrophoresis at 7.5 V/cm for 6 hr. The gel was stained and scanned as described under Fig. 1.

tated mechanism for the transport of this amino acid across the plasma membrane, sufficient leucine entered the cells when the extracellular concentration was relatively high. The incorporation of labeled leucine also was favored by an increase in the temperature of incubation (34). Figure 3A illustrates a pattern of radioactivity obtained after electrophoresis of total protein of greening cells, which were labeled during the first hour of exposure to light at 38°C. Peaks in the pattern corresponding to polypeptides *b* and *c*, identified by comparison with a pattern of radioactivity for the membrane fraction of such cells (Fig. 3B), were prominant and well resolved. The ability to resolve these polypeptides by electrophoresis of *total* cellular

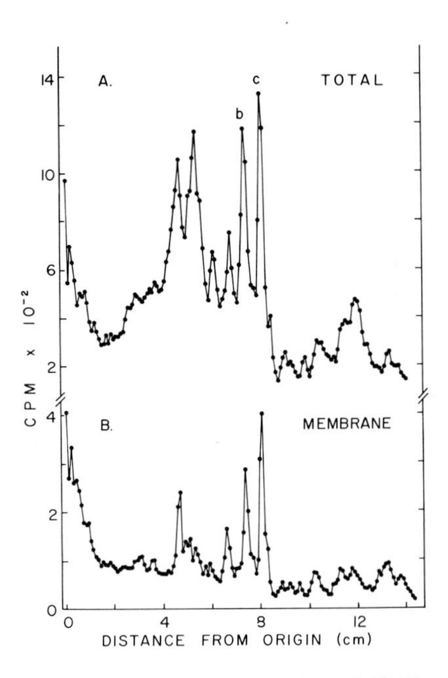

FIG. 3. Electrophoretic analysis of the incorporation of [^{3}H]leucine into total protein of greening *C. reinhardtii* y-1 cells **(A)** and into protein of the membrane fraction from greening cells **(B).** Etiolated cells, grown in the dark at 25°C, were suspended to a density of 6 × 10^6 cells/ml, incubated with shaking in the dark for 2 hr at 38°C and then exposed to light from incandescent lamps (about 80 J/m^2-sec after passage through a 7-cm H_2O filter). L-[^{3}H]leucine (0.7 Ci/mmole) was added to the cell suspension to a concentration of 3 × 10^{-5} M at the time light was given, and the incubation was continued for another hour. **A:** For preparation of total protein, cells were collected in 10% trichloroacetic acid and washed 2 times with 5% trichloroacetic acid and 1 time with H_2O. The sample (2.4 × 10^7 cells) was dissolved as described in Fig. 2. A 30-μl portion was applied to an 8% polyacrylamide gel and subjected to electrophoresis at 6.5 V/cm for 6 hr. After staining, the gel was sliced into 1-mm sections, each of which was digested with 0.1 ml 30% H_2O_2 at 60°C overnight and counted with 5 ml of toluene-Triton X100-Omnifluor scintillation fluid (53). **B:** For preparation of the membrane fraction, 2.4 × 10^7 cells were washed 2 times with 20 mM Tris-HCl (pH 7.6), and then broken by sonication. The sample was centrifuged for 40 min at 80,000 × *g*. The pellet was prepared for and subjected to electrophoresis as described for total protein.

proteins and the relatively extensive incorporation of [^{3}H]leucine into *b* and *c* have enabled analysis of the synthesis of these polypeptides without resorting to purification of the membrane. Electrophoresis of total protein for the analysis also eliminated potential difficulties in interpretation of results in those instances in which synthesis of the polypeptides, but not formation of membranes, occurs (see ref. 71).

SITE OF SYNTHESIS OF THE MAJOR POLYPEPTIDES OF THYLAKOID MEMBRANES

Cells of *C. reinhardtii* have, as do other eukaryotic plant cells, chloroplastic ribosomes distinct in character from cytoplasmic ribosomes (74–76). Figure 4 shows the results of an analysis of total ribosomes of *C. reinhardtii* y-1 by sedimentation through a sucrose gradient (74). Two types of ribosomes are present, a 70S population contributed by the chloroplast, and an 82S population from the cytoplasmic matrix. The differences in these ribosomes have been exploited in the manner illustrated in Fig. 4. The 70S chloroplastic ribosomes are inhibited by chloramphenicol and other inhibitors of prokaryotic-type ribosomes, whereas the 82S cytoplasmic ribosomes are sensitive to cycloheximide. The inhibitory actions of the antibiotics were observed in this experiment as a decrease in the amount of monomers in the samples, resulting from a "freezing" of polysomes as the inhibitors halted movement of ribosomes along mRNA strands. The selective nature of each antibiotic is clearly demonstrated. This specificity in the

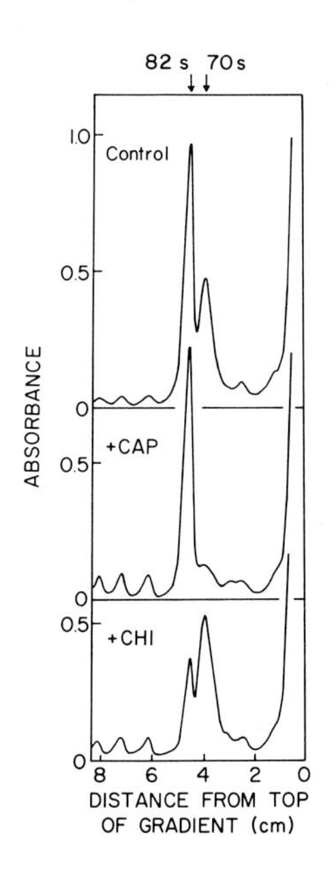

FIG. 4. Effects of chloramphenicol and cycloheximide on sedimentation of ribosomes from *C. reinhardtii* y-1. Etiolated cells were suspended to a density of 6×10^6 cells/ml, divided into three 100-ml portions, and exposed to light at 25°C. One portion served as a control, the second received chloramphenicol (final concentration, 100 μg/ml) after 4.5 hr in the light, while the third received cycloheximide (final concentration, 10 μg/ml) after 5.5 hr in the light. At 20 min after the addition of cycloheximide, the cells were collected and broken by passage through a French pressure cell. The ribosomes in a $10{,}000 \times g$ supernatant fluid of the broken cell preparation of each portion were analyzed on exponential sucrose gradients as described previously (74). CAP, chloramphenicol; CHI, cycloheximide.

action of each antibiotic has provided the basis for an investigation of the ribosomes involved in the synthesis of specific polypeptide components of the thylakoid membrane.

To determine which type of ribosome was involved in synthesizing polypeptides *b* and *c*, selective inhibitors of chloroplastic or cytoplasmic ribosomes were added to cultures of etiolated *C. reinhardtii* y-1 cells, which then were exposed to light. As the concentration of chloramphenicol was increased in increments from 10 to 200 μg/ml, the synthesis of chlorophyll was progressively inhibited (15,47,71). Synthesis of the large subunit of ribulose 1,5-diphosphate carboxylase, a convenient marker for the activity of chloroplastic ribosomes (72,77,78), was strongly inhibited by chloramphenicol (71,72). But no inhibition of the synthesis of *b* or *c* was observed with chloramphenicol or with streptomycin, a more effective inhibitor of chloroplastic ribosomes *in vitro* than chloramphenicol (76). As shown in Fig. 5, cells treated with streptomycin synthesized fully as much *b* and *c* after exposure to light as did control cells (compare patterns A and B).

On the other hand, cycloheximide completely blocked the accumulation of chlorophyll when added at concentrations of 2–5 μg/ml (15,47). And, as Fig. 5 shows, cycloheximide also inhibited the synthesis of polypeptides *b* and *c* (compare patterns A and C). In the presence of cycloheximide, in-

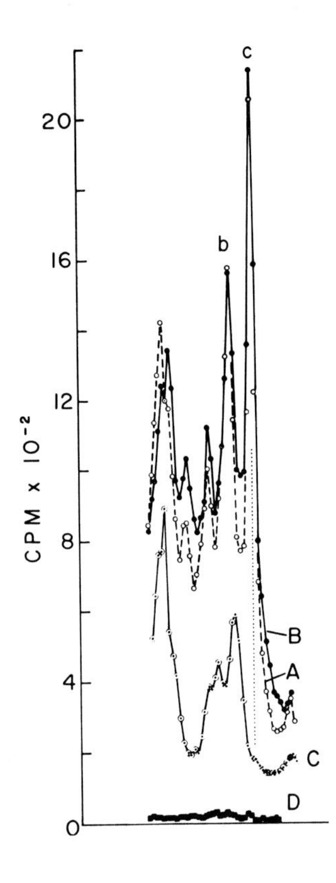

FIG. 5. Effects of streptomycin and cycloheximide on the synthesis of the major polypeptides of thylakoid membranes during greening of etiolated *C. reinhardtii* *y*-1 cells at 38°C. Dark-grown cells were suspended to 6×10^6 cells/ml and 4-ml portions were incubated with shaking in the dark at 38°C for 2 hr. Streptomycin (final concentration, 120 μg/ml) and cycloheximide (final concentration, 20 μg/ml) were added 1.5 hr and 10 min, respectively, before the end of the dark period. [^{3}H]leucine (0.7 Ci/mmole) then was added to a concentration of 3×10^{-5} M, and the cells were exposed to light as described under Fig. 3. After 1 hr of incubation in the light, the cells were collected and the total protein was subjected to electrophoresis on 10% polyacrylamide gels as described in Fig. 3. The patterns of radioactivity are for the regions of the gels encompassing the positions of the major membrane polypeptides (refer to Fig. 3). Pattern A, control cells; pattern B, streptomycin-treated cells; pattern C, cycloheximide-treated cells; and pattern D, cells treated with both streptomycin and cycloheximide.

corporation of [^{3}H]leucine into several polypeptides was observed, one of which migrated in 11% polyacrylamide gels at about the same rate during electrophoresis as did polypeptide *b*. However, in 8% gels, this labeled polypeptide migrated more rapidly than *b* and near the rate at which *c* moved through the gel. But the relative rates of migration of *b* and *c* during electrophoresis were not affected by a change in gel concentration. Therefore, on the basis of this analysis, the inhibition of synthesis of polypeptides *b* and *c* by cycloheximide was essentially complete. The cycloheximide-resistant incorporation of [^{3}H]leucine (Fig. 5, pattern C) was abolished when cells were treated with streptomycin in addition to cycloheximide (pattern D). This cycloheximide-resistant protein synthesis apparently had occurred on chloroplastic ribosomes.

The conclusion from the effects of these antibiotics was that the major polypeptides of thylakoid membranes in *C. reinhardtii* are synthesized on cytoplasmic ribosomes (34,53). Confirming evidence which supports a cytoplasmic site of synthesis of the major polypeptide of the membrane in higher plants was obtained by Machold and Aurich (79), also with antibiotics, and by Eaglesham and Ellis (80), who observed that the major membrane polypeptide was not synthesized by isolated chloroplasts *in vitro*.

KINETICS OF CHLOROPHYLL ACCUMULATION DURING GREENING OF *C. REINHARDTII* Y-1

When etiolated cells of *C. reinhardtii* y-1, grown in the dark at 25°C, were transferred to the light at the same temperature, a lag of 1–2 hr in duration was observed before significant accumulation of chlorophyll began (Fig. 6, curve A). The lag was considerably shortened, and a higher maximal rate of chlorophyll synthesis was achieved, if the temperature of the culture was increased to 37–40°C at the same time light was given (Fig. 6, curve B). But, as Fig. 6 also shows, the lag was eliminated entirely if the cells were incubated in the dark at the higher temperature for 2 hr prior to exposing the cells to light. Under the latter conditions, the accumulation of chlorophyll began immediately and proceeded at a linear rate during the first hour in the light (see Fig. 6, inset). The initial rate of chlorophyll synthesis under these conditions was 6–7 μg chlorophyll/10^{7} cells-hr, or about 18 μg chlorophyll/mg protein-hr. This rate is an order of magnitude greater than the rate observed during greening of most other etiolated plant cells (Table 1).

KINETICS OF THE SYNTHESIS OF MEMBRANE PROTEINS DURING GREENING OF *C. REINHARDTII* Y-1

Eytan and Ohad (47,48) first observed that synthesis of the major polypeptides of thylakoid membranes occurred in parallel with the accumulation of chlorophyll at 25°C. I have confirmed these results and have found that

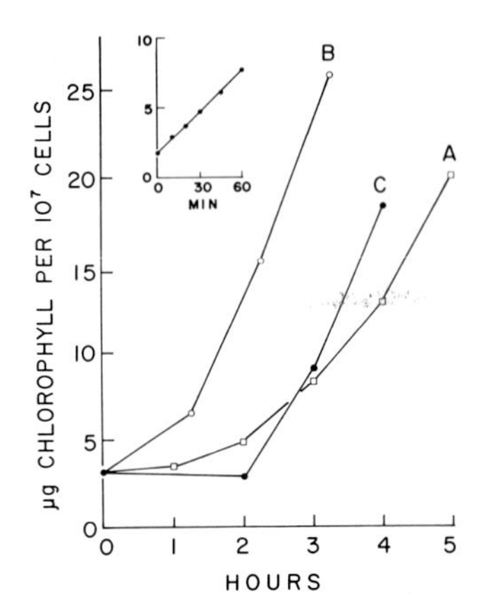

FIG. 6. Effect of temperature on the rate of accumulation of chlorophyll. Etiolated cells, grown in the dark at 25°C, were suspended to 4×10^6 cells/ml and treated as follows. Sample A: cells were exposed to light (about 15 J/m^2-sec from daylight fluorescent lamps) at 25°C throughout. Sample B: cells were placed in a water bath at 37°C at zero time and simultaneously exposed to the light. Sample C: cells were placed in a water bath at 37°C at zero time and maintained in the dark; at 2 hr they were exposed to the light until the end of the experiment. *Inset:* as in C, the cells had been preincubated at 37°C for 2 hr; zero time in this plot represents the beginning of the subsequent light exposure. Aliquots of each sample were removed periodically, and the chlorophyll content of the cells was determined spectrophotometrically (81).

coordinate synthesis of these membrane components also prevails at higher temperatures (34). The results of an experiment in which etiolated *C. reinhardtii* y-1 cells were incubated at 38°C for 2 hr in the dark before exposure to [^{3}H]leucine and to light are shown in Fig. 7. The incorporation of [^{3}H]leucine into cellular proteins, *including* that into *b* and *c*, was linear with time.

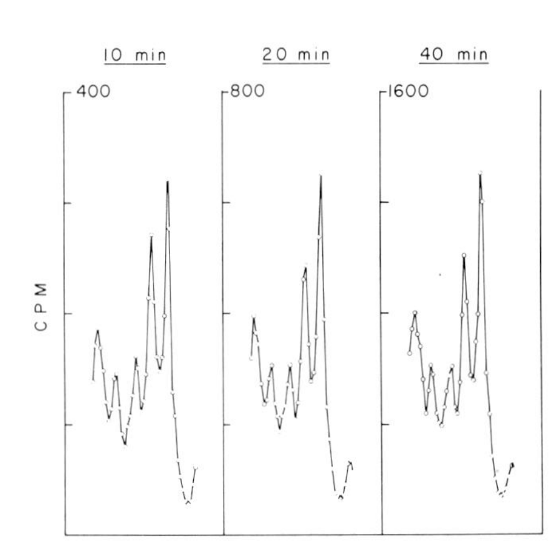

FIG. 7. Kinetics of the synthesis of polypeptides *b* and *c* during greening of etiolated cells of *C. reinhardtii* y-1 at 38°C. Dark-grown cells were suspended to 6×10^6 cells/ml and incubated with shaking at 38°C. After 2 hr in the dark, [^{3}H]leucine was added and the cells were exposed to light as described under Fig. 3. At the times indicated, an aliquot of the cell suspension was removed, the cells were collected and the total protein was subjected to electrophoresis as described in Fig. 3. Patterns of radioactivity for the region of the gels containing the major membrane polypeptides were plotted on ordinates corresponding to the length of time cells were in the light. Thus, identical patterns (relative peak heights) indicate linear incorporation of [^{3}H]leucine with time.

The sample taken at 10 min of illumination contained an amount of radioactivity in *b* and *c* one-half that found after 20 min in light. Experiments of this type revealed that at the higher temperature, synthesis of *b* and *c*, as of chlorophyll, began immediately when cells were exposed to light and proceeded at a linear rate.

REGULATION BY LIGHT OF THE SYNTHESIS OF THE MAJOR POLYPEPTIDES OF THYLAKOID MEMBRANES

It has been amply demonstrated in *C. reinhardtii* y-1, as well as in other plant cells which become etiolated in the dark, that light is required for the formation of thylakoid membranes. Not only is light required for synthesis of chlorophyll in these cells (27,82–85), it is also required for synthesis of the major polypeptides of the membrane (47,48,71–73). The involvement of light in the synthesis of chlorophyll is in general understood as the photoreduction of protocholorophyllide (24,82–84), although the mechanism of this reaction is not known. But little information is available regarding the role of light in the synthesis of the membrane polypeptides.

The most effective wavelength of light in the red region of the visible spectrum for the production of polypeptides *b* and *c* in *C. reinhardtii* y-1 at 38°C is about 650 nm, the wavelength also most effective in the synthesis of chlorophyll (Fig. 8). The reason for this coincidence, whether chlorophyll synthesis is required for the production of the polypeptides or vice versa, is not known. In cells that were exposed briefly to light and then returned to the dark, chlorophyll synthesis stopped abruptly, but synthesis of the polypeptides declined slowly over a period of 2 hr (48,52). In contrast, when greening cells were treated with cycloheximide to inhibit synthesis of the polypeptides in the light, the accumulation of chlorophyll again stopped abruptly (15), even though there apparently is no direct inhibition of chlorophyll synthesis by cycloheximide (87). These results suggest that the accumulation of chlorophyll may be dependent upon the synthesis of membrane polypeptides. Nevertheless, the spectral requirements suggest that the absorption of light by chlorophyll or its precursors may play a role in the synthesis of the polypeptides.

Investigations into the role of light and chlorophyll in the synthesis of the membrane polypeptides have attempted to draw conclusions from the kinetics of greening. The lag in the appearance of polypeptides *b* and *c* at 25°C suggested that light may be involved at the transcriptional level, and that a period of 1–2 hr is required before the mRNA becomes available for translation (34,48,72). The photoreduction of protochlorophyllide, an event which apparently happens in the chloroplast (88), was implicated in earlier studies as a possible factor in the control mechanism (72). However, it has not been established whether the genes for *b* and *c* are part of chloroplastic or nuclear DNA. But for translation to occur on cytoplasmic ribosomes, the

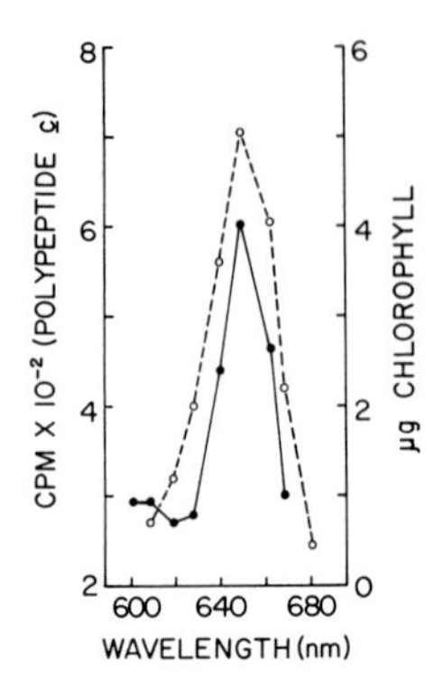

FIG. 8. Spectral requirements in the red region of the spectrum for the synthesis of chlorophyll and of polypeptide *c*. Portions of a suspension of etiolated cells (6×10^6 cells/ml) were placed into 50-ml beakers wrapped with black tape. The cells were incubated 2 hr in the dark at 38°C, and then illuminated with light transmitted by three-cavity interference filters (11 nm half-width; Ditric Optics, Inc., Marlboro, Mass.) mounted on top of the beakers. The intensity of light transmitted by the filters was approximately 1.4 J/m^2-sec. Chlorophyll in 10-ml portions of the cell suspension was analyzed spectrophotometrically (86) after 1 hr in the light and the values were corrected for the initial level present in the cells (0.6 μg of chlorophyll/10^7 cells). To assay synthesis of membrane polypeptides, 4-ml portions were treated as above. The cells were labeled with [^{3}H] leucine for 1 hr in the light as described in Fig. 3, and total protein of each sample was subjected to electrophoresis. The amount of radioactivity in the position of polypeptide *c* was determined from the patterns of radioactivity for illuminated samples and corrected for radioactivity in the same position in a dark-labeled sample. Chlorophyll, (●); polypeptide *c*, (○).

mRNA must exit from the responsible organelle. Therefore, the kinetics of greening at 25°C were consistent with a hypothesis for control exercised at the level of transcription.

A completely different situation emerged from experimental results obtained at 38°C. At the higher temperature synthesis of *b* and *c* began immediately upon illumination (Fig. 7). Because of the apparent physical separation of the transcriptional and translational events, it again was expected that a lag of some length would occur if light affected transcription of the mRNA for *b* and *c*. Even in bacterial cells, in which transcription and translation are coupled, at 37°C a lag of about 3 min ensues between the addition of an inducer and attainment of the maximal rate of synthesis of the induced protein (89–91). The absence of any lag in the synthesis of *b* and *c* in *C. reinhardtii* y-1 indicated that the mRNA for these polypeptides was already available at the time cells were exposed to light. Therefore, the higher temperature apparently permitted synthesis of this mRNA in the dark.

The kinetic evidence for the availability of the mRNA at the time cells were exposed to light was confirmed by experiments with actinomycin D at 38°C. The concentration of actinomycin D used in these experiments, 30 μg/ml, was sufficient to lower, within 10 min, total RNA synthesis to

15–20% of that observed in untreated cells (34). If the drug was added near the end of the 2-hr incubation period in the dark, i.e., 10 min before light was given, synthesis of *b* and *c* was not diminished during an hour of further incubation in the light. Treated cells made as much *b* and *c* as did control cells, even though some depression of the synthesis of most cellular proteins was observed in the presence of actinomycin D. However, as an indication of the effectiveness of the drug, if actinomycin D was added near the beginning of the 2-hr incubation in the dark, general protein synthesis was markedly diminished by the time cells were illuminated. Under these latter conditions, synthesis of *b* and *c* did not occur when cells were exposed to light (34).

These data prompted the conclusion that, at the higher temperatures, light is required, not for transcription of the mRNA for *b* and *c,* but for a later step in the synthetic process. Because the evidence indicated that the mRNA was transcribed in the dark, is there evidence that the mRNA was translated? An electrophoretic analysis of the polypeptides synthesized in the dark at 37–40°C revealed a component not observed in cells incubated at 25°C (Fig. 9, arrow). However, neither *b* nor *c* was synthesized in the dark. The characteristics of the synthesis of the new polypeptide, about 27,-000 daltons in mass, suggested that a relationship may exist between it and polypeptides *b* and *c,* and that its appearance was a consequence of transcription of the mRNA for *b* and *c* in the dark. From an analysis similar to that described in Fig. 5, it was concluded that the site of synthesis of this polypeptide also was on cytoplasmic ribosomes. Synthesis of this new polypeptide exhibited a lag, becoming detectable only after an hour of incu-

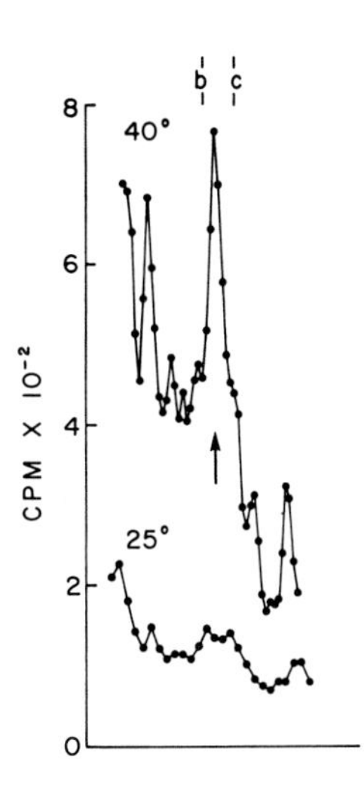

FIG. 9. Effects of temperature on the incorporation of [^{3}H]leucine in the dark. Etiolated cells were suspended to 5×10^6 cells/ml and incubated in the dark for 2 hr either at 25°C or at 40°C. [^{3}H]leucine (0.7 Ci/mmole) was added to a concentration of 2.5×10^{-5} M and the incubations were continued for an additional hour. Samples of total protein were subjected to electrophoresis as described under Fig. 3. The arrow points to a peak in the pattern for cells incubated at 40°C that was not observed in cells kept at 25°C. The positions of polypeptides *b* and *c,* determined from gels of a companion sample labeled in the light, are indicated. The gel containing the 40°C sample was cut into 1-mm slices, whereas the gel containing protein from cells incubated at 25°C was cut into 2-mm slices. Only midportions (about 5–9 cm from the origin) of the patterns of radioactivity are shown.

bation in the dark at the higher temperatures. Moreover, a trend was apparent in these experiments, with greater amounts of this polypeptide synthesized with increasing temperatures. Although a considerable amount of the 27,000-dalton component was synthesized in the dark at 40°C (Fig. 9), synthesis of this polypeptide was not detected if [^{3}H]leucine was added at the time cells were exposed to light (34).

If this apparent reciprocal relationship between synthesis of the 27,000-dalton polypeptide and of polypeptides *b* and *c* can be supported by further experimental evidence, it would indicate that light is not necessary for translation of the mRNA for *b* and *c*. In addition, a lack of involvement of light in the translational stage is consistent with an apparent physical separation between the chloroplastic pigments which presumably absorb the light and the cytoplasmic ribosomes on which the mRNA is translated. There is, therefore, no evidence that would support, nor any logical reason to invoke, an involvement of light at the translational level during synthesis of the major membrane polypeptides.

It is clear from a comparison of Figs. 7 and 9 that production of *b* and *c* requires light. Since the evidence does not favor a role of light in transcription or translation of the mRNA for polypeptides *b* and *c*, are there post-translational events that possibly could require light? It is not known if polypeptides *b* or *c* are primary products of translation, or if they are derived from other, perhaps larger, precursor molecules. However, the results of several types of experiments provide hints that proteolysis may be involved in processing these major membrane polypeptides, in particular, polypeptide *c*. First, Table 2 shows the amino acid compositions of the polypeptides after purification by electrophoresis (34). The compositions are expressed as residues of each amino acid per molecule, assuming that each fraction contains predominantly polypeptides homogeneous in type. The remarkable similarity in the compositions is consistent with the suggestion that the two fractions are related, either as both being derived from a common presursor or with polypeptide *c* originating as a result of proteolysis of polypeptide *b*. Second, when etiolated cells were exposed to light in the presence of chloramphenicol (200 μg/ml) at 25°C, formation of thylakoid membranes was strongly inhibited, but synthesis of *b* and *c* on cytoplasmic ribosomes was not affected by the antibiotic. In such experiments an excess of polypeptide *c* accumulated in the soluble fraction of the cell, but a commensurate accumulation of *b* was not observed (71,72). A possible explanation of these results is that, since polypeptide *b* was not rescued by incorporation into membranes, it was degraded to yield a soluble form of polypeptide *c*. Third, the extensive synthesis of a single, new polypeptide, about 27,000 daltons in mass, in the dark at higher temperatures (Fig. 9) suggested that a single type of mRNA is transcribed for the membrane polypeptides, if indeed these polypeptides are related. Because production of this new polypeptide decreased immediately upon exposure of cells to

TABLE 2. *Amino acid composition of chloroplast membrane polypeptides* b *and* c

	(residues/molecule)	
Amino acid	*b*	*c*
Gly	31	28
Ala	30	25
Leu	26	25
Asx	24	20
Glx	23	20
Pro	17	17
Phe	16	15
Thr	15	10
Lys	15	10
Val	13	9
Ser	12	8
Ile	9	9
Tyr	9	8
Arg	8	7
Met	4	4
Trp	4	4
His	3	3
½ Cys	1	1

Data from Ref. 34.

light (34), it may represent an intermediate in the processing of precursors or an abortive product of proteolysis produced only in the absence of light.

A PUZZLE

Considerably more data must be obtained before definite conclusions can be drawn for a proteolytic step in the production of the major membrane polypeptides. Furthermore, a mechanism by which a protease could be controlled by light is not obvious. For light to be effective, it must be absorbed by a pigment, which presumably is located in the chloroplast. The spectrum of the effective wavelengths is relatively narrow, with a maximum at about 650 nm at 38°C (Fig. 8). But at least two candidates are present in the chloroplast which could respond to light of this wavelength. The first of these is protochlorophyllide.

In higher plants, photoreduction of the protochlorophyllide holochrome exhibits an action spectrum with a maximum at 650 nm (82). The accumulation of chlorophyll during greening also shows the same maximum (24). This action spectrum matches the absorption properties of the protochlorophyllide holochrome. Under some conditions a second, minor, form of photoconvertible protochlorophyllide also is present, which absorbs maximally at about 635 nm (92).

Variable results have been obtained on the spectral requirements for greening of etiolated algae. McLeod et al. (93) obtained a maximum at 650 nm in the action spectrum for the initial accumulation of chlorophyll in an etiolated mutant strain of *C. reinhardtii.* An etiolated strain of *Chlorella* also responded maximally to light of 650 nm (83). However, Ohad (94) found that, if greening *C. reinhardtii* y-1 cells were exposed to light transmitted by filters after a period of exposure to white light at 25°C, continued accumulation of chlorophyll and of membrane proteins exhibited a maximum at about 630 nm. In *Euglena,* the absorption spectrum of etiolated cells has a maximum at 635 nm (95), and the action spectrum for the accumulation of chlorophyll in this algae also has a maximum between 630 and 635 nm (96). Possibly, the variant maxima for the action spectra for greening of these algae result from variations in the absorption properties of the same pigment, protochlorophyllide holochrome, in different environments.

A second candidate for the agent capable of producing the spectra shown in Fig. 8 is chlorophyll *b,* of which small amounts remain in etiolated *C. reinhardtii* y-1 cells (13,45). Chlorophyll *b* maximally absorbs red light of 650–652 nm *in vivo* (97,98). Most, if not all, the chlorophyll *b* in thylakoid membranes is bound, along with an equal amount of chlorophyll *a,* to the major membrane polypeptide (60,61). The functional roles of chlorophyll *b* have not been elucidated, but it is generally designated as an accessory, light-harvesting pigment to photosystem II (3). In this or perhaps another capacity, it is reasonable to expect that chlorophyll *b* would be capable of rapid, photosensitized reduction–oxidation reactions, the same as chlorophyll *a.* By this means, chlorophyll *b* possibly could perform work as the result of the absorption of light. However, the characteristics of the spectral requirements (Fig. 8) would seem to indicate a specialized form of chlorophyll *b,* distinct from cooperative reactions with chlorophyll *a.* No evidence has been obtained for such a role for chlorophyll *b* in the process of greening, but there also is no conclusive evidence to the contrary.

Assuming that reduction of protochlorophyllide most likely is the photoevent that controls production of polypeptides *b* and *c* in *C. reinhardtii,* it remains difficult to reconcile these two processes. Phototransformable protochlorophyllide is bound to a protein, the holochrome, and is itself changed as it absorbs light. From the kinetics of synthesis of polypeptides *b* and *c,* this change must immediately be translated into activation of other reactions. Not enough is known of the properties of the holochrome to determine if this protein is directly involved in production of polypeptides *b* and *c*. But if it were, it would be expected that production of polypeptides *b* and *c* would stop abruptly when illumination is discontinued. Instead, at both 25°C and 38°C a gradual decrease in the production of polypeptides *b* and *c* occurs over a period of about 2 hr after light is removed (48,52, and *unpublished results*).

An alternate possibility is that the product of the photoreaction, chlorophyllide *a,* promotes a reaction involved in the production of polypeptides *b* and *c*. As these polypeptides apparently bind both chlorophylls *a* and *b*, control of the production of the polypeptides may be related to this binding and to the subsequent synthesis of chlorophyll *b*. Such a suggestion is supported by the existence of mutants of barley and of *C. reinhardtii*, which are deficient in the major polypeptides of the thylakoid membranes as well as in chlorophyll *b* (54,70).

Thus we conclude that at 37–40°C light is involved only in the final stage in the synthesis of the major polypeptides of thylakoid membranes. But the mechanism of this reaction is unknown, and the available data resemble pieces of a puzzle for which the central pattern-forming piece is missing. Further work is necessary to determine with certainty the effective light receptor in this process and how the absorption of light by this receptor influences the activity of other reactions.

It should be pointed out that additional control mechanisms might be operative at 25°C. At the lower temperature, mRNA for the major membrane polypeptides seems not to be made in the dark. A lag of 1–2 hr is found before synthesis of these polypeptides can be detected in the light. Therefore, a mechanism for transcriptional control also might exist (52,72), which becomes nonfunctional as the temperature is raised.

Because polypeptides *b* and *c* are stable and undergo little degradation within the membrane (73), if a proteolytic event is involved, it necessarily would have to occur prior to incorporation of the polypeptides into the membrane. Nothing is known of the process by which these polypeptides enter the chloroplast during or after translation on cytoplasmic ribosomes, but the actions of proteases have been observed in other instances in which proteins are transferred across membranes during synthesis (99,100), or are involved in assembly of specific structures (101). Numerous proteolytic enzymes have been found in plant cells (102), to which useful functions have not been ascribed.

The major polypeptides of thylakoid membranes apparently exist as a complex with chlorophyll in the membranes, as mentioned earlier. This complex might form as soon as the polypeptides enter the chloroplast. Specific aggregation of this complex with other lipids and proteins synthesized within the chloroplast then results in the development of a functional membrane structure.

SUMMARY

Thylakoid membranes of *C. reinhardtii* y-1 contain two major polypeptides, whose production requires a light-dependent reaction. With the development of conditions for linear and rapid kinetics of greening of *C. reinhardtii* y-1, an examination of these light-dependent events can now be launched. The chlorophyll *b* in thylakoid membranes is apparently complexed with these major polypeptides, along with an equal amount of chlorophyll *a*. The synthesis of chlorophyll seems to be involved in controlling the production of the membrane polypeptides. Yet the accumulation of chlorophyll under normal greening conditions appears to be dependent upon the synthesis of membrane polypeptides. Therefore, a situation may exist in which the components of a complex association reciprocally affect production of each other. Assembly of the membrane may follow in due course from the specific interactions of these components, as their syntheses are triggered by light.

ACKNOWLEDGMENT

This investigation was supported by grants from the National Science Foundation, GB-32411 and BMS 71–01550 A02.

REFERENCES

1. Ke, B. (1973): The primary electron acceptor of photosystem I. *Biochim. Biophys. Acta,* 301:1–33.
2. Trebst, A. (1974): Energy conservation in photosynthetic electron transport of chloroplasts. *Annu. Rev. Plant Physiol.,* 25:423–458.
3. Radmer, R., and Kok, B. (1975): Energy capture in photosynthesis: Photosystem II. *Annu. Rev. Biochem.,* 44:409–433.
4. Shapiro, S. L., Kollman, V. H., and Campillo, A. J. (1975): Energy transfer in photosynthesis: pigment concentration effects and fluorescent lifetimes. *FEBS Lett.,* 54:358–362.
5. Govindjee (editor) (1975): *Bioenergetics of Photosynthesis.* Academic Press, New York.
6. Branton, D. (1969): Membrane structure. *Annu. Rev. Plant Physiol.,* 20:209–238.
7. Goodenough, U. W., and Staehelin, L. A. (1971): Structural differentiation of stacked and unstacked chloroplast membranes. Freeze-etch electron microscopy of wild-type and mutant strains of *Chlamydomonas. J. Cell Biol.,* 48:594–619.
8. Kirk, J. T. O. (1971): Chloroplast structure and biogenesis. *Annu. Rev. Biochem.,* 40:161–196.
9. Kretzer, F. (1973): Molecular architecture of the chloroplast membranes of *Chlamydomonas reinhardi* as revealed by high resolution electron microscopy. *J. Ultrastruct. Res.,* 44:146–178.
10. Klein, S. M., and Vernon, L. P. (1974): Protein composition of spinach chloroplasts and their photosystem I and photosystem II subfractions. *Photochem. Photobiol.,* 19:43–49.
11. Anderson, J. M. (1975): Possible location of chlorophyll within chloroplast membranes. *Nature,* 253:536–537.
12. Bryan, G. W., Zadylak, A. H., and Ehret, C. F. (1967): Photoinduction of plastids and of chlorophyll in a *Chlorella* mutant. *J. Cell Sci.,* 2:513–528.
13. Ohad, I., Siekevitz, P., and Palade, G. E. (1967): Biogenesis of chloroplast membranes. I. Plastid dedifferentiation in a dark-grown algal mutant (*Chlamydomonas reinhardi*). *J. Cell Biol.,* 35:521–552.
14. Ohad, I., Siekevitz, P., and Palade, G. E. (1967): Biogenesis of chloroplast membranes. II. Plastid differentiation during greening of a dark-grown algal mutant (*Chlamydomonas reinhardi*). *J. Cell Biol.,* 35:553–584.
15. Hoober, J. K., Siekevitz, P., and Palade, G. E. (1969): Formation of chloroplast membranes in *Chlamydomonas reinhardi* y-1. Effects of inhibitors of protein synthesis. *J. Biol. Chem.,* 244:2621–2631.
16. Herron, H. A., and Mauzerall, D. (1972): The development of photosynthesis in a greening mutant of *Chlorella* and an analysis of the light saturation curve. *Plant Physiol.,* 50:141–148.

17. Senger, H., and Bishop, N. I. (1972): The development of structure and function in chloroplasts of greening mutants of *Scenedesmus,* I. Formation of chlorophyll. *Plant Cell Physiol.,* 13:633–649.
18. Smith-Johannsen, H., and Gibbs, S. P. (1972): Effects of chloramphenicol on chloroplast and mitochondrial ultrastructure in *Ochromonas danica. J. Cell Biol.,* 52:598–614.
19. Schiff, J. A. (1973): The development, inheritance and origin of the plastid in *Euglena. Adv. Morphogen.,* 10:265–312.
20. Henningsen, K. W., and Boynton, J. E. (1974): Macromolecular physiology of plastids. IX. Development of plastid membranes during greening of dark-grown barley seedlings. *J. Cell Sci.,* 15:31–55.
21. Robertson, D., and Laetsch, W. M. (1974): Structure and function of developing barley plastids. *Plant Physiol.,* 54:148–159.
22. Wolff, J. B., and Price, L. (1960): The effects of sugars on chlorophyll biosynthesis in higher plants. *J. Biol. Chem.,* 235:1603–1608.
23. Miller, R. A., and Zalik, S. (1965): Effect of light quality, light intensity and temperature on pigment accumulation in barley seedlings. *Plant Physiol.,* 40:569–574.
24. Ogawa, T., Inoue, Y., Kitajima, M., and Shibata, K. (1973): Action spectra for biosynthesis of chlorophylls *a* and *b* and β-carotene. *Photochem. Photobiol.,* 18:229–235.
25. Alberte, R. S., Thornber, J. P., and Naylor, A. W. (1972): Time of appearance of photosystems I and II in chloroplasts of greening jack bean leaves. *J. Exp. Bot.,* 23:1060–1069.
26. Lagoutte, B., and Duranton, J. (1972): The action of light at the structural protein level on etiolated plastids from *Zea mays* L. *FEBS Lett.,* 28:333–336.
27. Gassman, M. L. (1973): The conversion of photoinactive $protochlorophyllide_{633}$ to phototransformable $protochlorophyllide_{650}$ in etiolated bean leaves treated with δ-aminolevulinic acid. *Plant Physiol.,* 52:590–594.
28. Rebeiz, C. A., Crane, J. C., Nishijima, C., and Rebeiz, C. C. (1973): Biosynthesis and accumulation of microgram quantities of chlorophyll by developing chloroplasts *in vitro. Plant Physiol.,* 51:660–666.
29. Remy, R. (1973): Pre-existence of chloroplast lamellar proteins in wheat etioplasts. Functional and protein changes during greening under continuous or intermittent light. *FEBS Lett.,* 31:308–312.
30. Alberte, R. S., Fiscus, E. L., and Naylor, A. W. (1975): The effects of water stress on the development of the photosynthetic apparatus in greening leaves. *Plant Physiol.,* 55: 317–321.
31. Smillie, R. M., Evans, W. R., and Lyman, H. (1963): Metabolic events during the formation of a photosynthetic from a nonphotosynthetic cell. *Brookhaven Symp. Biol.,* 16: 89–108.
32. Bishop, D. G., and Smillie, R. M. (1970): The effect of chloramphenicol and cycloheximide on lipid synthesis during chloroplast development in *Euglena gracilis. Arch. Biochem. Biophys.,* 137:179–189.
33. Oh-hama, T., and Senger, H. (1975): The development of structure and function in chloroplasts of greening mutants of *Scenedesmus.* III. Biosynthesis of δ-aminolevulinic acid. *Plant Cell Physiol.,* 16:395–405.
34. Hoober, J. K. (1976): Kinetics and regulation of synthesis of the major polypeptides of thylakoid membranes in *Chlamydomonas reinhardtii* y-1. *J. Cell Biol., in press.*
35. Mego, J. L., and Jagendorf, A. T. (1961): Effect of light on growth of Black Valentine bean plastids. *Biochim. Biophys. Acta,* 53:237–254.
36. Brawerman, G., Pogo, A. O., and Chargaff, E. (1962): Induced formation of ribonucleic acids and plastid protein in *Euglena gracilis* under the influence of light. *Biochim. Biophys. Acta,* 55:326–334.
37. Kempner, E. S., and Miller, J. H. (1965): The molecular biology of *Euglena gracilis,* I. Growth conditions and cellular composition. *Biochim. Biophys. Acta,* 104:11–17.
38. Carell, E. F., Johnston, P. L., and Christopher, A. R. (1970): Vitamin B_{12} and the macromolecular composition of *Euglena. J. Cell Biol.,* 47:525–530.
39. Woodward, J., and Merrett, M. J. (1975): Induction potential for glyoxylate cycle enzymes during the cell cycle of *Euglena gracilis. Eur. J. Biochem.,* 55:555–559.
40. McCullough, W., and John, P. C. L. (1972): A temporal control of the *de novo* synthesis

of isocitrate lyase during the cell cycle of the eucaryote *Chlorella pyrenoidosa. Biochim. Biophys. Acta,* 296:287–296.
41. Wanka, F., Moors, J., and Krijzer, F. N. C. M. (1972): Dissociation of nuclear DNA replication from concomitant protein synthesis in synchronous cultures of *Chlorella. Biochim. Biophys. Acta,* 269:153–161.
42. Pratt, L. H., and Bishop, N. I. (1968): Chloroplast reactions of photosynthetic mutants of *Scenedesmus obliquus. Biochim. Biophys. Acta,* 153:664–674.
43. Bishop, N. I., and Wong, J. (1971): Observations on photosystem II mutants of *Scenedesmus:* Pigments and proteinaceous components of the chloroplasts. *Biochim. Biophys. Acta,* 234:433–445.
44. Levine, R. P., and Goodenough, U. W. (1970): The genetics of photosynthesis and of the chloroplast in *Chlamydomonas reinhardi. Annu. Rev. Genet.,* 4:397–408.
45. Goldberg, I., and Ohad, I. (1970): Biogenesis of chloroplast membranes, IV. Lipid and pigment changes during synthesis of chloroplast membranes in a mutant of *Chlamydomonas reinhardi* y-1. *J. Cell Biol.,* 44:563–571.
46. Jacobson, B. S., Jaworski, J. G., and Stumpf, P. K. (1974): Fat metabolism in higher plants, LXII. Stearyl-acyl carrier protein desaturase from spinach chloroplasts. *Plant Physiol.,* 54:484–486.
47. Eytan, G., and Ohad, I. (1970): Biogenesis of chloroplast membranes. VI. Cooperation between cytoplasmic and chloroplast ribosomes in the synthesis of photosynthetic lamellar proteins during the greening process in a mutant of *Chlamydomonas reinhardi* y-1. *J. Biol. Chem.,* 245:4297–4307.
48. Eytan, I., and Ohad, I. (1972): Biogenesis of chloroplast membranes. VIII. Modulation of chloroplast lamellae composition and function induced by discontinuous illumination and inhibition of ribonucleic acid and protein synthesis during greening of *Chlamydomonas reinhardi* y-1 mutant cells. *J. Biol. Chem.,* 247:122–129.
49. Kirk, J. T. O., and Tilney-Bassett, R. A. E. (1967): *The Plastids. Their Chemistry, Growth and Inheritance.* Freeman, London.
50. Vernon, L. P., and Shaw, E. R. (1971): Subchloroplast fragments: Triton X-100 method. *Methods Enzymol.,* 23:277–289.
51. Beck, J. C., and Levine, R. P. (1973): Synthesis of chloroplast membrane lipids in *Chlamydomonas reinhardi. J. Cell Biol.,* 59:20a.
52. Hoober, J. K., and Stegeman, W. J. (1975): Regulation of chloroplast membrane synthesis. In: *Genetics and Biogenesis of Mitochondria and Chloroplasts,* pp. 225–251, edited by C. W. Birky, Jr., T. J. Byers, and P. S. Perlman. Ohio State University Press, Columbus, O.
53. Hoober, J. K. (1970): Sites of synthesis of chloroplast membrane polypeptides in *Chlamydomonas reinhardi* y-1. *J. Biol. Chem.,* 245:4327–4334.
54. Levine, R. P., and Duram, H. A. (1973): The polypeptides of stacked and unstacked *Chlamydomonas reinhardi:* Chloroplast membranes and their relation to photosystem II activity. *Biochim. Biophys. Acta,* 325:565–572.
55. Lagoutte, B., and Duranton, J. (1971): Physicochemical study of structural proteins of chloroplasts from *Zea mays* L. *Biochim. Biophys. Acta,* 253:232–239.
56. Menke, W., and Ruppel, H.-G. (1971): Molekulargewicht, Grosse und Gestalt von Proteinen der Thylakoidmembran. *Z. Naturforsch.,* 26b:825–831.
57. Henriques, F., Vaughan, W., and Park, R. (1975): High resolution gel electrophoresis of chloroplast membrane polypeptides. *Plant Physiol.,* 55:338–339.
58. Machold, O. (1975): On the molecular nature of chloroplast thylakoid membranes. *Biochim. Biophys. Acta,* 382:494–505.
59. Kung, S. D., and Thornber, J. P. (1971): Photosystem I and II chlorophyll–protein complexes of higher plant chloroplasts. *Biochim. Biophys. Acta,* 253:285–289.
60. Brown, J. S., Alberte, R. S., and Thornber, J. P. (1974): Comparative studies on the occurrence and spectral composition of chlorophyll–protein complexes in a wide variety of plant material. In: *Proceedings of the 3rd International Congress on Photosynthesis,* pp. 1951–1962, edited by M. Avron. Elsevier, Amsterdam.
61. Thornber, J. P. (1975): Chlorophyll-proteins: Light-harvesting and reaction center components of plants. *Annu. Rev. Plant Physiol.,* 26:127–158.
62. Levine, R. P., Burton, W. G., and Duram, H. A. (1972): Membrane polypeptides associated with photochemical systems. *Nature* [*New Biol.*], 237:176–177.

63. Anderson, J. M., and Levine, R. P. (1974): Membrane polypeptides of some higher plant chloroplasts. *Biochim. Biophys. Acta,* 333:378–387.
64. Jennings, R. C., and Eytan, G. (1973): Biogenesis of chloroplast membranes. XIV. Inhomogeneity of membrane protein distribution in photosystem particles obtained from *Chlamydomonas reinhardi* y-1. *Arch. Biochem. Biophys.,* 159:813–820.
65. Nolan, W. G., and Park, R. B. (1975): Comparative studies on the polypeptide composition of chloroplast lamellae and lamellar fractions. *Biochim. Biophys. Acta,* 375:406–421.
66. Arntzen, C. J., Dilley, R. A., and Crane, F. L. (1969): A comparison of chloroplast membrane surfaces visualized by freeze-etch and negative staining techniques, and ultrastructural characterization of membrane fractions obtained from digitonin-treated spinach chloroplasts. *J. Cell Biol.,* 43:16–31.
67. Hong, K., and Hubbell, W. L. (1972): Preparation and properties of phospholipid bilayers containing rhodopsin. *Proc. Natl. Acad. Sci. USA,* 69:2617–2621.
68. Segrest, J. P., Gulik-Krzywicki, T., and Sardet, C. (1974): Association of the membrane-penetrating polypeptide segment of the human erythrocyte MN-glycoprotein with phospholipid bilayers. I. Formation of freeze-etch intramembrane particles. *Proc. Natl. Acad. Sci. USA,* 71:3294–3298.
69. Henriques, F., and Park, R. B. (1975): Further chemical and morphological characterization of chloroplast membranes from a chlorophyll *b*-less mutant of *Hordeum vulgare. Plant Physiol.,* 55:763–767.
70. Thornber, J. P., and Highkin, H. R. (1974): Composition of the photosynthetic apparatus of normal barley leaves and a mutant lacking chlorophyll *b. Eur. J. Biochem.,* 41:109–116.
71. Hoober, J. K. (1972): A major polypeptide of chloroplast membranes of *Chlamydomonas reinhardi.* Evidence for synthesis in the cytoplasm as a soluble component. *J. Cell Biol.,* 52:84–96.
72. Hoober, J. K., and Stegeman, W. J. (1973): Control of the synthesis of a major polypeptide of chloroplast membranes in *Chlamydomonas reinhardi. J. Cell Biol.,* 56:1–12.
73. Beck, D. P., and Levine, R. P. (1974): Synthesis of chloroplast membrane polypeptides during synchronous growth of *Chlamydomonas reinhardtii. J. Cell Biol.,* 63:759–772.
74. Hoober, J. K., and Blobel, G. (1969): Characterization of the chloroplastic and cytoplasmic ribosomes of *Chlamydomonas reinhardi. J. Mol. Biol.,* 41:121–138.
75. Bourque, D. P., Boynton, J. E., and Gillham, N. W. (1971): Studies on the structure and cellular location of various ribosomes and ribosomal RNA species in the green alga *Chlamydomonas reinhardi. J. Cell Sci.,* 8:153–183.
76. Chua, N.-H., Blobel, G., and Siekevitz, P. (1973): Isolation of cytoplasmic and chloroplast ribosomes and their dissociation into active subunits from *Chlamydomonas reinhardtii. J. Cell Biol.,* 57:798–814.
77. Blair, G. E., and Ellis, R. J. (1973): Protein synthesis in chloroplasts, I. Light-driven synthesis of the large subunit of Fraction I protein by isolated pea chloroplasts. *Biochim. Biophys. Acta,* 319:223–234.
78. Bottomley, W., Spencer, D., and Whitfeld, P. R. (1974): Protein synthesis in isolated spinach chloroplasts: comparison of light-driven and ATP-driven synthesis. *Arch. Biochem. Biophys.,* 164:106–117.
79. Machold, O., and Aurich, O. (1972): Sites of synthesis of chloroplast lamellar proteins in *Vicia faba. Biochim. Biophys. Acta,* 281:103–112.
80. Eaglesham, A. R. J., and Ellis, R. J. (1974): Protein synthesis in chloroplasts, II. Light-driven synthesis of membrane proteins by isolated pea chloroplasts. *Biochim. Biophys. Acta,* 335:396–407.
81. Arnon, D. (1949): Copper enzymes in isolated chloroplasts. Polyphenol oxidases in *Beta vulgaris. Plant Physiol.,* 24:1–15.
82. Koski, V. M., French, C. S., and Smith, J. H. C. (1951): The action spectrum for the transformation of protochlorophyll to chlorophyll *a* in normal and albino corn seedlings. *Arch. Biochem. Biophys.,* 31:1–17.
83. Bogorad, L. (1966): The biosynthesis of chlorophylls. In: *The Chlorophylls,* pp. 481–510. Edited by L. P. Vernon and G. R. Seely. Academic Press, New York.
84. Gassman, M. L. (1973): A reversible conversion of phototransformable protochlorophyll(ide)$_{650}$ to photoinactive protochlorophyll(ide)$_{633}$ by hydrogen sulfide in etiolated bean leaves. *Plant Physiol.,* 51:139–145.

85. Nielsen, O. F., and Kahn, A. (1973): Kinetics and quantum yield of photoconversion of protochlorophyll(ide) to chlorophyll(ide) *a*. *Biochim. Biophys. Acta,* 292:117–129.
86. Vernon, L. P. (1960): Spectrophotometric determination of chlorophyll and pheophytins in plant extracts. *Anal. Chem.,* 32:1144–1150.
87. Gassman, M., and Bogorad, L. (1967): Studies on the regeneration of protochlorophyllide after brief illumination of etiolated bean leaves. *Plant Physiol.,* 42:781–784.
88. Rebeiz, C. A., and Castelfranco, P. A. (1973): Protochlorophyll and chlorophyll biosynthesis in cell-free systems from higher plants. *Annu. Rev. Plant Physiol.,* 24:129–172.
89. Pardee, A. B., and Prestidge, L. S. (1961): The initial kinetics of enzyme induction. *Biochim. Biophys. Acta,* 49:77–88.
90. Kepes, A. (1969): Transcription and translation in the lactose operon of *Escherichia coli* studied by *in vivo* kinetics. *Prog. Biophys. Mol. Biol.,* 19:201–236.
91. Piovant, M., and Lazdunski, C. (1975): Different cyclic adenosine 3′,5′-monophosphate requirements for induction of β-galactosidase and tryptophanase. Effect of osmotic pressure on intracellular cyclic adenosine 3′,5′-monophosphate concentrations. *Biochemistry,* 14:1821–1825.
92. Kahn, A., and Nielsen, O. F. (1974): Photoconvertible protochlorophyll(ide)$_{635/650}$ *in vivo:* a single species or two species in dynamic equilibrium? *Biochim. Biophys. Acta,* 333:409–414.
93. McLeod, G. C., Hudock, G. A., and Levine, R. P. (1963): The relation between pigment concentration and photosynthetic capacity in a mutant of *Chlamydomonas reinhardi.* In: *Photosynthetic Mechanisms in Green Plants,* pp. 400–408. Publication 1145, Nat. Acad. Sci.–Nat. Res. Council, Washington, D.C.
94. Ohad, I. (1975): Cytoplasm-chloroplast interrelationship during biogenesis of chloroplast membranes in *Chlamydomonas reinhardi.* In: *Nucleocytoplasmic Relationships during Cell Morphogenesis in Some Unicellular Organisms,* edited by S. Puiseu-Dao. Elsevier, Amsterdam.
95. Zeldin, M. H., Cohen, C. E., Ben-Shaul, Y., and Schiff, J. A. (1975): Measurement *in vivo* of light-induced spectroscopic changes of protochlorophyll(ide) and chlorophyll(ide) in *Euglena. Plant Physiol.,* 56 (suppl.):33.
96. Egan, J. M., Jr., Dorsky, D., and Schiff, J. A. (1975): Events surrounding the early development of *Euglena* chloroplasts. VI. Action spectra for the formation of chlorophyll, lag elimination in chlorophyll synthesis, and appearance of TPN-dependent triose phosphate dehydrogenase and alkaline DNase activities. *Plant Physiol.,* 56:318–323.
97. Cederstrand, C. N., Rabinowitch, E., and Govindjee. (1966): Analysis of the red absorption band of chlorophyll *a in vivo. Biochim. Biophys. Acta,* 126:1–12.
98. Brown, J. S. (1972): Forms of chlorophyll *in vivo. Annu. Rev. Plant Physiol.,* 23:73–86.
99. Steiner, D. F., Kemmler, W., Tager, H. S., and Peterson, J. D. (1974): Proteolytic processing in the biosynthesis of insulin and other proteins. *Fed. Proc.,* 33:2105–2115.
100. Schechter, I., McKean, D. J., Guyer, R., and Terry, W. (1975): Partial amino acid sequence of the precursor of immunoglobulin light chain programmed by messenger RNA *in vitro. Science,* 188:160–162.
101. Hershko, A., and Fry, M. (1975): Post-translational cleavage of polypeptide chains: role in assembly. *Annu. Rev. Biochem.,* 44:775–797.
102. Ryan, C. A. (1973): Proteolytic enzymes and their inhibitors in plants. *Annu. Rev. Plant Physiol.,* 24:173–196.

Biogenesis and Turnover of Membrane Macromolecules,
edited by John S. Cook. Raven Press, New York 1976.

The Properties and Biosynthesis of RNA Associated with Surface Membranes of L Cells

Mary Catherine Glick

Department of Pediatrics, University of Pennsylvania School of Medicine, Children's Hospital of Philadelphia, Philadelphia, Pennsylvania 19104

Specific populations of RNA associated with the surface membrane could be instrumental in providing a rapid mechanism for protein synthesis, either for membrane proteins or for proteins participating in cellular events (1). One can postulate that the RNA could be activated by specific hormone-like substances, thus participating in the cellular functions brought about by external stimuli. It has been suggested that the transcription of mRNA is linked to surface membrane changes induced by cyclic AMP (2).

Surface membranes from mammalian cells have been shown by many laboratories to contain 3–10% of the RNA of the cell. These results have been summarized (3). Most authors consider this RNA as a contaminant; however, others suggest a meaningful role and have examined it by various techniques. To summarize some of these studies, the RNA appears to be of two populations. First, ribosomal RNA, similar to that found in cytoplasmic ribosomes (4–7), and second, an RNA or polynucleotide associated with the outside of the cell (8–12). Whether or not some of the latter RNA represents viral RNA present at the cell surface, normally or artifactually, remains to be elucidated. The first population of RNA, ribosomal, appears to participate in protein synthesis as surface membranes from a number of cell types incorporate amino acids into proteins (1,12–14). Recently, platinum-pyrimidine complexes have been used to demonstrate the presence of nucleic acid material, probably DNA, on the outer surface of tumor cells (15), substantiating a previous finding of DNA in surface membranes from lymphoid cells (16).

It therefore seems pertinent to present data characterizing the RNA associated with the surface membrane of the L cell, a mouse fibroblast. The data suggest two populations: (*i*) a unique, rapidly labeled heterogeneous RNA, and (*ii*) ribosomal RNA similar to that found within the cells.

METHODS

Growth of L Cells

Unless otherwise stated, L cells were grown and harvested as described (1). Cultures were examined at routine intervals for *Mycoplasma* and were negative. In the experiments requiring radioactive RNA, the cells were grown in the presence of [^{3}H]-5-uridine (24–28 Ci/mmole) or [^{14}C]-2-uridine (25 mCi/mmole) obtained from New England Nuclear Corporation, Boston, Mass. The time of growth in the presence of the isotope was 24 hr, unless specified with the individual experiments. The amount of isotope used was 0.5 μCi/ml of culture medium for [^{3}H]uridine and 0.25 μCi/ml of culture medium for [^{14}C]uridine, unless otherwise specified.

Preparation of Cell Fractions

When comparisons were made, surface membranes, cytoplasmic ribosomes, nuclei, and supernatant fractions were obtained from the same culture of L cells. Proteins were determined on each preparation of all cell fractions by the method of Lowry et al. (17). In some experiments, RNA was determined as described for these fractions (1) after hydrolysis with LiOH (18). Phosphate was assayed by the method of Bartlett (19). All radioactivities were determined by dissolving the fractions or whole cells in NCS (Amersham/Searle, Des Plaines, Ill.) and after the addition of liquifluor-toluene scintillation fluid, counting in a Packard Tri Carb scintillation counter.

Surface Membranes

Surface membranes were prepared from L cells by the Zn ion procedure (20). The final membrane preparations contained few or no nuclei and very little of other particulate matter of the cytoplasm as observed in the phase-contrast microscope (Fig. 1). The purified surface membranes were counted in a hemocytometer and represented a 10–20% yield of membranes from the whole cells. The surface membrane preparation contained 2 pg RNA and 30 pg protein per membrane, representing 8–11% of the total cell protein.

Ribosomes

A ribosomal fraction was prepared from the same whole cells from which the surface membranes were isolated (1). Comparison of the cytoplasmic ribosomes prepared by this technique, that is, in the presence of 1 mM $ZnCl_2$, have been made with ribosomes prepared by regular techniques (1).

Supernatant Material

The supernatant material remaining after centrifugation of the ribosomal fraction at 110,000 × *g* in the Spinco ultracentrifuge for 2 hr was examined in some experiments, whereas in others the supernatant material was precipitated with trichloroacetic acid (1), and the precipitate is referred to as "110,000 × *g* Sup, TCA ppt."

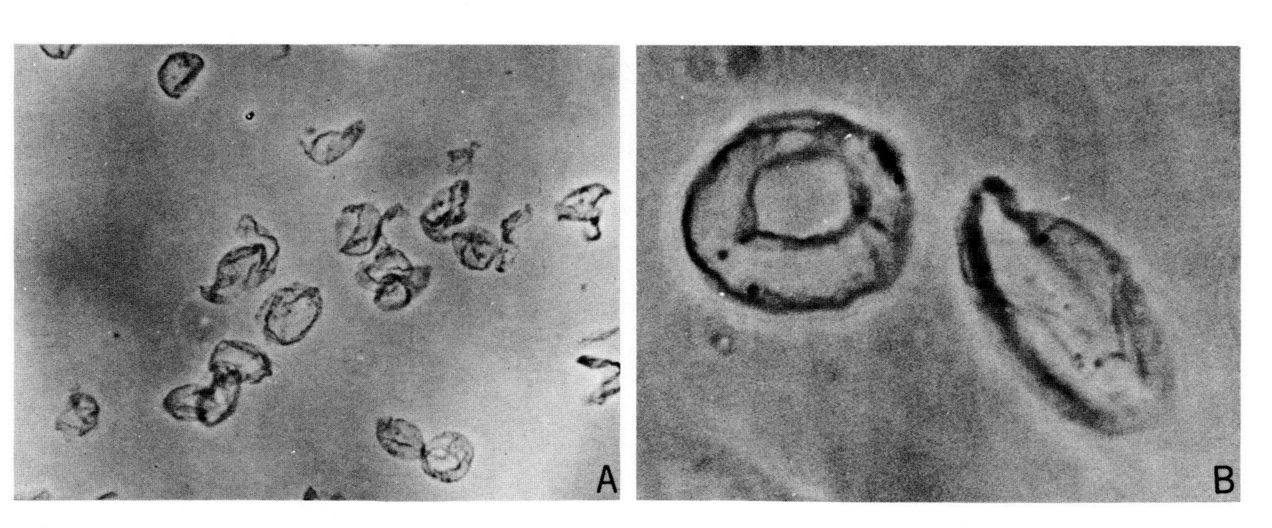

FIG. 1. Surface membranes isolated from L cells by the Zn ion procedure. Phase contrast **A:** ×350; **B:** ×1,400. [From Glick (3) by permission.]

Nuclei

The method for preparation of nuclei from the same L cells from which the surface membranes and microsomes were isolated has been described (21). These nuclei, which were counted in the hemocytometer, contained their outer membranes and were free from contaminating organelles as shown in the electron microscope (21).

Sucrose Gradient Centrifugation

Cells were grown in the presence of [^{3}H]uridine or [^{14}C]uridine and the surface membranes and cytoplasmic ribosomes were isolated. To compare the ribosomal particles the labeled cell fractions were combined, treated with sodium deoxycholate (DOC) and Brij 58, and centrifuged through gradients of sucrose solutions (22). To release the RNA, selected fractions from the gradient were made 0.2% with sodium dodecylsulfate (SDS) in the appropriate buffer (22) and centrifuged again through gradients of sucrose solutions (22). Nonlabeled *E. coli* ribosomes or rRNA, were included with each gradient as markers.

Polyacrylamide Gel Electrophoresis

Surface membranes, nuclei, and ribosomes were prepared from a 2 liter culture of L cells (4×10^5 cells/ml) labeled for 7 min with 1 mCi of [^{3}H]-uridine. The cell fractions were extracted with phenol at 55°C as described by Wagner et al. (23). After the addition of 2 vol ethanol and subsequent centrifugation, the precipitate was suspended in buffer containing 0.2% SDS for electrophoresis (24). Chick embryo fibroblasts, grown in the presence of 0.2 μCi of [^{14}C]uridine/ml of medium, were extracted with phenol in a similar manner, and the radioactive RNA was included in each gel as 28 and 18 S markers. Electrophoresis on 2.7% polyacrylamide gels (24) (0.6×9.0 cm) with 0.5% agarose (25) was for 2 or 3 hr at 3 mA/gel. The gels were cut and dispensed in 1 M NH_4OH with a Maizel apparatus (Savant Instruments, Inc., Hicksville, N.Y.) collecting fractions of 10 drops. The NH_4OH was evaporated and the gels suspended in 0.1 ml of 0.1 N NaOH and further processed for radioactive counting with NCS and liquifluor–toluene as described above.

Base Ratios and Specific Activities

For the experiments reporting the specific activities, the RNA was made radioactive by growing the L cells in the presence of [^{3}H]uridine for 24 hr. The surface membranes and ribosomal fraction were isolated from these cells. RNA was extracted from the surface membranes and cytoplasmic ribosomes and hydrolyzed with LiOH (18). After the removal of the lithium as the perchlorate, the recovery of radioactivity from the surface membranes and cytoplasmic ribosomes was 89 to 100%, respectively. The purine nucleotides present in the digest were further hydrolyzed with 1 N HCl for 1 hr at 100°C and evaporated to dryness. The hydrolysates were chromatographed on cellulose thin layer plates (25 μm, Analtech, Wilmington, Del.) in isopropanol : HCl : H_2O (65 : 16 : 19) for 15 hr. The spots, detected by UV absorption, corresponding to G, A, CMP, and UMP, were scraped from the plates and eluted for determination of radioactivity and UV absorption.

Incubation with RNase

Surface membranes or ribosomes were prepared from L cells made radioactive by growth in the presence of [^{3}H]uridine. Surface membranes (23,520 cpm) or ribosomes (13,480 cpm) were suspended in 1 ml of a sterile solution of 30% sucrose in 2 mM Tris-HCl, pH7.3, 0.8 mM $MgCl_2$, 200 μg of bovine serum albumin (3 × crystallized, Worthington Biochemical, Freehold, N.J.) and 10 μg RNase. Duplicate fractions, but without RNase, served as controls and were incubated for the specified times or precipitated immediately with 1 ml of 5% trichloroacetic acid. At the end of the incubation times, the incubates were precipitated with 5% trichloracetic acid in the cold and centrifuged for 20 min at 1,800 × *g*. Radioactivity was determined on aliquots of the supernatant solutions as described above.

Synthesis of RNA

A culture of L cells was grown to a concentration of 4–5 × 10^5 cells/ml of medium; 0.7 μCi of [^{3}H]uridine was added per milliliter of medium and incubation continued at 37°C. At the specified times an aliquot was removed and immediately chilled in a large volume of ice, centrifuged at 5°C and washed as described (1). The cells were suspended in 0.16 M NaCl and a portion was removed for counting in the hemocytometer and determination of radioactivity. The remaining cells were used for the preparation of surface membranes, ribosomes, nuclei, and 110,000 × *g* supernatant material. The final preparation of each cell fraction was analyzed for protein content and radioactivity. The surface membranes and nuclei were counted in a hemocytometer.

The following procedure was used for the treatment of the L cells with low or high concentrations of actinomycin D (Merck, Sharp and Dohme, Rahway, N.J.). Cells were grown to a concentration of 3.5–4.5 × 10^5 per ml of culture medium. One-half of the cells were removed to another spinner flask and served as control. The other half was treated for 25 min with the appropriate concentration of actinomycin D. After this time [^{3}H]uridine was added to both cultures. Aliquots containing 5 × 10^7 cells were removed from both cultures to ice at the times specified for each experiment. The cells were processed and the surface membranes, ribosomes, nuclei, and "110,000 × *g* sup, TCA ppt" were prepared.

RESULTS

Synthesis of RNA

Kinetics of Incorporation of [³H]Uridine into Cell Fractions

L cells were incubated with [³H]uridine and the distribution of radioactivity was determined in the various cell fractions. After 15 min the nucleus contained 27% of the total radioactivity taken up by the whole cell, while the surface membrane contained 3.5%.

The kinetics of incorporation at early times is shown in Fig. 2, and experiments pertinent to later times are shown in Figs. 3 and 6. The most

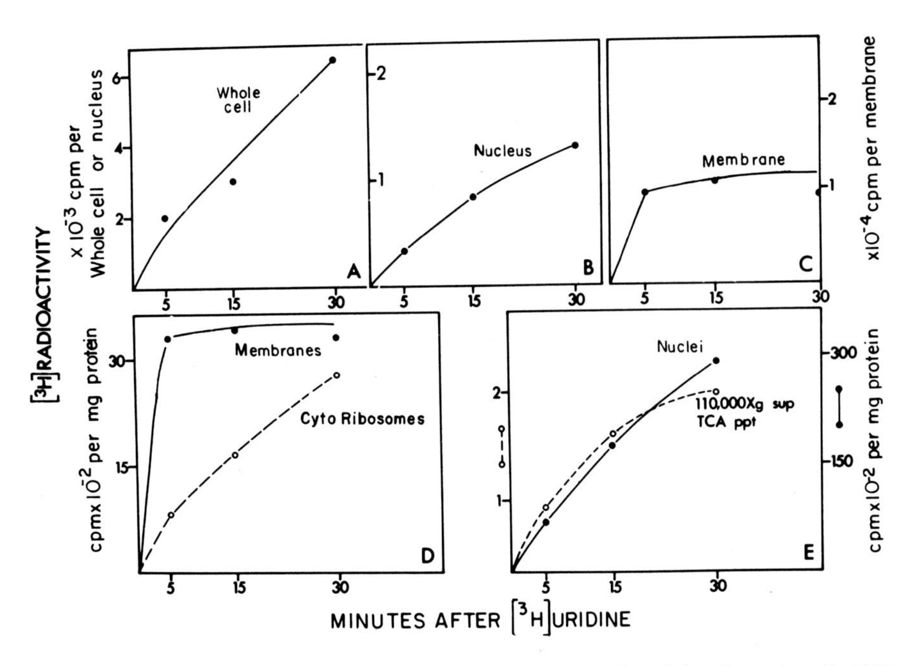

FIG. 2. Kinetics of incorporation of [³H]uridine into various L cell fractions. L cells (4 × 10⁵/ml of culture medium) were incubated at 37°C in the presence of [³H]uridine (0.7 μCi/ml of medium). At 5, 15, and 30 min, 19, 8.8, and 9.6 × 10⁷ cells, respectively, were removed to ice and harvested. Cell fractions were prepared for each interval. Radioactivity was expressed as counts per minute for the whole cell **(A)**; nucleus **(B)**; or surface membrane **(C)**; or was expressed as cpm/mg of protein **(D)** for the surface membranes (●) and ribosomes (○), and **(E)** for the nuclei (●) and "110,000 × *g* sup, TCA ppt" (○).

rapid rate of incorporation at early times was into the surface membrane fraction (Fig. 2C). The fast initial incorporation of [^{3}H]uridine into the surface membranes was even more striking when expressed as cpm/mg protein (Fig. 2D,E) rather than cpm/cell particle (Fig. 2B,C). Figure 2 also shows that after 5–10 min the rapid incorporation into surface membranes is followed by a much slower rate of incorporation. However, with the other cell fractions, the rate of synthesis did not slow down to that extent. It should be noted that after 60 min the synthesis of the surface membrane-associated RNA increased gradually with time (see Fig. 6).

Kinetics of Incorporation into RNA Fractions

In order to show that the rapid initial uptake of radioactivity into the surface membranes did not represent nonspecific absorption to the cell surface, the distribution of radioactivity into trichloroacetic acid-precipitable material was examined.

Within 5 min, 54% of the radioactivity incorporated into the surface membranes was acid-insoluble rising to 73% within 30 min. For the same time intervals only 17% of the radioactivity in the ribosomes was recovered in the acid-insoluble fraction, rising to 55% within 30 min. At later times 90% or more of the radioactivity of both cell fractions was found in the material which had been precipitated with trichloroacetic acid. Thus, the higher radioactivity found in the surface membrane fractions was not due to nonspecific absorption.

The rate of [^{3}H]uridine incorporation into the RNA of the surface membranes (mem-RNA) and cytoplasmic ribosomes (cyto-RNA) is shown in Fig. 3. In order to demonstrate that these differences were in RNA, the specific activities (cpm μmole of RNA-P) of the RNA in the two fractions were determined and the results are represented by the ratio of these specific activities (Fig. 3). The initial high ratio is the result of the more rapid incorporation into the mem-RNA. This ratio falls to a minimum by 90 min. From 90 min on, a constant amount of [^{3}H]uridine was incorporated into both mem-RNA and cyto-RNA.

In this same experiment, aliquots of the cells which were pulsed for 45 min with [^{3}H]uridine were chased with nonradioactive medium for 30 and 120 min. Following the 30-min chase, more radioactivity was chased from the cyto-RNA than from the mem-RNA. The incorporation of radioactivity into the surface membranes continued at an undiminished rate. On the other hand, following the 120-min chase, a marked drop in the radioactivity was observed. This striking difference in the amount of radioactivity incorporated by mem-RNA when compared to the cyto-RNA during the chase with cold medium is shown by a large increase and subsequent fall in the ratio of the specific activities (Fig. 3).

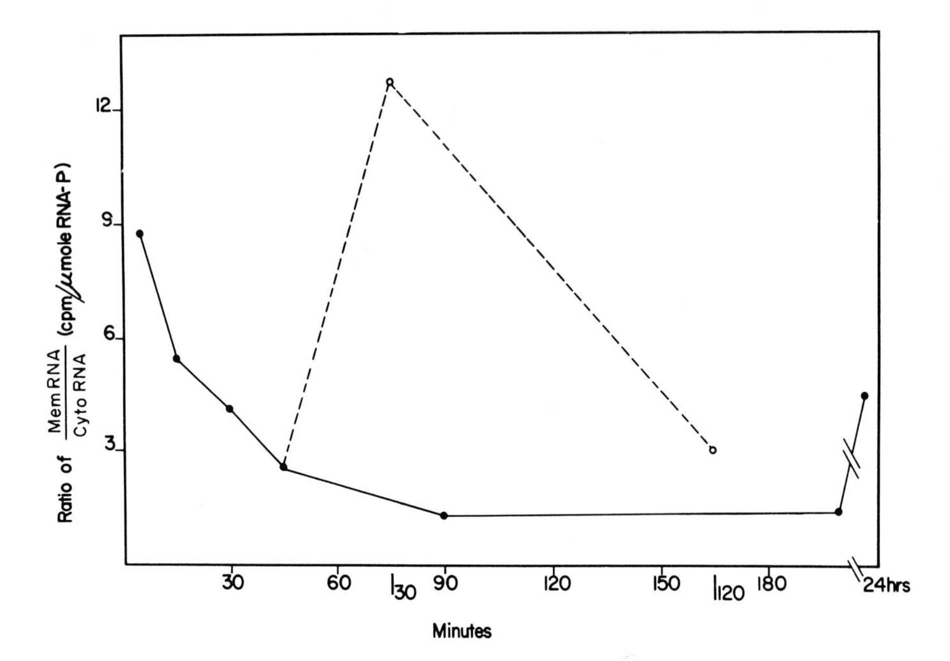

FIG. 3. Ratio of the specific activities of mem-RNA and cyto-RNA. The experiment was similar to that described in Fig. 2 with the exception that 7.7×10^7 cells were removed at each time and only surface membranes and ribosomes were prepared. In addition, at the 45-min interval, 1.5×10^8 cells were removed, washed sterilely with 0.16 M NaCl at 37°C, and resuspended in 300 ml of fresh medium with no radioactivity. 7.5×10^7 cells, respectively, were removed after 30 and 120 min of additional incubation, chilled, washed, and surface membranes and ribosomes were prepared. The extraction of RNA and phosphate determination were as described in the text. The specific activity (cpm/μmole of RNA-P) of the mem-RNA was compared to the specific activity of the cyto-RNA and plotted to give the ratios for each time interval (●), and the 30 and 120 min chase of the 45 min pulse (○).

Size of RNA Species of Surface Membranes

Mem-RNA labeled with [^{3}H]uridine at various times was examined by centrifugation through gradients of sucrose solutions in the presence of SDS (Fig. 4). The RNA from the 45-min pulse (Fig. 4A) was heterogeneous with respect to isotope incorporation, whereas by 90 min most of the radioactivity was found in the 28 and 18S RNA (Fig. 4B). The heterogeneity appeared to be completely lost by 200 min (Fig. 4C). The RNA labeled for 24 hr showed a pattern similar to Fig. 4C. The RNA of the 45-min pulse followed by the 30-min chase was hardly heterogeneous (Fig. 4D) and showed a pattern which resembled more the 90-min pulse of RNA (Fig. 4B) than that obtained with the 45-min pulse alone (Fig. 4A). After the 120-min chase (Fig. 4E), the 18 and 28S RNA again represented most of the radioactivity.

To examine further the RNA obtained with a short pulse, the L cells were exposed to [^{3}H]uridine for 7 min and surface membranes, ribosomes, and nuclei were prepared. The RNA from these fractions was extracted with phenol and examined by electrophoresis on polyacrylamide gels. mem-RNA (Fig. 5A), when labeled for this short period, was heterogeneous and showed prominent peaks between the 18 and 28S markers. Similar patterns were obtained for 8- to 10-min pulsed mem-RNA and always showed the two peaks between the 18 and 28S markers. The profile of the RNA obtained from the nuclei (nuc-RNA) showed the presence of 45 and 28S peaks as well as an 18S component (Fig. 5B). As 18S RNA is not normally seen in pulse-labeled nuc-RNA (26), it is assumed that in our method of preparation, the nucleus, which has the outer membrane intact retained 18S RNA. In any event, the mem-RNA profile is markedly different from the nuc-RNA and rules out the possibility of nuc-RNA contamination in the surface membrane preparation.

Effect of Actinomycin D on RNA Synthesis

The effect of actinomycin D on the uptake of [^{3}H]uridine into the various cell fractions was examined. Cells were exposed to actinomycin D (10 μg/ml of medium) for 25 min, after which [^{3}H]uridine was added and the time course of incorporation of the label into the cell fractions was determined. The kinetics of incorporation was different for each cell fraction incubated with or without actinomycin D. In contrast to the other cell fractions (Fig. 6A–C), there was no inhibition of the incorporation into the surface membrane fraction for the first 60 min (Fig. 6D); after this period the incorporation declined rapidly.

The synthesis of nonribosomal RNA was examined by preincubating the L cells for 25 min with low concentrations of actinomycin D (0.08 μg/ml of medium). This concentration has been used to inhibit specifically the synthesis of ribosomal RNA (26). Again, as with the high concentrations of actinomycin D, the surface membranes showed little inhibition in the amount of [^{3}H]uridine incorporated within 60 min, and never more than 12% (Fig. 7). The other cell fractions showed more inhibition, 30–40% for the nuclear fraction and 20–30% for the ribosomal fraction (Fig. 7). The difference between the membrane fractions was even more pronounced when the pulse of labeled uridine was examined after a shorter period. Thus, incubation for 30 min showed no inhibition by actinomycin D in mem-RNA, whereas all the other fractions were significantly inhibited. When the pulse was prolonged for 120 min, all of the cell fractions showed marked inhibition in the amount of radioactivity incorporated in the presence of actinomycin D.

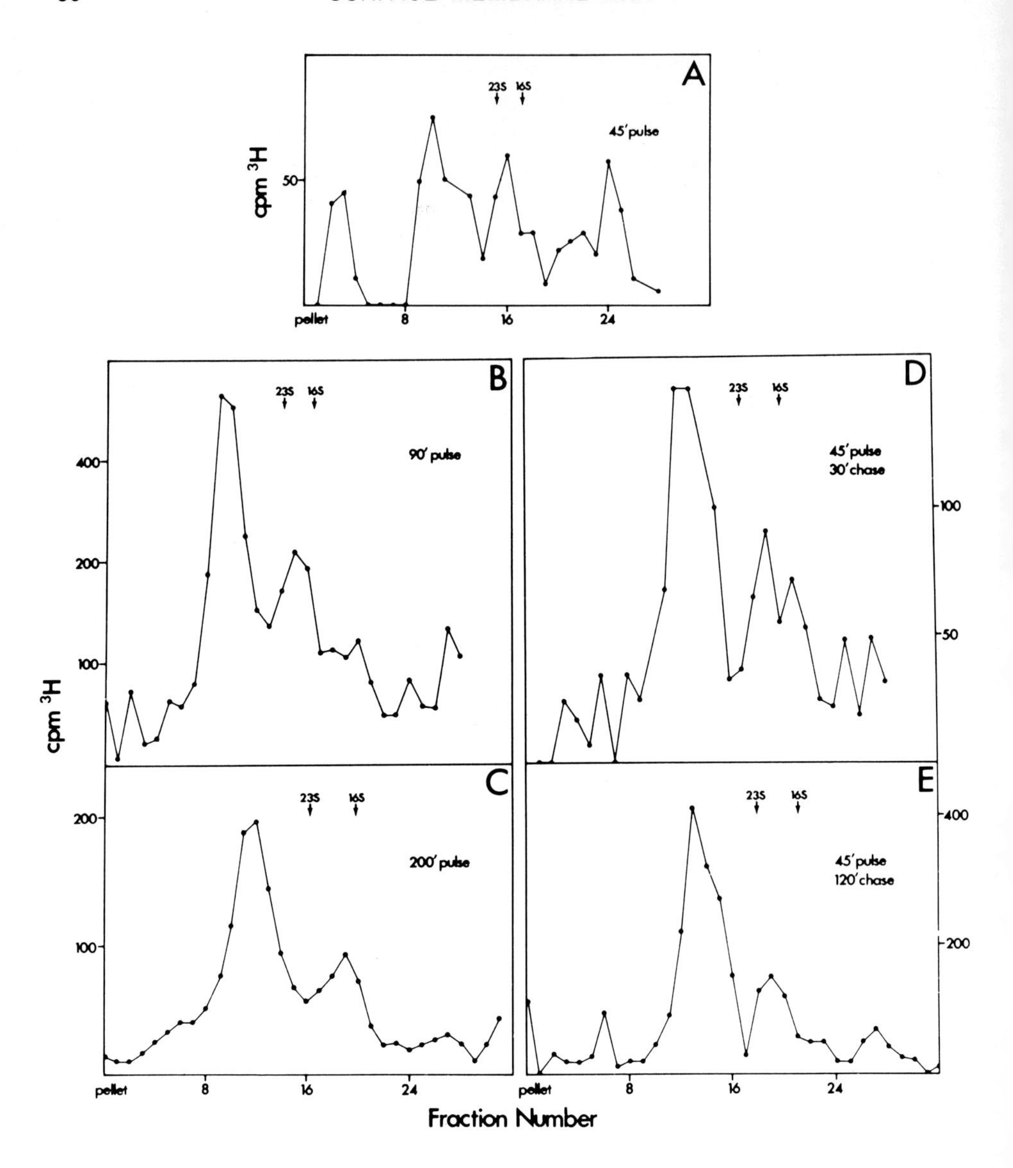

FIG. 4. Comparison by sucrose gradient centrifugation of the RNA of the surface membranes synthesized at various time intervals. Surface membranes, obtained from the experiment described in Fig. 3 were made 0.2% with SDS and centrifuged through 15–30% sucrose solutions with 0.01 M Tris (ph 7.4), 0.1 M NaCl, and 0.01 M EDTA for 16 hr at 53,000 × *g*. *Escherichia coli* ribosomes were included as 23 and 16S markers. The mem-RNA preparations were from the pulse intervals of 45 min **(A),** 90 min **(B),** 200 min **(C),** and 45 min with a 30-min **(D),** and 120-min **(E)** chase.

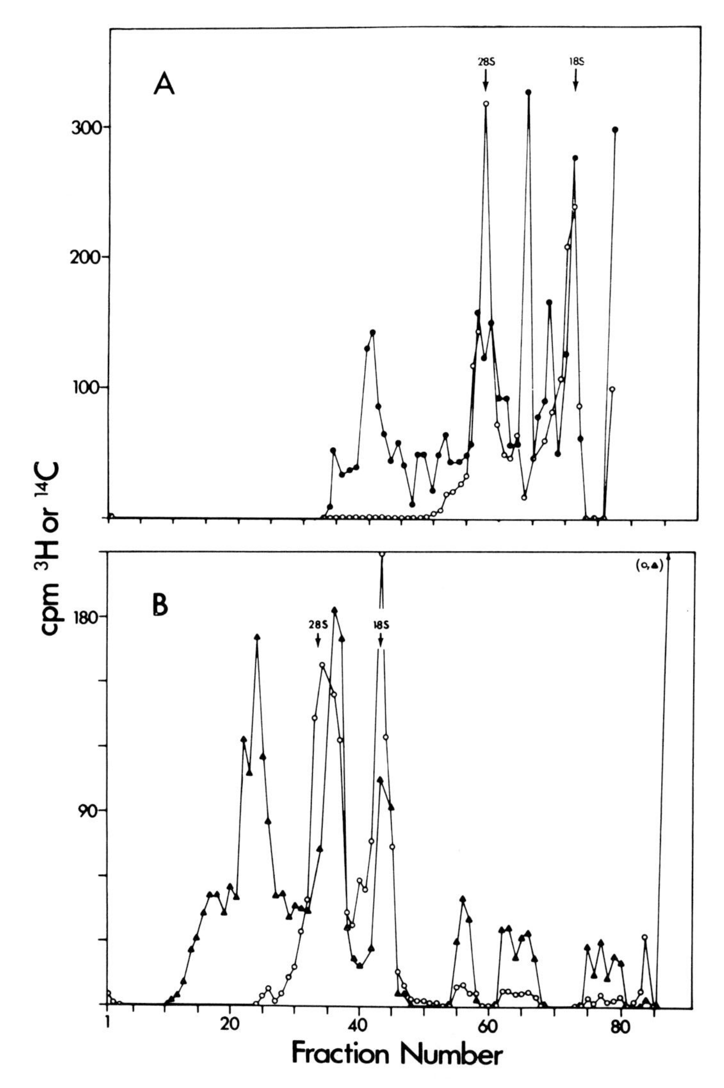

FIG. 5. Polyacrylamide gel electrophoresis of RNA obtained from surface membranes (●) and nuclei (▲) after a short interval pulse. A culture (2 liters) of L cells (4×10^5 cells/ml) was pulsed with 1 mCi of [^{3}H]uridine for 7 min. Cells were harvested and surface membranes **(A)** and nuclei **(B)** were prepared. Prior to electrophoresis, the RNA was extracted from both fractions with phenol, precipitated with ethanol and suspended in buffer for electrophoresis containing 0.2% SDS. Chick embryo fibroblasts (○) made radioactive with [^{14}C]uridine were treated in a similar manner and the radioactive RNA was included in each gel as 28 and 18S markers. Electrophoresis on 2.7% polyacrylamide gels with 0.5% agarose was for 3.5 hr **(A)** or 2 hr **(B)** at 3 mA/gel.

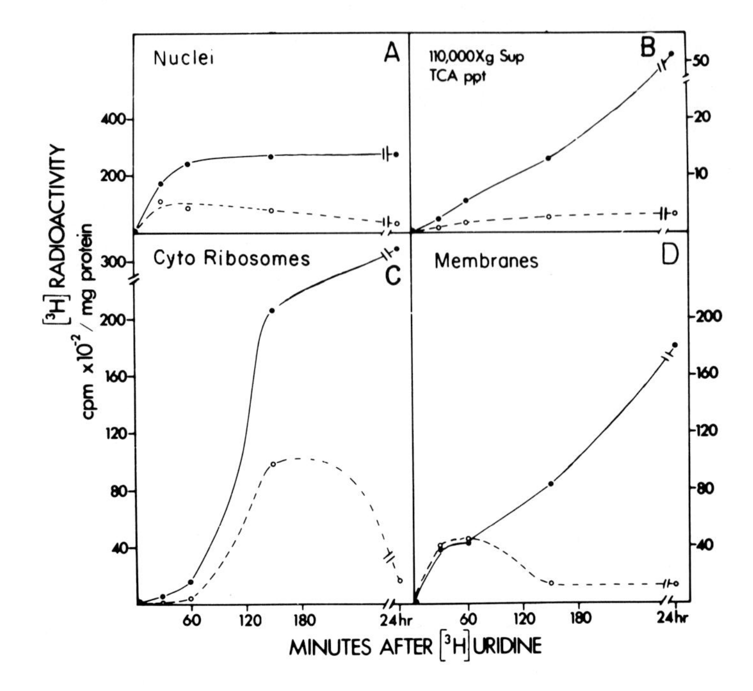

FIG. 6. RNA synthesis in the presence of actinomycin D. L cells, grown to a concentration of 4.5×10^5 cells/ml, were divided into two cultures. Actinomycin D (10 μg/ml of medium) was added to one culture (○) while the other culture served as a control (●). After incubation for 25 min, [^{3}H]uridine (0.43 μCi/ml of medium) was added to each and incubation continued, with aliquots removed at 30, 60, 150 min, and 24 hr. The cells were harvested from both cultures and nuclei **(A),** "110,000 $\times$ *g* sup, TCA ppt" **(B),** cytoplasmic ribosomes **(C),** and surface membranes **(D),** were prepared at the stated times. The amount of radioactivity incorporated into each fraction is expressed as cpm/mg of protein.

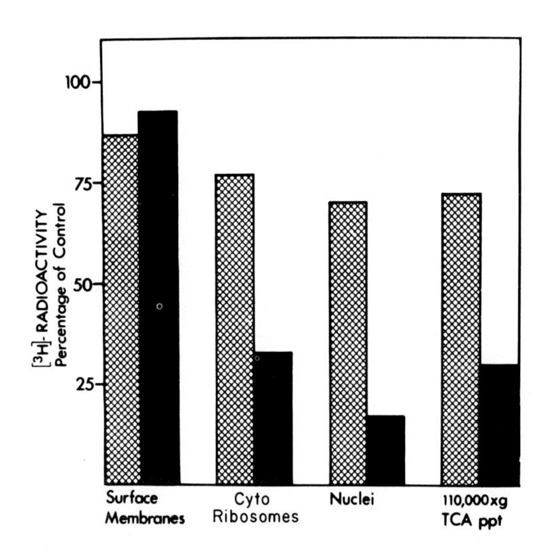

→

Comparison of Ribosomal RNA Derived from Surface Membranes and Cytoplasm

Sucrose Gradient Centrifugation

Surface membranes were isolated from L cells grown in the presence of [^{3}H]uridine, whereas cytoplasmic ribosomes were obtained from [^{14}C]-uridine-containing cultures. The two radioactive-labeled fractions were combined and treated with DOC and Brij 58 to liberate the bound ribosomes, and subsequently were centrifuged through a sucrose gradient. The ^{3}H-labeled RNA of the surface membrane preparation sedimented as 60 S and 80 S ribosomes and as polysomes (Fig. 8A). The ^{14}C-labeled cytoplasmic fraction showed a similar sedimentation profile, but a shift from the 60S toward the 50S region of the gradient as compared to the ribosomes from the surface membranes (Fig. 8A).

To analyze the sedimentation profile of the RNA, selected fractions from the ribosome gradient (Fig. 8A) were treated with SDS and further applied to a new sucrose gradient. The pellet as shown in Fig. 8A, representing the more rapidly sedimenting region of the ribosome gradient and fraction 6, representing the 80S RNA region, were found in the second gradient, to be composed of 18 and 28S RNA for both the ^{3}H- and ^{14}C-labeled preparations. The 60S region of the ribosome gradient (fraction 13, Fig. 8A) was composed of 16–18 and 28–30S RNA for the surface-membrane-derived [^{3}H]RNA. The 60S cytoplasmic ribosomes, although composed of 16–18S RNA, showed only a small amount of RNA in the 28–30S region of the gradient. The 50S region of the ribosome gradient (fraction 15, Fig. 8A) showed that the mem-RNA was composed of 28–30, 16–18, and 10S RNA (Fig. 8B). A similar profile was observed for the cyto-RNA, with the exception that little 28–30S RNA was present (Fig. 8B). The RNA profiles of the cytoplasmic fractions suggest some measure of RNA degradation, which could account for this lack of the 28–30S RNA obtained from the 50 and the 60S regions of the gradient.

FIG. 7. RNA synthesis in the presence of low and high concentrations of actinomycin D. The experimental details were as described for the experiment in Fig. 6 with the exception that the starting culture was divided into four cultures. Two cultures were treated with actinomycin D, 0.08 and 10 μg/ml of medium, respectively, for 25 min, and the other two cultures served as controls. After this time, all cultures were made radioactive with [^{3}H]uridine (0.5 μCi/ml of medium) and incubated for 60 min. Cell fractions were prepared as described in the text. The amount of radioactivity incorporated into each cell fraction in the presence of 0.08 μg (*hatched bar*) or 10 μg (*solid bar*) of actinomycin D is expressed as percentage of the radioactivity incorporated into the cell fractions prepared from the two control cultures.

Base Composition

The base composition of mem-RNA was found to be similar to that of cyto-RNA (Table 1). In addition, no significant differences were found for the (A + U)/(G + C) ratio, as well as the specific activities of UMP and CMP (Table 1).

Sensitivity to RNase

The mem-RNA was found to be more resistant to RNase digestion than the cyto-RNA. Only 57% of the mem-RNA was solubilized under condi-

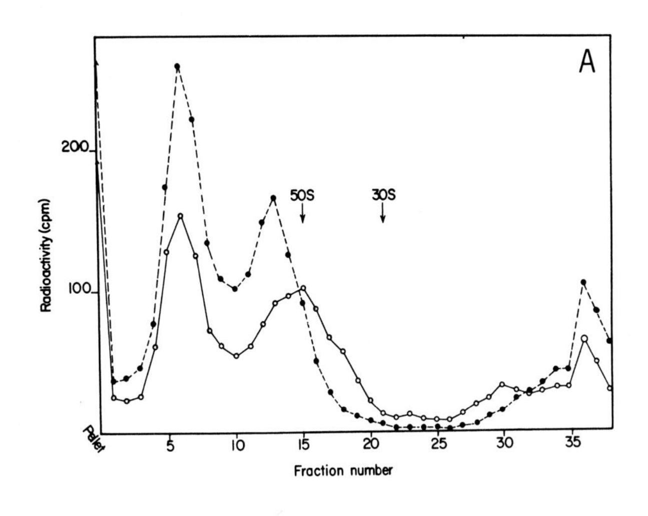

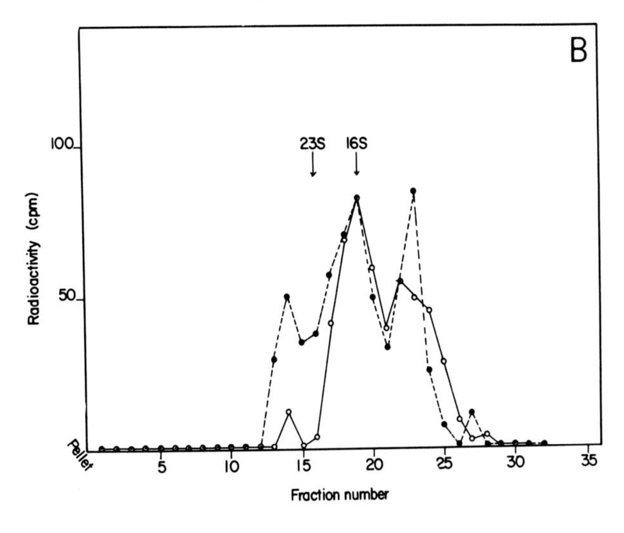

TABLE 1. *Comparison of the base composition and specific activities of mem-RNA and cyto-RNA*

Source of RNA	G	A	CMP	UMP	$\frac{A+U}{G+C}$
	Base composition (% of total)				
Surface membranes	41.1	15.3	27.1	16.5	0.47
Cytoplasmic ribosomes	40.6	17.4	24.8	17.1	0.52
	Specific activity (cpm/nmole)				
Surface membranes	<0.05	73	3,650	4,460	
Cytoplasmic ribosomes	<0.07	65	4,740	4,370	

The base composition of the RNA extracted from surface membranes or cytoplasmic ribosomes was expressed as percentage of the total nanomoles of the individual bases and were means of two determinations. The standard deviations were all <1.0. The specific activity (cpm/nmole) of the individual bases were determined on mem-RNA or cyto-RNA prepared from L cells grown in the presence of [^{3}H]uridine (26.9 Ci/mmole) for 24 hr.

tions (37°C/30 min) which released 97% of the cyto-RNA (Table 2). Under more mild conditions (25°C/10 min) 21% of the mem-RNA was solubilized as compared to 34% of the cyto-RNA.

Nature of the Association of the Ribosomes to the Surface Membranes

Ribosomal-like particles associated with the surface membranes were visualized in the electron microscope. These particles appeared on the inner surface of the isolated membranes (Fig. 9). Because the possibility exists

---

FIG. 8. Sucrose gradient centrifugation of mem-RNA and cyto-RNA. Surface membranes and ribosomes were prepared from L cells made radioactive by growth in the presence of [^{3}H]- or [^{14}C]uridine. **A:** The isolated surface membranes (●) and cytoplasmic ribosomal fraction (○) were combined, made 0.5% with DOC and Brij 58 and centrifuged through 15–30% sucrose solutions with 0.01 M Tris (pH 7.4), 0.01 M NaCl, and 1.5 mM $MgCl_2$ for 16 hr at 53,000 × *g*. **B:** Fraction 15 from **A** was made 0.2% with SDS, 0.01 M Tris (pH 7.4), 0.01 M EDTA, and 0.1 M NaCl and centrifuged as described for **A.** *E. coli* ribosomes were included in the gradients marking 50 and 30S ribosomal RNA and subsequently marking 23 and 16S RNA.

that these ribosomes are nonspecifically bound to the surface membranes during the isolation procedures, several experiments were performed to examine this. L cells were grown in the presence of [^{3}H]uridine for 24 hr, harvested, and a ribosomal fraction prepared. The fresh radioactive ribosomal preparation (45,650 cpm) was added to washed, nonradioactive L cells (4.75×10^7 cells) and surface membranes were prepared by the Zn ion procedure (20). After partial purification, less than 1% of the added radioactivity remained with the isolated surface membranes, indicating that nonspecific binding of the ribosomes probably did not occur under these conditions. Whole cells, treated with 1 mM $ZnCl_2$, used as a control with the same amount of added ^{3}H-labeled ribosomes, retained 2% of the radioactivity after five washings with 0.16 M NaCl. A similar control of whole

TABLE 2. *Degradation of mem-RNA and cyto-RNA by RNase*

Source of RNA	Assay	RNA degraded by RNase 25°C/10 min	RNA degraded by RNase 37°C/30 min
		% of total radioactivity	
Surface membranes	Control not incubated	0	0.4
	Control incubated	0	2.5
	RNase	21	57
Cytoplasmic ribosomes	Control not incubated	0	6
	Control incubated	6	5
	RNase	34	97

Surface membranes and cytoplasmic ribosomes were prepared from L cells grown for 24 hr in the presence of [^{3}H]uridine (26.9 Ci/mmole). Surface membranes (23,520 cpm) cytoplasmic ribosomes (13,480 cpm) were incubated under the specified conditions as described in Methods and were subsequently precipitated with 1 ml of 5% trichloroacetic acid. The control, not incubated, was precipitated immediately with 5% trichloroacetic acid. All fractions were centrifuged at $1,800 \times g$ for 20 min. The radioactivity released into the supernatant solutions was determined and expressed as percentage of the total radioactivity in the surface membrane or cytoplasmic ribosomal fractions.

cells not treated with 1 mM $ZnCl_2$ retained 1% of the total radioactivity.

In other experiments, the free and bound ribosomes were separated according to the procedures of Blobel and Potter (27). L cells were grown in the presence of [^{3}H]uridine for 24 hr and harvested. The L cells were homogenized either with 1 mM $ZnCl_2$ by the procedure for preparing surface membranes or by the procedure of Blobel and Potter. The homogenates were subjected to centrifugation under conditions which were reported to separate free and bound ribosomes. The percentage of the total radioactivity which was isolated from the homogenate as free ribosomes was similar (11–19%) whether the ribosomes were prepared by the Zn^{2+} procedures or the procedures of Blobel and Potter. In addition, if 1 mM $ZnCl_2$ was added to the L cells after homogenization by the latter procedure, there was no decrease in the free ribosomes. These results again indicated that there was no nonspecific binding of the Zn^{2+}-treated ribosomes to the surface membranes.

Using low-temperature spectroscopy, the surface membranes were found to contain less than 5% of the concentration of cytochrome b_5 found in the endoplasmic reticulum, indicating very little contamination. In addition, no stretches of endoplasmic reticulum were seen with the electron microscope.

A few experiments were performed in attempting to remove the ribosomes from the surface membranes. Sonication fragmented the membranes, but

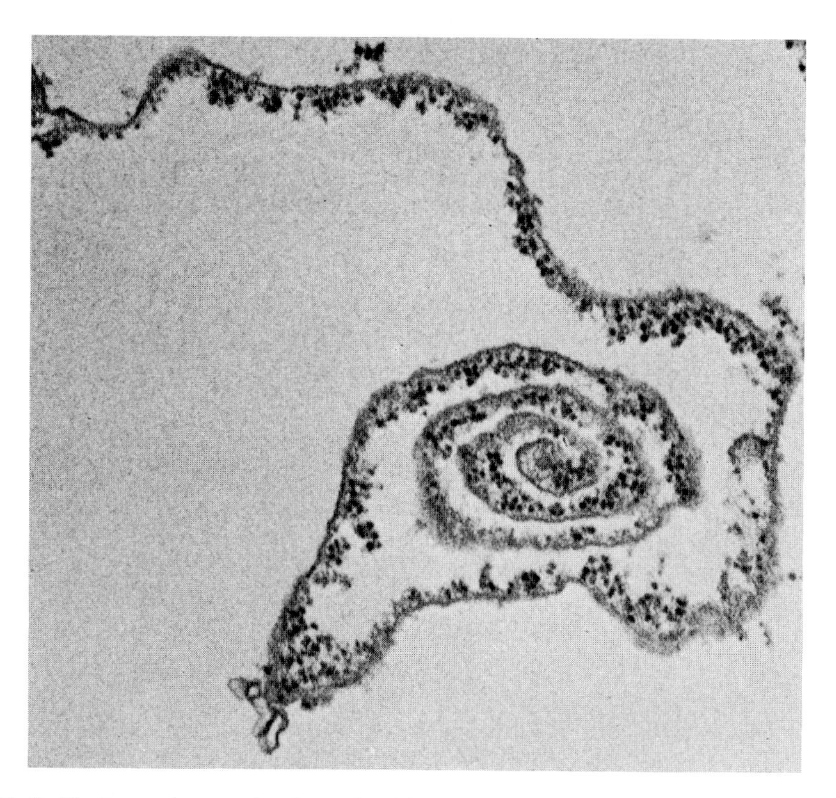

FIG. 9. Electron micrograph of an ultrathin section of an isolated surface membrane from an L cell. ×40,800.

the pieces of membranes as viewed in the electron microscope retained the ribosome-like particles. Treatment with 6 M urea did not remove the ribosome-like particles.

DISCUSSION

The results presented in this chapter suggest that the surface membrane associated RNA (mem-RNA) has some unique properties. The mem-RNA is more rapidly labeled than the cyto-RNA (Fig. 2). Early labeled mem-RNA is not sensitive to actinomycin D under conditions wherein synthesis of cyto-RNA or nuc-RNA is inhibited (Fig. 7).

Yudkin and Davis (28) observed that the labeling of mem-RNA in *Bacillus megatherium* was not inhibited, and, in fact, was stimulated by treatment with actinomycin D (10 μg/ml). With this concentration, other RNA species were 70% inhibited. They suggest that the rapidly synthesized RNA flows from the cytoplasm to the membrane during treatment with actinomycin D. The results in Fig. 6 show that in the presence of actinomycin D, nuc-RNA synthesis is inhibited within 30 min, while mem-RNA synthesis continues for 60 min. The results reported here agree with the notion that either the RNA flows from the nucleus to the membrane, or that this RNA is being transcribed in the membrane itself. In addition, when the radioactivity was chased by cold medium, incorporation of radioactivity into the mem-RNA continued at early times, whereas incorporation into cyto-RNA decreased, again suggesting a movement of RNA to the surface membranes.

Another difference noted was that mem-RNA, when associated with the surface membranes was less sensitive to degradation by RNase than the cyto-RNA (Table 2). However, other structural and chemical properties were similar. That is, the base compositions (Table 1) and sedimentation properties were similar (Fig. 8). Such a similarity has also been indicated in the ribosomal RNA associated with surface membranes and cytoplasm of human diploid lymphocytes (4) and Ehrlich ascites tumor cells (5).

The incorporation of [^{3}H]uridine into the surface membranes appears to be a three-step process—a rapid initial incorporation, followed by a slower rate, and then a slow steady synthesis to 13 hr when saturation is reached. The meaning of this is not apparent at the present time, although the rapid initial incorporation could represent an affinity of mRNA for membrane-bound ribosomes. On the other hand, it could represent RNA associated with the external side of the surface membrane (7–10). The subsequent incorporation could represent the slower movement of ribosomes from the cytoplasm to the cell surface. Regardless of the meaning, it does appear that at least a portion of the mem-RNA is metabolically distinct from cyto-RNA and nuc-RNA. That mem-RNA probably fulfills a specialized function is suggested by the fact that the polypeptides synthesized *in vitro* by the surface membranes isolated from L cells are different from those synthesized by the cytoplasmic ribosomes (1). Similar differences in polypeptide synthesis have been found for Ehrlich ascites tumor cells and baby hamster kidney fibroblasts (Glick, *unpublished observations*). It remains to be demonstrated whether or not these findings are common to surface membranes from all cell types.

SUMMARY

RNA associated with isolated surface membranes from L cells was compared with cytoplasmic RNA from the same cells. The profiles obtained by centrifugation through gradients of sucrose solutions and the base composition were similar for RNA from both sources.

A number of differences were noted: (*i*) when the synthesis of RNA was examined, the specific activity (cpm/nmole of RNA-P) of the RNA of the surface membranes was higher than that of the cytoplasmic RNA from the same cells after a short exposure to [^{3}H]uridine; (*ii*) concentrations of actinomycin D which were inhibitory to the synthesis of RNA found in nuclei, cytoplasmic ribosomes, and supernatant material were not inhibitory at early times to the formation of RNA found in the surface membranes; (*iii*) the RNA associated with the surface membranes was less sensitive to RNase than was the RNA of the ribosomal fractions. These differences and others indicate that, although the ribosomal RNA associated with the surface membranes may be chemically and structurally similar to that from the cytoplasm, at least a portion of the surface membrane RNA appears to be distinct metabolically from either the cytoplasmic or nuclear RNA.

ACKNOWLEDGMENTS

The electron micrograph was taken by Dr. Herbert Blough, Department of Ophthalmology. I wish to thank Dr. Clayton Buck, Kansas State University, for the preparation of *E. coli* ribosomes and Dr. Leonard Hayflick, Stanford University, for examining the cultures for the presence of *Mycoplasma*. I am particularly grateful to Professor U. Z. Littauer, Weizmann Institute, Rehovot, Israel, for his helpful discussions during the preparation of this manuscript. The excellent technical assistance of Ms. Roberta Koser, Annamarie Klein, and Fred Henretig is gratefully acknowledged. The investigation was supported by U.S. Public Health Service Grant 5P01-A107005-04 and American Cancer Society Grant PRA-68.

REFERENCES

1. Glick, M. C., and Warren, L. (1969): Membranes of animal cells. III. Amino acid incorporation by isolated surface membranes. *Proc. Natl. Acad. Sci. USA*, 63:563–570.

2. Korinek, J., Spelsberg, T. C., and Mitchell, W. M. (1973): mRNA transcription linked to the morphological and plasma membrane changes induced by cyclic AMP in tumor cells. *Nature,* 246:455–458.
3. Glick, M. C. (1976): Isolation of surface membranes from mammalian cells. In: *Mammalian Cell Membranes,* edited by G. A. Jamieson and D. M. Robinson. Butterworths, London.
4. Del Villano, B. C., and Lerner, R. A. (1972): Further studies on plasma membrane-associated nucleic acids in cultured human lymphocytes. *Fed. Proc.,* 31:804Abstr.
5. Juliano, R., Ciszkowski, J., Wait, D., and Mayhew, E. (1972): RNA associated with plasma membranes of Ehrlich ascites carcinoma cells. *FEBS Let.,* 22:27–30.
6. Levitan, I. B., Mushynski, W. E., and Ramirez, G. (1972): Highly purified synaptosomal membranes from rat brain: Preparation and characterization. *J. Biol. Chem.,* 247:5376–5381.
7. Van Blitterswijk, W. J., Emmelot, P., and Feltkamp, C. A. (1973): Studies on plasma mambranes. XIX. Isolation and characterization of a plasma membrane fraction from calf thymocytes. *Biochim. Biophys. Acta,* 298:577–592.
8. Weiss, L., and Mayhew, E. (1967): Ribonucleic acid within the cellular peripheral zone and the binding of calcium to iogenic sites. *J. Cell. Physiol.,* 69:281–291.
9. Davidova, S. Ya., and Shapot, V. S. (1970): Liporibonucleoprotein complex as an integral part of animal cell plasma membranes. *FEBS Let.,* 6:349–351.
10. Rieber, M., and Bacalao, J. (1974): An "external" RNA removable from mammalian cells by mild proteolysis. *Proc. Natl. Acad. Sci. USA,* 71:4960–4964.
11. Chipowsky, S., Schnaar, R., and Roseman, S. (1975): Interactions of fibroblasts with insoluble analogues of cell surface carbohydrates. *Fed. Proc.,* 34:614 (*Abstr.*).
12. Ramirez, G., Levitan, I. B., and Mushynski, W. E. (1972): Highly purified synaptosomal membranes from rat brain, incorporation of amino acids into membrane protein *in vitro. J. Biol. Chem.,* 247:5382–5390.
13. Booyse, F. M., and Rafelson, M. E. (1969): In: *Dynamics of Thrombus Formation and Dissolution,* p. 149, edited by S. A. Johnson and M. M. Guest. Lippincott, Philadelphia.
14. Hendler, R. W. (1968): *Protein Biosynthesis and Membrane Biochemistry.* Wiley, New York.
15. Aggarwal, S. K., Wagner, R. W., McAllister, P. K., and Rosenberg, B. (1975): Cell-surface associated nucleic acid in tumorigenic cells made visible with platinum–pyrimidine complexes by electron microscopy. *Proc. Natl. Acad. Sci. USA,* 72:928–932.
16. Lerner, R. A., Meinke, W., and Goldstein, P. A. (1971): Membrane-associated DNA in the cytoplasm of diploid human lymphocytes. *Proc. Natl. Acad. Sci. USA,* 68:1212–1216.
17. Lowry, D. H., Rosebrough, N. J., Farr, A. L., and Randall, R. J. (1951): Protein measurement with the folin phenol reagent. *J. Biol. Chem.,* 193:265–275.
18. Gebicki, J. M., and Freed, S. (1966): Microdetermination of nucleotides in hydrolyzates of RNA. *Anal. Biochem.,* 14:253–257.
19. Bartlett, G. R. (1959): Phosphorus assay in column chromatography. *J. Biol. Chem.,* 234:466–468.
20. Warren, L., and Glick, M. C. (1969): Isolation of surface membranes of tissue culture cells. In: *Fundamental Techniques in Virology,* p. 66, edited by K. Habel and N. P. Salzman. Academic Press, New York.
21. Glick, M. C., Comstock, C. A., Cohen, M. A., and Warren, L. (1971): Membranes of animal cells. VIII. Distribution of sialic acid, hexoamines and sialidase in the L cell. *Biochim. Biophys. Acta,* 233:247–257.
22. Knight, E., Jr., and Darnell, J. E. (1967): Distribution of 5s RNA in HeLa Cells. *J. Mol. Biol.,* 28:491–502.
23. Wagner, E. K., Katz, L., and Penman, S. (1967): The possibility of aggregation of ribosomal RNA during hot phenol SDS deproteinization. *Biochem. Biophys. Res. Commun.,* 28:152–159.
24. Bishop, D. H. L., Claybrook, J. R., and Spiegelmen, S. (1967): Electrophoretic separation of viral nucleic acids on polyacrylamide gels. *J. Mol. Biol.,* 26:373–387.
25. Peacock, A. C., and Dingman, C. W. (1967): Resolution of multiple ribonucleic acid species by polyacrylamide gel electrophoresis. *Biochemistry,* 6:1818–1827.

26. Perry, R. P., and Kelly, D. E. (1970): Inhibition of RNA synthesis by actinomycin D: Characteristic dose–response of different RNA species. *J. Cell Physiol.,* 76:127–139.
27. Blobel, G., and Potter, V. R. (1967): Studies on free and membrane-bound ribosomes in rat liver. I. Distribution as related to total cellular RNA. *J. Mol. Biol.,* 26:279–292.
28. Yudkin, M. D., and Davis, B. (1965): Nature of the RNA associated with the protoplast membrane of *Bacillus megaterium. J. Mol. Biol.,* 12:193–204.

Biogenesis and Turnover of Membrane Macromolecules,
edited by John S. Cook. Raven Press, New York 1976.

Genetic and Cellular Properties of Ouabain-Resistant Mutants

Raymond M. Baker

Department of Biology and Center for Cancer Research, Massachusetts Institute of Technology, Cambridge, Massachusetts 02139

Genetic methods are being employed with increasing facility in studies of cultured mammalian cells (1,2). The isolation and characterization of particular mutants altered in the plasma membrane should contribute substantially to our understanding of the functions and organization of that organelle (3). For that purpose, one direct approach is to examine the occurrence and properties of cells with altered responses to membrane-active agents. The steroid drug ouabain has proved to be a suitable selective agent for this purpose. It is known to specifically inhibit the plasma membrane Na^+-, K^+-, Mg^{2+}-activated ATPase (EC 3.6.1.3) which mediates active Na/K exchange (4), and the properties of this system have been studied extensively at the physiological and molecular levels (5,6). For the past several years my laboratory has been involved in the isolation and characterization of cultured mouse, hamster, and human cells that are resistant to ouabain (7–9).

My purpose in this chapter is to outline our findings to date concerning the phenotypes of ouabain-resistant (Oua^R) rodent and human cells and to indicate a unifying view of the nature of the alteration and of the dominance shown by this trait. The cell types utilized for these studies differ with respect to their degree of aneuploidy, but a more apparent distinction relevant to the Oua^R character is the difference between the species in sensitivity to ouabain. The wild-type (WT) human cells respond to drug concentrations approximately 10^4-fold lower than those required to comparably affect the WT mouse and hamster cells (Fig. 1). Both types of cells give rise to variants that are relatively ouabain-resistant, the properties of which are described below.

OUABAIN-RESISTANT MOUSE AND HAMSTER CELLS

Our studies of ouabain-resistance were initiated with heteroploid mouse L cells, and were subsequently pursued with the Chinese hamster ovary (CHO) cell line, because of its more stable "pseudodiploid" chromosome constitution (10) and the feasibility of utilizing additional genetic markers

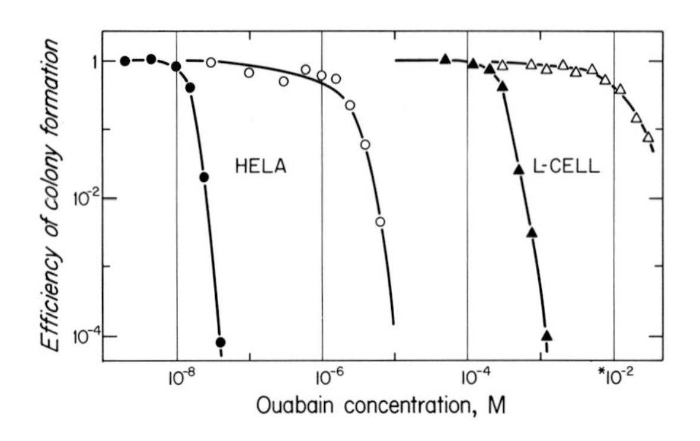

FIG. 1. The relative plating efficiencies of human HeLa cells, mouse L cells, and spontaneous ouabain-resistant variants of each, as a function of ouabain concentration in standard culture medium (5–6 mM K^+). The wild-type L-cell line (▲) and its mutant subclone Oua^RL-A7 (△) have been described previously (7). The wild-type HeLa clone (●) was derived from strain CCL2 of the American Type Culture Collection, and its mutant subclone Oua^RH-B1 (○) was selected in 2×10^{-7} ouabain (9). [The data points in the range 3×10^{-3} M to 3×10^{-2} M ouabain are from assays at doses comparable in effect to those shown, but in fact tenfold lower in both K^+ and ouabain concentrations (7).]

in this material (1,2). The main findings from these studies have been reported in detail (7):

1. Selections of large populations of CHO or L cells by single-step plating for viability in the presence of ouabain yielded variant clones that are substantially more resistant to the drug than WT (Fig. 1). Various isolates showed 3- to 200-fold increased drug resistance in terms of cytotoxic dose. The phenotypes bred true and were stable over prolonged periods of cultivation in the absence of ouabain.

2. Luria-Delbruck fluctuation analyses indicated that the Oua^R cells arise spontaneously from WT in the absence of selective conditions at rates of $\sim$5–6 $\times 10^{-8}$ per cell generation for both L cells and CHO cells. Oua^R cells could be induced at frequencies of about 10^{-4} per surviving WT cell by treatment with the mutagen ethylmethane sulfonate (EMS), compared to typical background frequencies on the order of 10^{-6} spontaneous variants per WT cell. Efficient induction of Oua^R hamster V79 cells by several other mutagens also has been described recently (11). No gross karyotypic abnormalities are associated with the phenotype.

3. Oua^R L cells examined were found to be relatively resistant to ouabain inhibition of K^+ influx into whole cells (cf. ref. 12 for Oua^R Ehrlich ascites rat cells) and to inhibition of Na-K-ATPase activity in isolated plasma membranes. Membrane preparations from the wild type and a resistant clone showed the same requirements for optimal enzyme activity in the absence of ouabain (13,7).

4. Somatic cell hybrids formed from Oua^R and WT CHO cells expressed

the ouabain-resistance trait in dose–response assays for cytotoxicity, although the hybrids were typically less resistant than the Oua^R parental cell type. This result is exemplified by the data shown in Fig. 2.

These findings imply that ouabain resistance is a consequence of specific mutations affecting the binding of the inhibitor to Na-K-ATPase, or conceivably the response of the transport enzyme to bound ouabain. In the section below, concerning Oua^R human cells, I will discuss further data bearing directly on a mechanism of resistance. First, let us consider here recent results with CHO cells which detail the gene dosage effect associated with dominance of the Oua^R trait.

We have examined various Oua^R CHO cells and hybrids to determine full dose–response curves for ouabain inhibition of their K^+ transport function, measured in terms of influx of ^{86}Rb tracer (8). The results for a WT and two independent mutant CHO clones are shown in Fig. 3. Influx into the WT cells declines steadily with drug concentrations increasing from about 3×10^{-6} M to 1×10^{-4} M. Influx into the Oua^RC-C11 cells declines in this dose range, but at a substantially reduced rate, while transport by the Oua^RC-A1 cells is not affected at all until very high ouabain concentrations are employed. In terms of the ouabain concentration necessary to

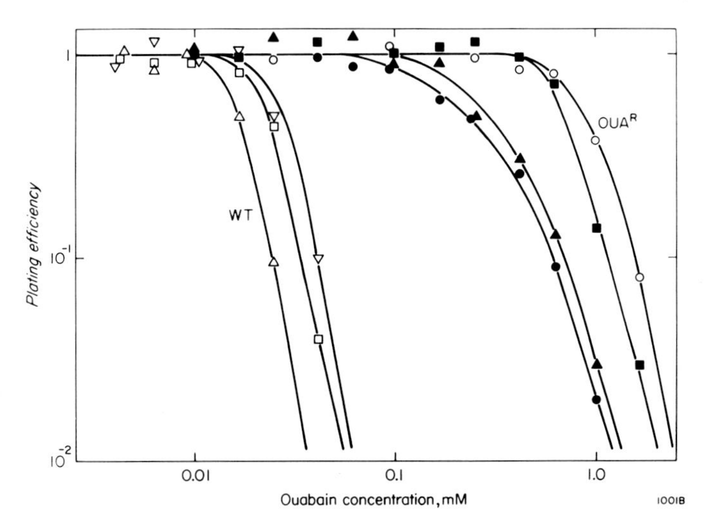

FIG. 2. The relative plating efficiencies as a function of ouabain concentration, 0.5 mM K^+, of parental and hybrid CHO cell clones: (▽) pseudodiploid Ade⁻ cells, WT for ouabain sensitivity (Oua^S) with an auxotrophic requirement for adenine (17,18): (△) pseudodiploid GAT⁻ Oua^S cells, with an auxotrophic requirement for glycine, adenosine, and thymidine (7,19); (□) pseudotetraploid Ade⁻ Oua^S × GAT⁻ Oua^S hybrid clone; (○) pseudodiploid GAT⁻ Oua^RC-C11 mutant cells, selected from the GAT⁻ Oua^S line; (●, ▲, ■) pseudotetraploid Ade⁻ Oua^S × GAT⁻ Oua^RC-C11 hybrid clones of independent origin. Isolation of the GAT⁻ Oua^RC-C11 clone and derivation of the hybrids, by fusion and selection for complementation of the auxotrophy markers, were carried out as previously described (7).

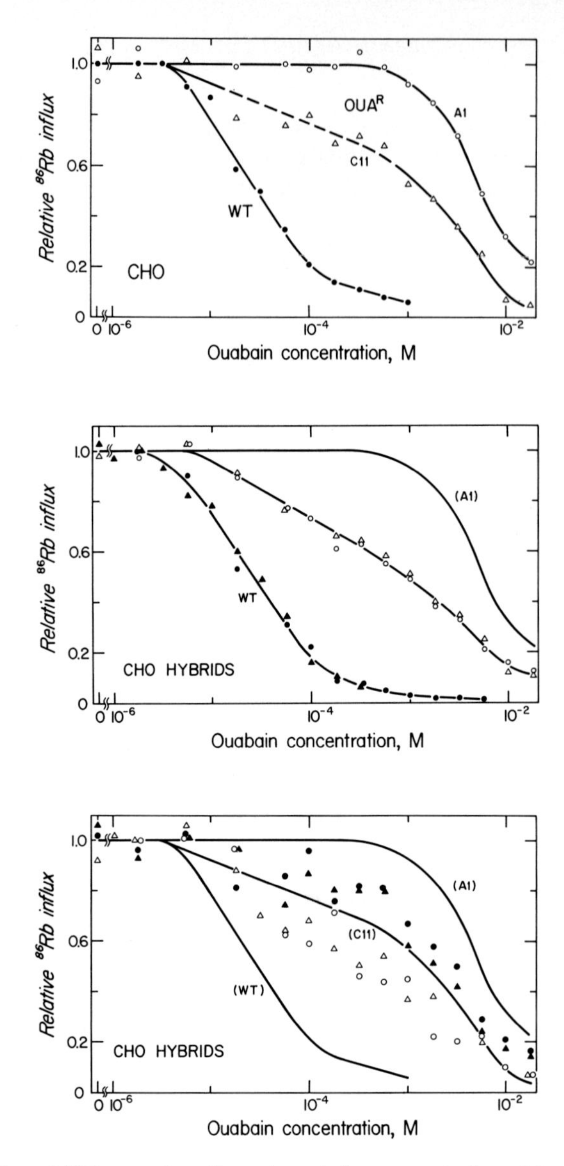

FIG. 3 (*top*). Influx of ^{86}Rb as a function of ouabain concentration, 0.5 mM K^+, for three pseudodiploid CHO cell clones: (●) Pro⁻ Oua^S cells, with WT ouabain sensitivity and an auxotrophic requirement for proline (1,18); (○) Pro⁻ Oua^{R}C-A1 cells derived from the latter (7); (△) GAT⁻ Oua^{R}C-C11 cells (cf. Fig. 2). The influx assays were conducted as previously described (8) except that the concentration of K^+ in the medium was 0.5 mM.

FIG. 4 (*middle*). Dose responses of ^{86}Rb influx for pseudotetraploid hybrid CHO cells: (●, ▲) Pro⁻ Oua^S × GAT⁻ Oua^S hybrids; (○, △) Pro⁻ Oua^{R}C-A1 × GAT⁻ Oua^S hybrids. The circles represent assays on pooled populations of around 10 hybrid clones and the triangles indicate assays on a clonal hybrid isolate of each type. The upper curve, without data points, indicates the result for the Pro⁻ Oua^{R}C-A1 parental cells shown in Fig. 3.

FIG. 5 (*bottom*). Dose responses of ^{86}Rb influx for pseudotetraploid hybrid CHO cells: (○, △) Pro⁻ Oua^S × GAT⁻ Oua^{R}C-C11 hybrids; (●, ▲) Pro⁻ Oua^{R}C-A1 × GAT⁻ Oua^{R}C-C11 hybrids. The three curves represent the results shown in Fig. 3 for the Oua^S, Oua^{R}C-C11, and Oua^{R}C-A1 parental cells.

reduce influx to 50% the level observed in the absence of drug, the Oua^{R}C-C11 and Oua^{R}C-A1 clones show, respectively, approximately 50- and 200-fold increases in the drug resistance.

The analogous results for *hybrids* formed from the WT and Oua^{R}C-A1 cells are shown in Fig. 4, in comparison to control data for hybrids between the two appropriate WT cell lines. The latter are nearly identical to the results for the pseudodiploid WT represented in Fig. 3. The WT × A1 hybrid cells show a susceptibility to ouabain intermediate to that of the two parental cell types and quite similar to that observed for the Oua^{R}C-C11 cells (Fig. 3). Figure 5 shows the results for hybrids of the Oua^{R}C-C11 cells with either WT or with Oua^{R}C-A1 cells. In each case the hybrids exhibit a dose–response intermediate to those of the two parental cell types, implying that all the parental alleles are being expressed, and the resulting phenotype is determined by their relative dosage.

The generality of the results illustrated in Figs. 3, 4, and 5 was tested by examining in analogous fashion the flux responses for a matrix of parental and hybrid cells that included two other independent mutants in addition to Oua^{R}C-A1 and Oua^{R}C-C11. The results are summarized in Table 1. The shapes of the dose–response curves for the Oua^{R}C-D1 and Oua^{R}C-E1 clones were similar to that shown for the Oua^{R}C-A1 cells in Fig. 3. The flux responses of the hybrid cells were nearly always intermediate to those of the two parental cell types, the two exceptions for hybrids of Oua^{R}C-D1 with other Oua^R lines lying within the range of experimental error.

Another point documented by these data is that cell viability is not lost until the transport function has been reduced below a threshold value. Comparison of the dose–responses in terms of viability and in terms of ^{86}Rb influx (Figs. 3 and 5 compared to Fig. 2 for the Oua^{R}C-C11 and the WT × C11 hybrid cells, and Fig. 4 compared to Fig. 11 of ref. 7 for the WT × A1

TABLE 1. *The relative ouabain doses required to reduce ^{86}Rb influx by 50% for various CHO cells and their hybrids*

Parental cells		Hybrids: GAT$^-$ ~1×	GAT$^-$ Oua^R-C11 50×	GAT$^-$ Oua^R-D1 120×
Pro$^-$	1×	1×, *1×	13×, *10×	20×
Pro$^-$ Oua^R-A1	200×	30×, *20×	70×, *80×	120×
Ade$^-$	1×	1.7×, *0.5×	10×, *7×	*7×
Ade$^-$ Oua^R-E1	170×	*5×	100×, *80×	*100×

Each of the Oua^R clones was isolated from the indicated parental auxotroph, and the hybrids were derived and assayed as indicated above. For ease of comparison, the results are expressed in terms of a ouabain concentration of 3×10^{-5} M (0.5 mM K$^+$) as a reference dose = 1×. Values with asterisks are from assays on cell populations containing a mixture of hybrid clones and those without asterisks refer to results for clonal hybrid isolates.

hybrid cells) indicates that the ouabain concentration necessary to reduce transport to 50% corresponds closely to the concentration required to reduce colony-forming efficiency to 10%.

OUABAIN-RESISTANT HUMAN CELLS

Investigations of ouabain resistance in human cells were undertaken with the primary objectives of taking advantage of their high drug sensitivity in characterization experiments and of employing the euploid experimental material afforded by explanted normal human fibroblasts. A number of Oua^R clones were isolated from five different strains of diploid human fibroblasts in two different laboratories (8). They were detected at frequencies of approximately 10^{-6} per viable cell in EMS-mutagenized cultures, but occur at frequencies $<10^{-7}$ in untreated cultures. Most selected clones retained their drug resistance upon subsequent cultivation.

The Oua^R diploid fibroblast isolates examined showed 20- to 180-fold increased resistance with respect to ouabain inhibition of growth, and also with respect to ouabain inhibition of ^{86}Rb influx into whole cells, demonstrating a correspondence here also between affects of the drug upon transport and viability (8). The shapes of the dose–response curves obtained for inhibition of transport were generally analogous to that shown for the CHO Oua^RC-A1 cells in Fig. 3, little decrease of flux into the mutant cells being discernible at some doses completely inhibitory for WT cells. For one Oua^R diploid fibroblast clone examined in some detail, no differences from WT cells in K^+ activation of transport or equilibrium Na:K balance were found, but at equilibrium with low ouabain concentrations the number of ouabain molecules bound per unit cell surface area was evidently half or less the number for WT cells (8,9). Study of these strains has, of course, been limited because of their senescence in culture.

We have also assembled a collection of established Oua^R human clones from the heteroploid HeLa cell line and have surveyed some of their properties (9,3). The results from these experiments furnish an enlightening supplement to observations with the other cell types.

Dose–response curves for ouabain inhibition of flux into the Oua^R HeLa isolates give clear indication of components of transport activity with differing ouabain susceptibility. The data in Fig. 6 illustrate this point for three independent HeLa mutants in comparison to the WT. For each of the clones there is a reduction of influx with increasing dose through the range 3×10^{-8} to 3×10^{-7} M, but whereas for WT the flux declines continuously to ouabain-insensitive background level, for the mutants an intermediate level is sustained until a second decline occurs for doses 1 to 2 orders of magnitude greater than those effective for WT. The resistant component in the mutants appears to represent 50 to 70% of the influx responsive to ouabain, a level sufficient for the cells to maintain viability

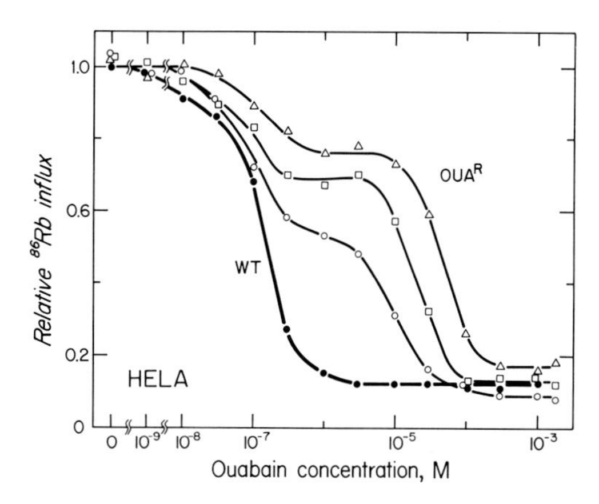

FIG. 6. Influx of ^{86}Rb as a function of ouabain concentration for WT HeLa cells (●) and the mutant HeLa clones Oua^{R}H-A3 (○), Oua^{R}H-B1 (□), and Oua^{R}H-E1 (△). The assays were conducted as previously described (8), with 5–6 mM K^+ in the medium and 2-hr incubation of the cells with drug prior to addition of the isotope and determination of the rates of uptake.

in these selective drug concentrations (compare Figs. 1 and 6 for the Oua^{R}H-B1 line).

Equilibrium levels of [^{3}H]ouabain binding for these HeLa mutants are 65 to 75% that for WT. There are no large differences in the rates at which equilibrium is approached, but with a similar isolate E. T. Brake and J. S. Cook (*personal communication*) found evidence for two types of binding sites with very different dissociation rates. In assays at 1.6×10^{-8} M to 7.5×10^{-7} M ouabain (0.1 mM K^+), we observed a gradual rise in equilibrium binding—which is specific in the sense that cation transport is inhibited in this dose range, and that a minimum of 80–90% of bound molecules can be competed out by very large excess of unlabeled drug (9). Scatchard plots of these data, shown in Fig. 7 for the WT and Oua^{R}H-B1 clones, are con-

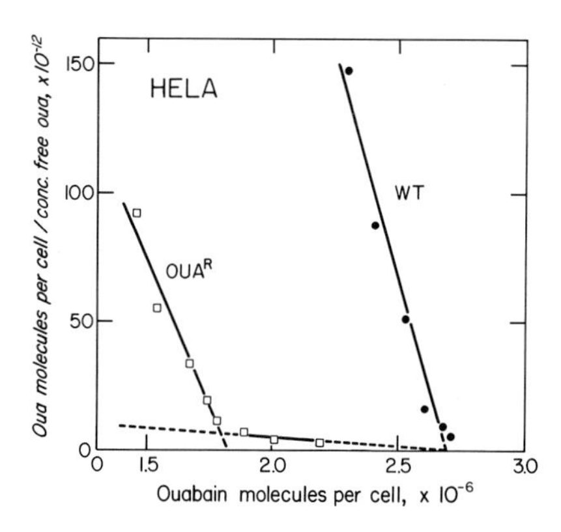

FIG. 7. Scatchard plots of data for ouabain-binding to the WT (●) and to the Oua^{R}H-B1 (□) HeLa clones. These determinations were made in medium containing 0.1 mM K^+; hence, the binding constants measured here are lower than the inhibitory concentrations observed in the experiments of Fig. 6.

sistent with the concept of two different populations of ouabain-binding sites in the resistant cells where only one population is indicated for the WT. These results imply that the WT binding is attributable to 2.7×10^6 sites/cell with $K_d \sim 3 \times 10^{-9}$ M, while one component of the Oua^{R}H-B1 binding is attributable to 1.8×10^6 sites with $K_d \sim 4 \times 10^{-9}$ M. The plot as drawn (assuming the same total number of sites per cell for each cell type, as they exhibit virtually the same cell volumes and influxes in the absence of drug) suggests that a second component of binding in the mutant may be due to 0.9×10^6 sites with $K_d \sim 9 \times 10^{-8}$ M.

It thus appears that in the Oua^R cells a fraction of the high-affinity binding sites have been supplanted by low-affinity sites, as would be expected if the isolates were heterozygous for an allele specifying the reduced affinity. For the Oua^{R}H-B1 line described in Fig. 7, there appears to be a single mutant allele in cells triploid for the locus. The fact that, according to the data in Fig. 6, Oua^R transport can maintain the influx at approximately two-thirds of the normal level is not inconsistent with this notion. The influx assays were performed after the cells had been preincubated in drug for 2 hr; by analogy to the case for partially inhibited WT cells (14), the unbound low-affinity sites might be expected to function with around two-fold increased efficiency as a consequence of elevated internal sodium after inhibition of the high-affinity sites.

Evidence of a different kind that the alteration in Oua^R HeLa cells is highly specific, probably affects the binding site, and that the changes in various mutants are different comes from tests of their cross resistances to compounds structurally and functionally (15,16) closely related to ouabain. We have assayed a number of independent isolates for their responses to the drugs digoxin (Dg) and digitoxin (Dt), which differ only with respect to a hydroxyl group at the steroid C-14 position that is thought to be important to binding (15). Representative data (9) are shown in Table 2, to support the following inferences: (*i*) All Oua^R mutants are cross-resistant to some extent. (*ii*) The relative degrees of cross resistance vary, as indexed by the ratios of increments in resistances shown in the last column. (*iii*) By these criteria, two broad categories of mutants at least can be distinguished. Some Oua^R isolates are no less resistant to Dt than to ouabain, but do show substantially less cross resistance to Dg. The differential in resistances to Dg and Dt here probably reflects the importance of the configuration at the C-14 position of the steroid to its activity against ATPase altered in a particular region of the binding site. Other clones are most resistant to ouabain and show lesser and approximately equal increments in resistances to Dg and Dt, possibly indicating changes affecting other components of the binding sites. (*iv*) As would be predicted, mutants with similar cross-resistance patterns can be obtained by selecting for resistance to Dg or to Dt (Dg^R and Dt^R isolates, respectively) instead of to ouabain.

TABLE 2. *Relative resistances of Oua^R HeLa cells to ouabain (Oua) and to the related drugs digitoxin (Dt) and digoxin (Dg)*

	Relative doses for 1% PE			Ratios of increments in resistances	
HeLa clone	Oua	Dt	Dg	Oua:Dt	Dt:Dg
WT-HI control	(5×10^{-8} M) = 1×	(1×10^{-7} M) = 1×	(5×10^{-8} M) = 1×	1	1
Oua^R H-A2	10×	2.4×	5.0×	4.2	0.48
Oua^R H-A3	60×	24×	20×	2.5	1.2
Oua^R HI-B1	100×	24×	20×	4.2	1.2
Oua^R HI-C1	240×	600×	20×	0.40	30
Oua^R HI-E1	240×	600×	20×	0.40	30
Dt^R HI-G1	60×	600×	20×	0.10	30
Dg^R HI-H1	240×	600×	20×	0.40	30

For each cell line and drug, the dose required for reduction of colony-forming efficiency to 1% the value without drug was estimated from a series of small cultures inoculated with 200 cells each and graded drug concentrations. The isolation and properties of the various HeLa clones will be detailed elsewhere (9).

CONCLUSIONS

The findings outlined above imply that some cells carrying a single mutation to ouabain resistance exhibit two kinds of Na/K transport sites, with normal and reduced affinity for ouabain, respectively. The Oua^R HeLa mutants are evidently heterozygous for a codominant allele that principally affects ouabain binding. The term codominant is applied here in an operational sense to indicate that properties attributable to the different alleles could be distinguished in the phenotype of the heterozygote, as in the influx and binding assays with Oua^R HeLa described in Figs. 6 and 7. In contrast, in some instances with other cell types where the Oua^R marker is known to be heterozygous, as in CHO cell hybrids (Figs. 4 and 5), resistance was expressed at an intermediate level, but components due to different alleles have not been resolved by any of the assays yet applied. Here, expression of the marker would more appropriately be described as "incompletely" dominant. There is at present no good evidence as to whether the different types of result are due only to the different cell types (which, after all, require very different selective drug doses), or instead reflect the phenotypes of different mutant loci.

Some types of Oua^R isolates not known a priori to be heterozygous, including the CHO Oua^{R}C-A1 (Fig. 3) and also human diploid fibroblasts with around 50% normal binding (8), showed no significant reduction of influx at all over a range of ouabain doses inhibitory to WT transport activity. Although possibilities such as a completely dominant mode of resistance or X-linkage of the locus cannot yet be excluded, it seems most likely that failure to discern intermediate dominance in these experiments was a consequence of the conditions utilized for the influx assays. If half or less of the transport activity were inhibited by preincubation with selective drug doses, the resulting increase in internal sodium and pump activity due to resistant sites could have reestablished influx at an apparently normal level (14; cf. above). The fact that resistant and sensitive transport activities in the HeLa isolates could be distinguished might thus be a lucky consequence of the aneuploidy in these cells, since the relevant locus appears to be triploid and even at maximum efficiency the drug-resistant sites alone might be unable to sustain influx at the level usual in the absence of drug.

These studies with ouabain-resistant mutants demonstrate that good genetic markers for membrane-associated properties can be obtained and utilized in cultured cells. The mutants offer an interesting example of dominant expression of a membrane trait, and should prove valuable to cellular studies concerning the effects of transport changes and to molecular investigations of the target enzyme.

ACKNOWLEDGMENTS

Most of the experimental work described in this chapter was performed at the Ontario Cancer Institute and The Hospital for Sick Children, Toronto, Canada, with support from the National Cancer Institutes of Canada and the United States and the Medical Research Council of Canada. Recent support has been from the U.S. National Institutes of Health through grants GM-21665 and CA-14051. I am grateful to James Kao, Theresa Lam, and Lynette Spencer for excellent technical assistance, and to Frank Solomon and John Barry for discussions during the preparation of this manuscript.

REFERENCES

1. Puck, T. T. (1972): *The Mammalian Cell as a Microorganism.* Holden-Day, San Francisco, Calif.
2. Siminovitch, L. (1976): On the nature of hereditable variation in cultured somatic cells. *Cell,* 7:1–11.
3. Baker, R. M., and Ling, V. (1976): Membrane mutants of mammalian cells in culture. In: *Methods in Membrane Biology,* edited by E. D. Korn. Plenum Press, New York (*in press*).
4. Glynn, I. M. (1964): The action of cardiac glycosides on ion movements. *Pharmacol. Rev.,* 16:381–407.
5. Schwartz, A., Lindenmayer, G. E., and Allen, J. C. (1972): The Na^+, K^+-ATPase membrane transport system: importance in cellular function. In: *Current Topics in Membranes and Transport,* Vol. 3, pp. 1–82, edited by F. Bronner and A. Kleinzeller. Academic Press, New York.
6. Dahl, J. L., and Hokin, L. E. (1974): The sodium–potassium adenosinetriphosphatase. *Annu. Rev. Biochem.,* 43:327–356.
7. Baker, R. M., Brunette, D. M., Mankovitz, R., Thompson, L. H., Whitmore, G. F., Simino-

vitch, L., and Till, J. E. (1974): Ouabain-resistant mutants of mouse and hamster cells in culture. *Cell,* 1:9–21.
8. Mankovitz, R., Buchwald, M., and Baker, R. M. (1974): Isolation of ouabain-resistant human diploid fibroblasts. *Cell,* 3:221–226.
9. Baker, R. M. (1976): Properties of ouabain-resistant mutants of cultured human cells. (*in preparation*).
10. Deaven, L. L., and Petersen, D. F. (1973): The chromosomes of CHO, an aneuploid Chinese hamster cell line: G-band, C-band, and autoradiographic analyses. *Chromosoma,* 41:129–144.
11. Arlett, C. F., Turnbull, D., Harcourt, S. A., Lehmann, A. R., and Colella, C. M. (1976): Mutation induction by gamma rays, ultraviolet light, ethyl methanesulphonate, methyl methanesulphonate and the fungicide captan: A comparison of the 8-azaguanine and ouabain-resistance systems in Chinese hamster cells. *Mutat. Res.* (*in press*).
12. Mayhew, E. (1972): Ion transport by ouabain-resistant and -sensitive Ehrlich ascites carcinoma cells. *J. Cell. Physiol.,* 79:441–451.
13. Brunette, D. M. (1972): Membrane properties of mouse L-cells. Ph.D. thesis, University of Toronto.
14. Cook, J. S., Vaughan, G. L., Proctor, W. R., and Brake, E. T. (1975): Interaction of two mechanisms regulating alkali cations in HeLa cells. *J. Cell. Physiol.,* 86:59–70.
15. Wilson, W. E., Sivitz, W. I., and Hanna, L. T. (1970): Inhibition of calf brain membranal sodium- and potassium-dependent adenosine triphosphatase by cardioactive sterols—a binding site model. *Mol. Pharmacol.,* 6:449–459.
16. Rosenberg, H. M. (1975): Variant HeLa cells selected for their resistance to ouabain. *J. Cell. Physiol.,* 85:135–142.
17. Ness, D. (1973): Auxotrophic mutants of the Chinese hamster ovary cell line. M.S. thesis, University of Toronto.
18. Thompson, L. H., and Baker, R. M. (1973): Isolation of mutants of cultured mammalian cells. *Methods Cell Biol.,* 6:209–281.
19. McBurney, M. W., and Whitmore, G. F. (1974): Isolation and biochemical characterization of folate-deficient mutants of Chinese hamster cells. *Cell,* 2:173–182.

Biogenesis and Turnover of Membrane Macromolecules,
edited by John S. Cook. Raven Press, New York 1976.

Reorganization of the Sea Urchin Egg Surface at Fertilization and Its Relevance to the Activation of Development

David Epel and James D. Johnson

Marine Biology Research Division, Scripps Institution of Oceanography, University of California, San Diego, La Jolla, California 92093

This chapter focuses on the nature of the plasma membrane of the unfertilized egg and the changes that occur in this membrane upon fertilization. This membrane fuses with the sperm plasma membrane at fertilization, which fusion then triggers embryonic development. The interaction has especially been studied in the sea urchin egg, wherein one of the immediate responses is a massive reorganization of the egg surface. The greater part of this chapter will concern the significance of these plasma membrane changes of the sea urchin egg in the development of new transport systems and the activation of development.

The interactions of sperm and egg are diagrammatically depicted in Fig. 1. In the sea urchin, the receptor on the sperm is the acrosomal process which binds to a peripheral proteinaceous coat of the egg, known as the vitelline layer. One sperm, the fertilizing sperm, passes through this peripheral coat, presumably through the action of hydrolytic enzymes, and then fuses with the plasma membrane. This fusion touches off a rapid and massive secretory phase, resulting from the secretion or exocytosis of "cortical granules." These 1-μm-diameter granules are embedded in the cortex of the egg; the 75-μm diameter egg of *Stronglyocentrotus purpuratus* contains about 15,000 of these granules, and they account for 7% of the total egg protein. The secretion, which begins at the point of sperm fusion and propagates around the egg in 20 sec, results in the release of proteolytic and glycosidic enzymes to the outside. Action of these enzymes elevates the vitelline layer away from the egg surface and alters the sperm receptors of this layer as a part of the block to polyspermy (reviewed in refs. 1 and 2).

More importantly, in terms of this chapter, this exocytosis also results in a massive reorganization of the egg plasma membrane; electron microscope studies indicate that the cortical granule membrane fuses with the plasma membrane and a part of the granule membrane becomes incorporated into the new egg surface (3). The result is that the surface of the fertilized egg can be considered a mosaic of membrane, part derived from the cortical granule membrane and part derived from the original egg surface.

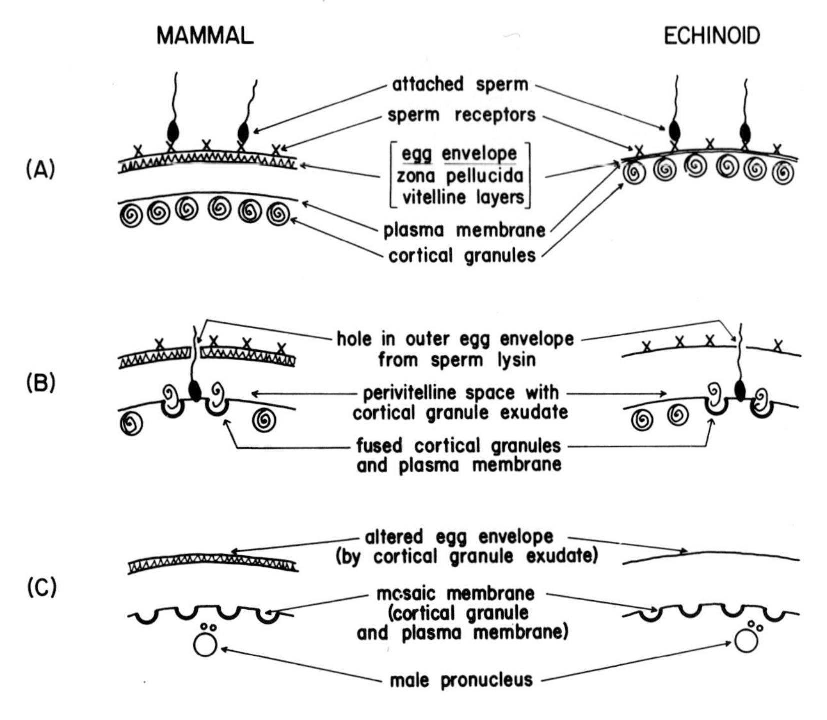

FIG. 1. A diagrammatic view of fertilization in the sea urchin and mammal egg. **A:** The phase of sperm attachment. **B:** Sperm fusion and the resultant secretory phase or exocytosis of the cortical granules (and elevation of the fertilization membrane in sea urchins). **C:** The surface of the fertilized egg, showing the mosaic nature of the plasma membrane and cortical granule membrane.

Accompanying these ultrastructural alterations are numerous physiological changes (see refs. 1 and 2). At the level of the membrane these include the development of Na^+-dependent amino acid transport, K^+ conductance, and transport systems for inorganic phosphate and nucleosides. At the level of the cytoplasm, protein synthesis increases markedly through increased translation of cytoplasmic mRNA. DNA synthesis is initiated, and there follows a period of rapid cell division and embryogenesis.

The major question we will address in this chapter is how these changes at the cell surface result in cytoplasmic changes responsible for initiating development. The conclusions we and our colleagues have reached, and which are developed here, are that the contact of the sperm with the egg results in increased intracellular calcium, which directly results in the cortical granule exocytosis and the reorganization of the egg plasma membrane. This exocytosis results in insertion of new plasma membrane necessary for the subsequent development of a Na^+-dependent amino acid transport system. The establishment of transport also requires an additional proteolytic alteration of the surface, an input of metabolic energy, and is sensitive to cytochalasin B. The second effect of the membrane reorganization is the loss of a regulatory surface glycoprotein. Loss of this protein can be related to the regulation of protein synthesis and may also regulate K^+ conductance and DNA synthesis as a part of the activation of the egg.

EXPERIMENTAL DISSECTION OF THE FERTILIZATION CHANGES

The changes at fertilization are an embarrassment of riches which are summarized in Fig. 2. They are depicted in temporal order and represent the sequence of changes in the eggs of *S. purpuratus* at 17°C (reviewed in ref. 1). These changes include early increases in intracellular Ca^{2+}, influx of Na^+ ion, initiation of the cortical exocytosis and the resultant block to polyspermy, activation of the enzyme NAD kinase, and a large increase in respiration. These changes all occur within the first minute after fertilization, are referred to as the "early" reactions, and are dependent on the increase of intracellular free calcium. These changes appear inexorably linked to each other and, as yet, cannot be experimentally dissociated. There then

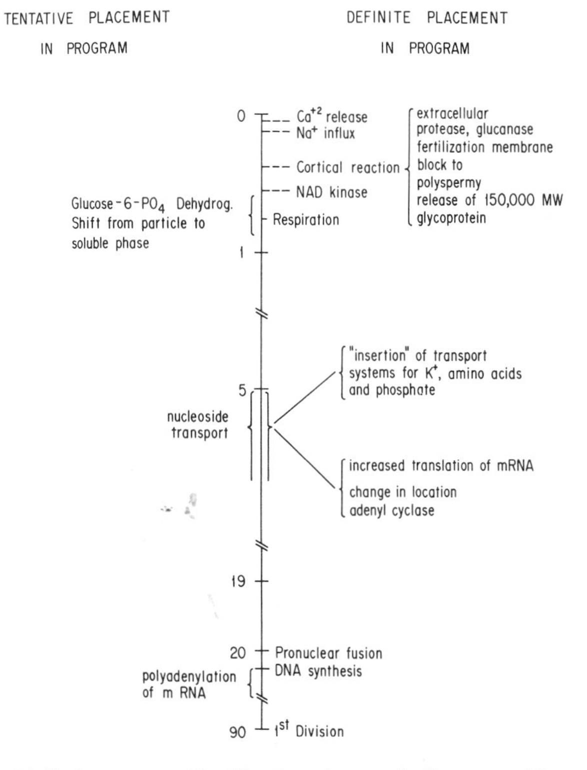

FIG. 2. A "timetable" of program of fertilization changes in the eggs of *Strongylocentrotus purpuratus. Right,* Well-defined changes; *left,* changes whose exact timing is unclear.

occur a series of "late" changes beginning at 5 min, which include the establishment of K^+ conductance in the cell membrane, activation of a Na^+-dependent amino acid transport system, increased protein synthesis, and initiation of DNA synthesis. Of these late changes, only the establishment of Na^+-dependent amino acid transport appears to be dependent or linked to the "early" changes; the other "late" changes all act as if they were independent of each other, i.e., they are not causally connected to each other and can be experimentally dissociated (4). As will be discussed, these changes probably result from loss of a regulatory surface protein, which acts in a suppressive fashion.

CALCIUM RELEASE—A PRIMARY EVENT

The evidence that a release of intracellular calcium is close to the primary event of fertilization was first suggested by Mazia in 1937 (5). He found an increase in dialyzable calcium after fertilization of *Arbacia* eggs. There was no change in total calcium; rather there appeared to be a translocation from the bound to the soluble form. That this increase in calcium was not coincidental, but causal, came from a recent experimental analysis by Steinhardt and Epel (6) and Chambers et al. (29). They found that increasing the intracellular Ca^{2+} content with the divalent cation ionophore A23187 would parthenogenetically activate eggs. The activation was independent of extracellular Ca^{2+} or Mg^{2+}, suggesting that the ionophore was releasing a divalent cation from intracellular stores. This is probably Ca^{2+}, since the bulk of the Mg^{2+} in the unfertilized sea urchin egg is in a free form, whereas the bulk of the Ca^{2+} is in a bound form (6).

This activation by ionophore has been observed in a wide variety of eggs, including mollusks, tunicates, amphibians, and mammals (7). In all cases except one, the activation was independent of exogenous divalent cations, which is consistent with the hypothesis that the ionophore acts by releasing divalent cations from intracellular stores. The exception, which supports the premise that Ca^{2+} is the activating agent, is the egg of the mollusk, *Spisula*. It requires exogenous Ca^{2+} for activation by ionophore A23187 (8).

Given that an increase in intracellular calcium is close to the primary event of fertilization, a major question is how the contact of the sperm acrosomal membrane with the egg plasma membrane results in an increase of intracellular calcium. A prerequisite to answering this question is identification of the store of intracellular calcium in the egg. Nakamura and Yasumasu (9) have presented evidence for a calcium-binding protein in the egg. Schuel and his colleagues (10), using pyroantimonate localization, claim that the plasma membrane itself binds calcium. If so, a sperm-induced perturbation in the plasma membrane could release calcium directly into the cytosol.

A second site for calcium store is the cortical granule itself. Vacquier (11) has made the important observation that adding Ca^{2+} to isolated cortical granules results in the loss of membrane integrity and the equivalent of an exocytosis, i.e., the contents of the granules become externalized. Making "lawns" of a cortical granule–plasma membrane preparation, he has presented evidence that the destabilization of the cortical granules by calcium is self-propagating. This suggests that the granules themselves may contain stores of labilizing agent (presumably calcium) which can propagate from granule to granule. This is an extremely intriguing hypothesis, although it is at variance with other experimental analyses showing that *in vivo* the propagation of the cortical granule exocytosis does not require adjacent granules (12).

The above experiment of Vacquier also provides an important clue as to what the increase in free Ca^{2+} is doing; by itself it can induce the cortical granule membrane to lose its integrity and release its contents. This release may be equivalent to exocytosis, but there has not yet been any evidence that in this *in vitro* system the granule membrane fuses with the overlying plasma membrane. However, given the general association of Ca^{2+} and secretion in many systems, Ca^{2+} is probably acting here, too, as an inducer of the cortical exocytosis.

CHANGES IN TRANSPORT PROPERTIES OF THE EGG

The next question we wish to consider is how the new mosaic membrane formed after cortical granule exocytosis is related to two transport changes that occur after fertilization. Experimental analysis indicates that one (K^+ conductance) develops independently of the cortical granule exocytosis, and hence must occur by modifications of the original plasma membrane, and the other (Na^+-dependent amino acid transport) develops only after the cortical granule exocytosis, suggesting that it depends on either the new mosaic membrane or some product of the exocytosis.

K^+ Conductance

First let us consider K^+ conductance, which apparently develops on the original plasma membrane. Work of Steinhardt and Mazia (13) has shown that one can activate K^+ conductance and several other aspects of egg metabolism by incubating eggs in millimolar concentrations of ammonia. This does not initiate a cortical reaction (13), indicating that alteration of the original plasma membrane leads to the conductance change.

Na^+-Dependent Amino Acid Transport

Further analyzing the ammonia activation, we found that protein synthesis was also activated in ammonia, but that there was no development of amino acid transport (4). However, if such eggs were fertilized, the Na^+-

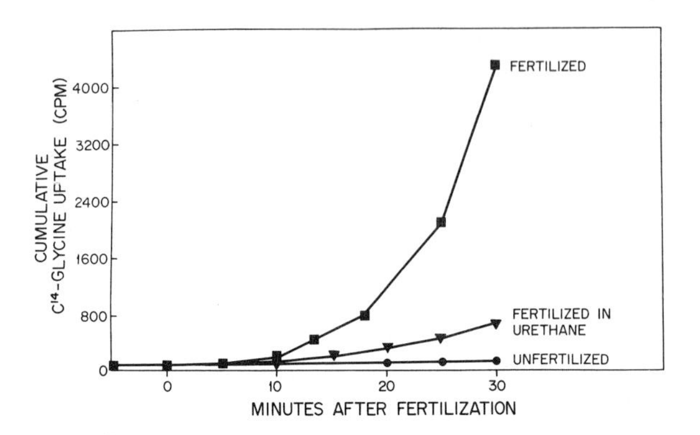

FIG. 3. Amino acid transport of *Arbacia punctulata* eggs following prevention of the cortical granule exocytosis by urethane [as described by Longo and Anderson (14)]. Eggs were inseminated in the presence of ^{14}C-glycine and cumulative uptake determined at the indicated times as described by Epel (18).

dependent amino acid transport system developed. This suggests that a cortical granule component might be needed, and this was affirmed by experiments in which *Arbacia punctulata* eggs were fertilized in the presence of urethane. Sperm entry occurs in urethane, but there is little cortical granule exocytosis (14). As can be seen (Fig. 3), very little transport develops in such eggs. Similar results have been obtained by Michael Garavito with *Strongylocentrotus purpuratus* in our laboratory. Garavito found that pretreatment of eggs in 40 mM NH_4Cl, pH 8, would prevent cortical granule breakdown upon a subsequent fertilization. Such eggs are activated and divide, but do not take up amino acid (Table 1). These results, then, are all consistent with the concept that the cortical exocytosis is a prerequisite for amino acid transport.

TABLE 1. ^{14}*C-glycine transport in Strongylocentrotus purpuratus*

Treatment	^{14}C-glycine transport (cpm) 60 min	120 min
Unfertilized, pH 8 seawater	278	358
Unfertilized, in 20 mM NH_4Cl, pH 8	297	365
Fertilized, in seawater	14,672	20,429
Fertilized, in ammonia (20% exocytosis)	1,354	2,814

^{14}C-glycine transport following fertilization in 20 mM NH_4Cl, pH 8, under conditions in which approximately 80% of the cortical granules do not undergo exocytosis.

Role of Exocytosis in Transport

Why is exocytosis required? One possibility is that some component of the amino acid permease system is "inserted" into the new surface resulting from the fusion of cortical granule and plasma membrane. Another is that the surface of the egg is altered by the enzymes contained in the cortical granules and which are normally released during the exocytosis.

Both hypotheses may be correct. So far, three enzymes have been described in the granules, a β-1,3-glucanase (15) and two trypsin-like proteases (16). The glucanase activity can be prevented by 2–10 mM threitol (D. Epel, *unpublished observations*); the protease activities by 10

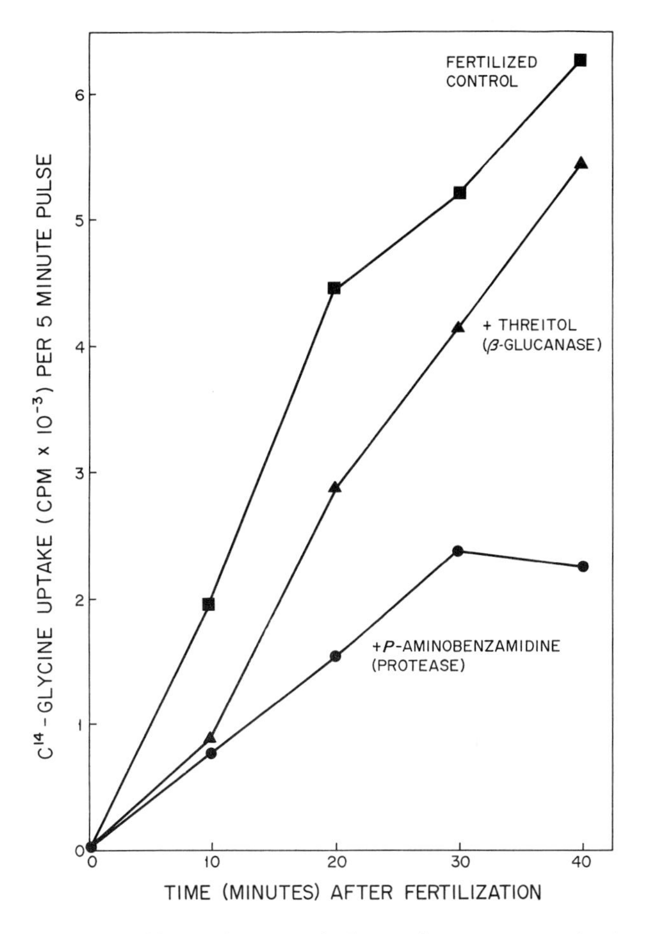

FIG. 4. The effects of inhibitors of two cortical granule enzymes on the development of amino acid transport. A 0.2% suspension of *Lytechinus pictus* eggs was incubated for 5 min in 2 mM threitol (β-glucanase inhibitor) or 10 mM *p*-aminobenzamidine (protease inhibitor). The eggs were then fertilized, pulsed for 5 min in ^{14}C-glycine at the indicated times, washed, and counts in the TCA-soluble supernatant determined.

mM *p*-aminobenzamidine (16). When eggs are fertilized in the presence of these inhibitors, only the protease inhibitors affect the development of transport (Fig. 4). The proteases apparently act to remove or alter some proteinaceous factor. If eggs are preincubated in bovine pancreatic trypsin and then fertilized in the presence of the trypsin inhibitors, there is no effect on transport. However, digestion of the egg surface with bovine trypsin, followed by incubation in ammonia, is insufficient to activate transport. A cortical exocytosis is still required, suggesting that the exposure of new surface may also be necessary.

Role of Amino Acid Transport in Development

The relevance of amino acid transport to normal development is unclear. The embryo normally uses endogenous protein reserves and embryos can develop through the pluteus stage in artificial seawater media which are initially devoid of the 10^{-7} M concentrations of dissolved amino acids that are present in coastal seawater. Also, embryonic development through the pluteus stage can occur when cortical granule exocytosis is prevented after fertilization, as with preincubation in high ammonia (Michael Garavito, *personal communication,* and Table 1) or by high hydrostatic pressure after insemination (17). Some transport eventually develops when cortical granule exocytosis is inhibited (Fig. 3), and perhaps this residual change in transport properties is enough to support development. Alternatively, the amino acid transport system may give the larvae a competitive edge in the race to feed, grow, and metamorphose in a plankton-poor environment. The transport system is a high-affinity one (K_m for glycine of 2×10^{-6} M, ref. 18). We have calculated that during the 30-day planktonic period of a sea urchin larva, these embryos would concentrate from the ambient levels an amino acid content equal to the protein content of the unfertilized egg.

Mechanisms in Development of Transport

Irrespective of its biological role, the change in Na^+-dependent transport provides an interesting experimental system for studying the development of an amino acid transport system. Earlier analysis showed that the development of this system could occur in the absence of new protein synthesis (18) indicating that transport results from an activation, possibly an insertion of preformed carriers. Some form of metabolic energy is required. Incubation of eggs in dinitrophenol after transport has been fully activated has no effect, but if dinitrophenol is added before the transport system has had a chance to develop, transport never appears (18). The nature of this energy-dependent event is unknown.

An additional clue as to how transport develops comes from the observation of Andrea Tenner and Epel that the *development,* but not the *function-*

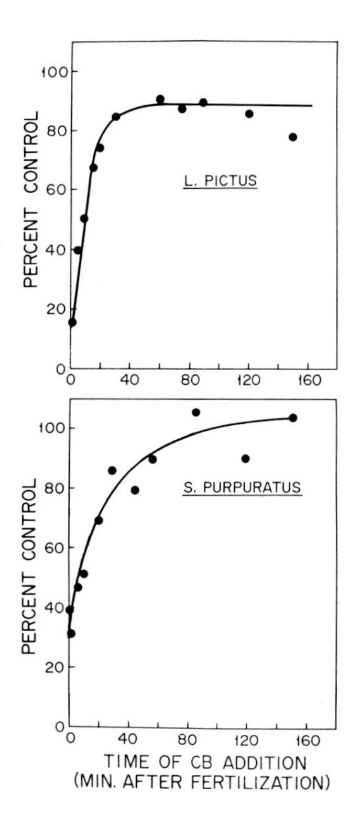

FIG. 5. Effects of cytochalasin B (CB) on the development of amino acid transport after fertilization of eggs of *Lytechinus pictus* and *Strongylocentrotus purpuratus.* A 0.2% suspension of eggs were fertilized and at the indicated times CB was added to a final concentration of 0.25 μg/ml. Thirty minutes after adding CB the eggs were pulsed for 5 min with ^{14}C-glycine, washed, and total uptake determined. The results depicted compare the percent transport of treated eggs to the control eggs. (Unpublished data of Andrea Tenner and David Epel.)

ing, of amino acid transport is also sensitive to small amounts (0.05–0.5 μg/ml) of cytochalasin B. The pattern of inhibition is remarkably similar to that seen with the energy inhibitors. This is shown in Fig. 5, which depicts the extent of amino acid transport that develops over a 2-hr period in relation to the time of cytochalasin B addition. As seen, addition of cytochalasin B after 2 hr has no effect on the fully developed transport system. However, addition of cytochalasin B very shortly after fertilization profoundly inhibits the development of transport. If transport develops from an insertion of preformed carriers into the plasma membrane, these experiments indicate that such insertion is energy-dependent and cytochalasin B-sensitive; their similar sensitivity suggests that both inhibitors may be acting at the same site.

Two Distinct Transport Domains in the Embryo Membrane

As noted above, the insertion of potassium conductance into the plasma membrane does not require a cortical exocytosis. Similarly, the insertion of potassium conductance is not sensitive to cytochalasin B (up to 10 μg/ml; R. Steinhardt, *personal communication*). It would thus appear that two separate transport systems become inserted into the membrane after

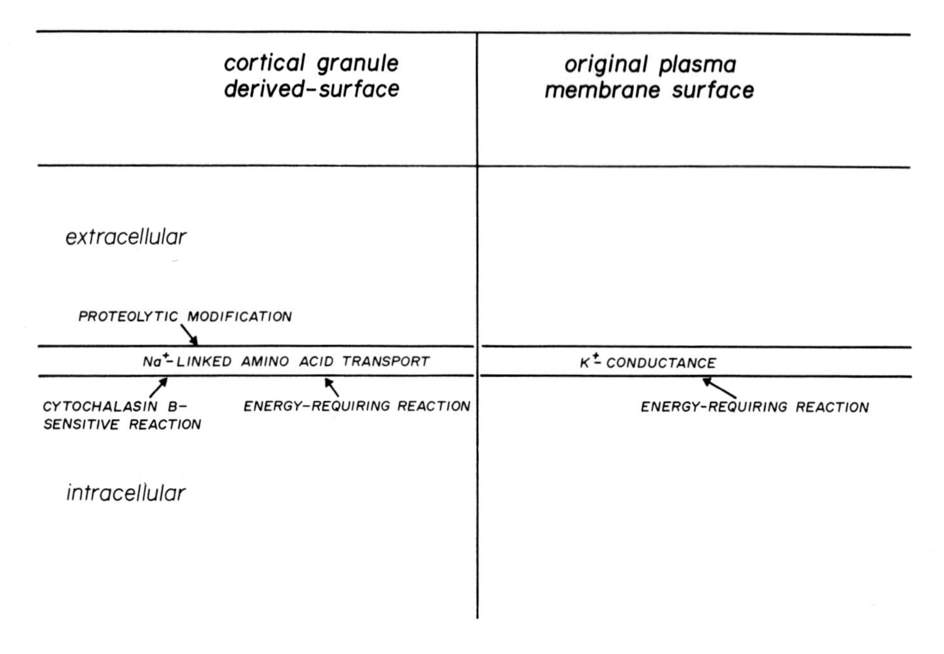

FIG. 6. A diagrammatic representation of the plasma membrane of the fertilized egg. As described in the text, the work of Steinhardt and Mazia (13) indicate that K^+ conductance develops from a modification of the original plasma membrane, possibly involving the loss of cell-surface protein. Na^+-dependent amino acid transport appears to develop in a different domain of the membrane, possibly that exposed by the cortical granule fusion. Proteolytic modification is required and development of transport is sensitive to cytochalasin B.

fertilization, and by different mechanisms. The amino acid transport system requires the cortical granule exocytosis and proteolytic activity, suggesting an insertion directly into an altered cortical granule membrane via an energy-requiring and cytochalasin-sensitive process. Conversely, the insertion of potassium conductance occurs in the original plasma membrane; energy is also required (19), but the process is unaffected by cytochalasin B. A model of the membrane depicting these separate domains is shown in Fig. 6.

METABOLIC ACTIVATION AND PLASMA MEMBRANE CHANGES

Exocytosis and Activation

We now consider another aspect of the cortical granule exocytosis—its relationship to the metabolic activation of the egg. As previously noted, exocytosis and activation seem unrelated to each other, since one can activate metabolism in the absence of a cortical reaction. Intimations that this was the case had been made earlier by Sugiyama and Kojima, but definitive evidence came from the aforementioned work of Steinhardt and Mazia on partial activation of eggs by ammonia (13). Subsequent work has shown that a number of agents will activate metabolism of eggs without a cortical reaction. These include nicotine (20), procaine (21), and ethylamine (22). That a cortical reaction is not needed suggests either that the initial increase in intracellular calcium activates metabolism by some second mechanism independent of the cortical reactions, or that the metabolic activation results from some second change which is independent of calcium. Our recent experimental analysis suggests that normally the cortical reaction results in loss of a surface glycoprotein, which acts as a suppressor of egg metabolism (20). Activating agents, such as ammonia, also cause loss of this surface glycoprotein in the absence of a cortical granule exocytosis, possibly by a membrane perturbation (20,22).

Regulation by Surface Glycoproteins

The evidence for this came from experiments we began at the Marine Biological Laboratory several summers ago (23). Using the lactoperoxidase procedure for labeling the egg surface, we found that there was a loss of a large amount of surface label when eggs were fertilized or activated by ionophore. This loss of label was primarily in a 150,000-MW glycoprotein. To our surprise, this loss was not dependent on a cortical reaction; it also occurred when eggs were activated with ammonia. Using other parthenogenetic agents such as nicotine or procaine, we found that there was a strong correlation between the extent of loss of labeled protein and the extent of activation of protein synthesis. This suggested that the loss of surface protein might be related to the activation of protein synthesis (20).

To test this hypothesis, we concentrated the macromolecules released from eggs when their metabolism was activated with ammonia, and added the dialyzed macromolecules back to activated eggs. We then activated unfertilized eggs with 10 mM ammonia, pH 8, and at 30 min removed the eggs from ammonia. The eggs were then resuspended in either seawater or in a seawater solution containing the mixture of macromolecules (at a concentration of about 30 μg protein/ml). We found that protein synthesis of eggs incubated in this cell surface protein mixture gradually began to decrease to the rate of the unfertilized egg (Fig. 7 and ref. 20). A similar correlation has been made by Mazia and his colleagues (22); they find that a number of treatments and agents perturbing the surface will activate egg metabolism.

A similar relationship between surface proteins and metabolic regulation has been suggested for both embryonic and adult vertebrate cells. Several workers have observed that the loss of a large molecular weight surface protein accompanies cell transformation (24,25). Koch (26) found that

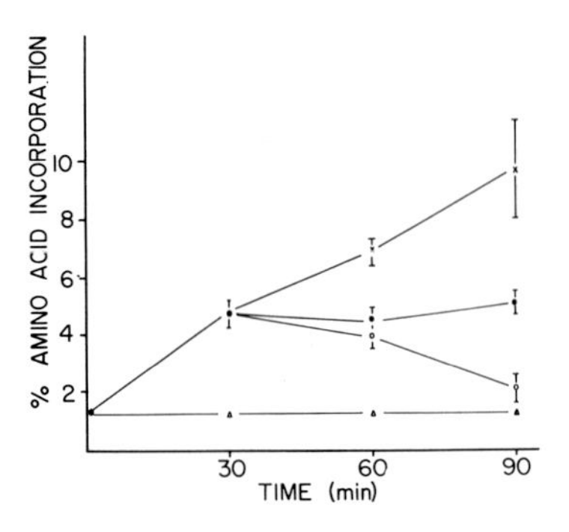

FIG. 7. The effects of the nondialyzable components released by activated eggs on protein synthesis. Unfertilized *L. pictus* eggs were incubated in 10 mM NH_4Cl, pH 8 for 30 min at which time a portion of the 0.2% suspension was pulsed with [^{3}H]valine for 5 min. The eggs were then washed twice in seawater and separate groups incubated in ammonia-containing seawater (**X**), regular seawater (●) or in seawater containing the nondialyzable cell surface components (○). Unfertilized controls (△) remained in seawater. 2.0-ml samples in duplicate from each group were pulsed for 5 min at 60 and 90 min. The error bars represent the standard deviation from 5 experiments [Data of Johnson and Epel (20)].

protein synthesis could be suppressed in HeLa-S_3 cells by glycopeptides isolated form the cell surface.

Mechanisms of Regulation by Surface Molecules

The most intriguing question raised by these findings is how the protein or proteins might be acting to suppress the metabolism of the egg. Our thinking is shaped by the observation that the effect is reversible, suggesting a rebinding to the cell, probably to a peripheral site. Also, these proteins must act in some pervasive fashion, since the various activating agents activate the late and independent changes of fertilization, which include such diverse and seemingly unrelated activities as K^+ conductance, DNA, and protein synthesis. Two possible models accounting for these activations are shown in Fig. 8. The first of these (model I) can be considered as an enzyme modification model, analogous to those proposed for hormone action, where the binding of a surface protein to the membrane results in the suppression of a regulatory cytoplasmic enzyme. Removal of this protein at fertilization or by activating agents results in a conformational change in the cytoplasmic enzyme and a resultant increase in enzyme activity. A likely candidate for this regulatory enzyme would be a nucleotide cyclase; however, there are no reported changes in cyclic AMP and cyclic GMP (reviewed in ref. 20). Alternatively, the cell-surface protein may be acting directly on a protein-modifying enzyme, such as a protein kinase. Such an enzyme is associated with the membrane of sea urchin eggs (27).

Model II (Fig. 8) can be considered a recompartmentation model. Such a

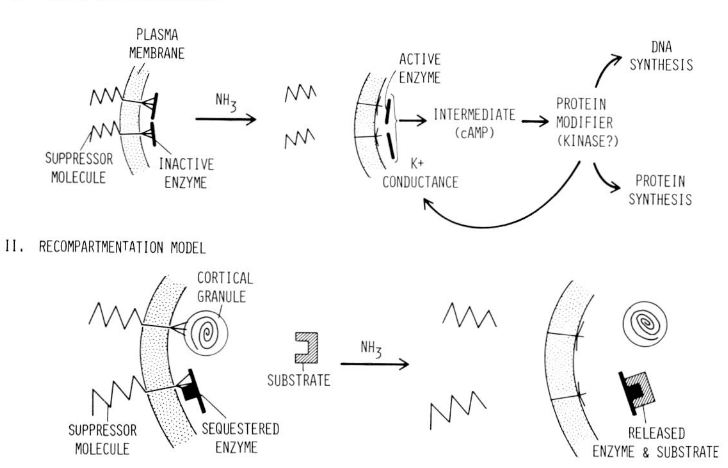

FIG. 8. Two models to account for the suppression of cell activity by surface macromolecules. The "protein modification model" assumes that the surface or suppressor molecule is somehow connected transmembrane to some regulatory enzyme, such as adenyl cyclase. Loss of the suppressor results in increased enzyme activity and the product (e.g., cyclic AMP) then activates a protein-modifying enzyme, such as a protein kinase. The "recompartmentation model" assumes that the cell-surface proteins sequester or mask critical enzymes from their substrate. Loss of the suppressor allows enzyme and substrate to react with each other and to promote cell synthesis.

model is appealing, since a change in cell structure could explain the pervasive effects that occur after fertilization. This recompartmentation could be in two dimensions, e.g., a change in membrane fluidity, or it could be in three dimensions within the cytoplasm, e.g., a change in cell actin or myosin. There is precedent for recompartmentation, since several enzymes, notably glucose 6-phosphate dehydrogenase, undergo a change from a particulate to a soluble form after fertilization (28).

CONCLUSIONS AND CAUTIONS

Fertilization results in a profound reorganization of the plasma membrane. This reorganization and its consequences are summarized in Fig. 9. A release of intracellular calcium, triggered by fusion of sperm and egg, initiates the cortical exocytosis and the resultant formation of a mosaic membrane. The exocytosis has other consequences, which are indicated as part of a cascade or dependent sequence. Na^+-dependent amino acid transport occurs in or on the domain contributed by the cortical granule surface. Development of transport requires a proteolytic modification of the surface and an energy input; it is also highly sensitive to cytochalasin B.

Other changes of fertilization, essentially the "late" changes such as K^+ transport and protein synthesis, appear to be related to a modification

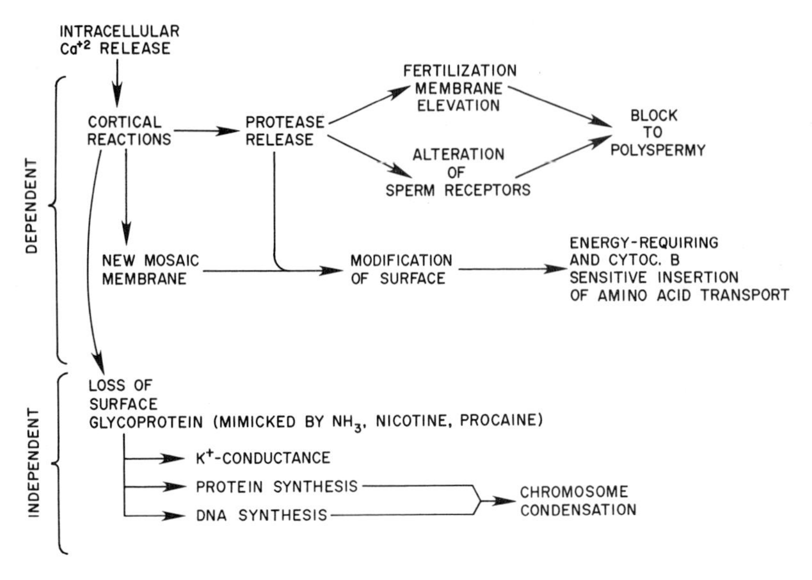

FIG. 9. Possible causal connections between various changes of fertilization. As described in the text, contact of sperm with egg releases intracellular calcium, which results in the cortical exocytosis and a chain of linked or dependent events. These include release of proteases, elevation of the fertilization membrane and the block to polyspermy. The exposed new surface is also modified by the proteases, and this surface in conjunction with an energy-requiring and cytochalasin-sensitive step becomes further modified to transport amino acids. The cortical exocytosis also results in loss of a surface glycoprotein. This protein can be released by ammonia, nicotine, procaine, etc. in the absence of the cortical exocytosis. This protein appears to regulate protein synthesis and may also regulate other metabolic events, such as K^+ conductance and DNA synthesis. These late changes are depicted as "independent," since they are not causally linked to one another.

of the original plasma membrane surface. Our findings indicate that protein synthesis is regulated by a surface glycoprotein, and it is tempting to speculate that this protein also controls K^+ conductance and DNA synthesis.

As depicted in Fig. 9, this protein release normally accompanies the exocytosis of fertilization. However, activating agents such as ammonia, nicotine, and procaine will release this protein with no cortical exocytosis. Once initiated, the metabolic changes turned on by these agents proceed independently of each other, and they are therefore considered as independent or parallel changes in Fig. 9.

A most interesting problem raised by these findings concerns the seemingly small role played by calcium in activation. So far, Ca^{2+} has only been implicated in the cortical exocytosis. As shown, a major accompaniment of this exocytosis is the release of the 150,000 MW glycoprotein. Yet, this glycoprotein can also be released by amine-type activating agents with no cortical exocytosis (ammonia and nicotine activate metabolism under conditions where the cortical exocytosis is not inhibited). Is the only linkage between Ca^{2+} and metabolism the release of the surface protein as a result of the membrane perturbations of exocytosis? Yet, eggs are not activated to divide by the amine-type activating agents, suggesting calcium initiates other critical changes. Also, the ionophore A23187 activates eggs which do not exhibit any apperent exocytosis, such as some mollusks (*Spisula*) and tunicate eggs.

Obviously, more information is required on the activation of these eggs. Are changes in Ca^{2+} compartmentalization also resulting in release of surface suppressor proteins in these eggs? Is calcium acting directly on some other cytoplasmic process? Or, are both types of changes a part of the activation of development? Answers to these questions can be obtained from comparative studies of different eggs.

So far, these studies on regulation at fertilization reveal that metabolic control of embryonic cells is remarkably similar to that operating in differentiated cells; fortunately, such regulation is more easily studied in eggs. An important question for the future will be to what extent such surface regulation is involved in embryonic determination and differentiation.

ACKNOWLEDGMENT

This work was supported by Grant BMS 73-07003-A02 from the National Science Foundation.

REFERENCES

1. Epel, D. (1975): The program of and mechanisms of fertilization in the echinoderm egg. *Am. Zool.*, 15:507–522.
2. Monroy, A. (1973): Fertilization and its biochemical consequences, *Addison-Wesley Module in Biology No. 7*. Addison-Wesley, Reading, Mass.
3. Millonig, G. (1969): Fine structural analysis of the cortical reaction in the sea urchin egg: after normal fertilization and after electric induction. *J. Submicrosc. Cytol.*, 1:69–84.
4. Epel, D., Steinhardt, R., Humphreys, T., and Mazia, D. (1974): An analysis of the partial metabolic derepression of sea urchin eggs by ammonia; the existence of independent pathways. *Dev. Biol.*, 40:245–255.
5. Mazia, D. (1937): The release of calcium in *Arbacia* eggs on fertilization. *J. Cell Comp. Physiol.*, 10:291–308.
6. Steinhardt, R. A., and Epel, D. (1974): Activation of sea urchin eggs by a calcium ionophore. *Proc. Natl. Acad. Sci. USA*, 71:1915–1919.
7. Steinhardt, R. A., Epel, D., Carroll, E. J., and Yanagimachi, R. (1974): Is calcium ionophore a universal activator for unfertilized eggs? *Nature*, 252:41–43.
8. Schuetz, A. W. (1975): Induction of nuclear breakdown and meiosis in *Spisula solidissima* oocytes by calcium ionophore. *J. Exp. Zool.*, 191:433–440.
9. Nakamura, M., and Yasumasu, I. (1974): Mechanism for increase in intracellular concentration of free calcium in fertilized sea urchin egg. A method for estimating intracellular concentration of free calcium. *J. Gen. Physiol.*, 63:374–388.
10. Cardasis, C., Schuel, H., and Herman, L. (1974): Ultrasturatural localization of calcium in unfertilized sea urchin eggs. *Cell Biol.*, 63:49a.
11. Vacquier, V. D. (1975): The isolation of intact cortical granules from sea urchin eggs: Calcium ions trigger granule discharge. *Dev. Biol.*, 43:62–74.
12. Allen, R. D. (1958): The initiation of development. In: *The Chemical Basis of Develop-*

ment, pp. 17–66, edited by W. D. McElroy and B. Glass. Johns Hopkins University Press, Baltimore, Md.
13. Steinhardt, R. A., and Mazia, D. (1972): Development of K^+-conductance and membrane potentials in unfertilized sea urchin eggs after exposure to NH_4OH. *Nature,* 241:400–401.
14. Longo, F. J., and Anderson, E. (1970): A cytological study of the relation of the cortical reaction to subsequent events of fertilization in urethane-treated eggs of the sea urchin *Arbacia punctulata. J. Cell Biol.,* 47:646–665.
15. Epel, D., Weaver, A. M., Muchmore, A. V., and Schimke, R. T. (1969): β-1,3-glucanase of sea urchin eggs. Release from particles at fertilization. *Science,* 163:294–296.
16. Carroll, E. J., and Epel, D. (1975): Isolation and biological activity of the proteases released by sea urchin eggs following fertilization. *Dev. Biol.,* 44:22–32.
17. Chase, D. (1967): Inhibition of the cortical reaction with high hydrostatic pressure and its effects on the fertilization and development of sea urchin eggs. Ph.D. thesis, University of Washington, Seattle.
18. Epel, D. (1972): Activation of an Na^+-dependent amino acid transport system upon fertilization of sea urchin eggs. *Exp. Cell Res.,* 72:74–89.
19. Steinhardt, R. A., Shen, S., and Mazia, D. (1972): Membrane potential, membrane resistance and an energy requirement for the development of potassium conductance in fertilization reaction of echinoderm eggs. *Exp. Cell Res.,* 72:195–203.
20. Johnson, J. D., and Epel, D. (1975): A relationship between release of surface proteins and the metabolic activation of sea urchin eggs at fertilization. *Proc. Natl. Acad. Sci. USA,* 72:4474–4478.
21. Vacquier, V. D., and Brandiff, B. (1975): DNA synthesis in unfertilized sea urchin eggs can be turned on and turned off by the addition and removal of procaine hydrochloride. *Dev. Biol.,* 47:12–31.
22. Mazia, D., Schatten, G., and Steinhardt, R. A. (1975): The turning on of activities in unfertilized sea urchin eggs; correlation with changes of the surface. *Proc. Natl. Acad. Sci. USA,* 72:4469–4473.
23. Johnson, J. D., Dunbar, B. S., and Epel, D. (1974): Fertilization-associated changes in the plasma membrane proteins of *Arbacia punctulata* eggs. *Biol. Bull.,* 147:485.
24. Yamada, K., and Weston, J. A. (1974): Isolation of a major cell surface glycoprotein from fibroblasts. *Proc. Natl. Acad. Sci., USA,* 71:3492–3496.
25. Hynes, R. O. (1974): Role of surface alterations in cell transformation: The importance of proteases and surface proteins. *Cell,* 1:147–156.
26. Koch, G. (1974): Reversible inhibition of protein synthesis in HeLa cells by exposure to proteolytic enzymes. *Biochem. Biophys. Res. Commun.,* 61:817–824.
27. Murofushi, H. (1974): Protein kinase in the sea urchin egg cortices. Its purification and characterization and some properties of an endogenous protein kinase in the cortices. *Biochem. Biophys. Acta,* 364:260–271.
28. Isono, N., Ishusuka, A., and Nakano, E. (1963): Studies on glucose-6-phosphate dehydrogenase in sea urchin eggs. I. *J. Fac. Sci. Univ. Tokyo,* Sect. 4, 10:55–66.
29. Chambers, E. L., Pressman, B. E., and Rose, B. (1974): The activation of sea urchin eggs by divalent ionophores A23187 and X-537A. *Biochem Biophys. Res. Commun.,* 60:126–132.

Biogenesis and Turnover of Membrane Macromolecules,
edited by John S. Cook. Raven Press, New York 1976.

Development of Chemical Excitability in Skeletal Muscle

Douglas M. Fambrough* and Peter N. Devreotes

*Department of Biophysics, Johns Hopkins University, Baltimore, Maryland 21210; and *Department of Embryology, Carnegie Institution of Washington, Baltimore, Maryland 21210*

The mechanisms involved in the turnover of the acetylcholine (ACh) receptor are of special interest to cell biologists since the receptor is an "integral" membrane protein. Therefore, the intracellular processing, incorporation into the surface membrane, and degradation of the acetylcholine receptor must include and probably typify the biogenesis and turnover of "integral" plasma membrane proteins. A unique advantage of the ACh receptor over other known membrane proteins is that a specific, sensitive, irreversible marker exists for the receptor – [^{125}I]monoiodo-α-bungarotoxin.

Beyond its interest as a "trace" component of the plasma membrane of embryonic myotubes, the receptor is a major, perhaps the only, protein component of the postsynaptic membrane at neuromuscular junctions. The details of receptor turnover in embryonic muscle cells will be relevant to the mechanism of formation of the neuromuscular junctions and of synapses between nerve cells.

ACCUMULATION OF ACH RECEPTORS IN PLASMA MEMBRANES

Figure 1 illustrates the accumulation of ACh receptors, α-bungarotoxin binding sites, on surfaces of primary chick myotubes through their development for 10 days in culture. During the first 2 days after plating, prior to the fusion of myoblasts into multinucleated myotubes, very little [^{125}I]monoiodo-α-bungarotoxin binding activity is associated with the myoblasts. At the time of the onset of fusion, at about 2 days, there is a rapid increase in the rate of accumulation of ACh receptors on newly formed myotubes. The accumulation continues as more myoblasts fuse into myotubes until, at about 10 days after the initial plating, the number of receptors accumulated reaches a plateau level. Qualitatively similar ACh receptor accumulation curves have been obtained for primary muscle cultures of chick, cow, and for several myogenic cell lines from rat (1–3). The quantitative aspects of the accumulation curves (rate of increase) obtained by various investigators differ, probably because the rate of receptor accumulation is related to the overall kinetics of muscle development which, in turn, depend upon

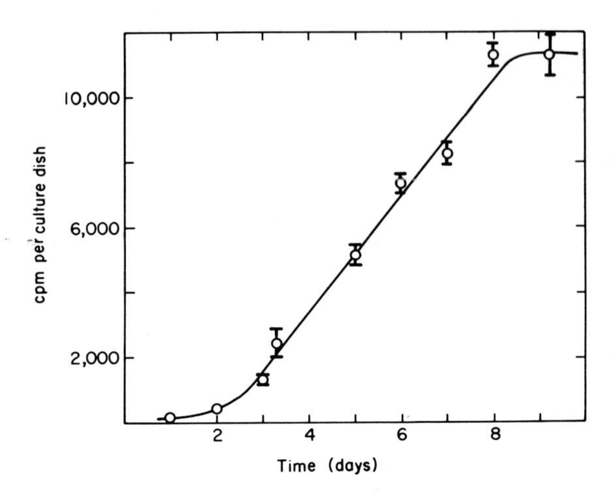

FIG. 1. Accumulation of ACh receptors as a function of culture age. A large set of chick muscle cultures, prepared by mechanical disruption of 11-day chick embryo leg muscle, and plating at 5×10^5 cells per 35-mm petri dish, were cultured in F12 at 37°C. At each of indicated times, [^{125}I]monoiodo-α-bungarotoxin was added to the culture medium so that the final concentration was 0.2 μg/ml. After 60 min, unbound [^{125}I]monoiodo-α-bungarotoxin was removed by three 10-min rinses in Hank's balanced salt solution containing 0.5% BSA. The amount of bound radioactivity was then determined by scintillation counting.

such factors as the individual lot of serum used in the medium, initial plating density, and frequency of feeding.

Figure 2 illustrates the accumulation of receptors into surface membranes of adult rat diaphragm following section of the phrenic nerve (4). Other investigators have obtained similar time courses for increases in α-bungarotoxin binding or in sensitivity to iontophoretically applied ACh in denervated rat diaphragm (5,6) and in denervated mouse and frog muscles (7,8).

There are many differences, of course, between myogenesis in tissue culture and the onset of denervation changes in adult muscle. Developing muscle in tissue culture rapidly gains mass and surface area, while the denervated muscle changes relatively little in these respects and eventually loses both mass and surface area as denervation atrophy sets in. Thus, in adult muscle, the increase in ACh receptors following denervation represents an increase in the number of receptors per unit area of surface. In tissue-cultured muscle the rate of increase in receptor density is less rapid than the increase in total receptors. Despite these obvious differences, however, there is a similarity. ACh receptor production and incorporation into plasma membranes is initiated and continues until the nonsynaptic (or extrajunctional) surface membrane contains approximately 500–1,000 functional ACh receptor sites per square micrometer (9) and becomes very responsive to extracellular ACh. The molecular mechanisms by which the

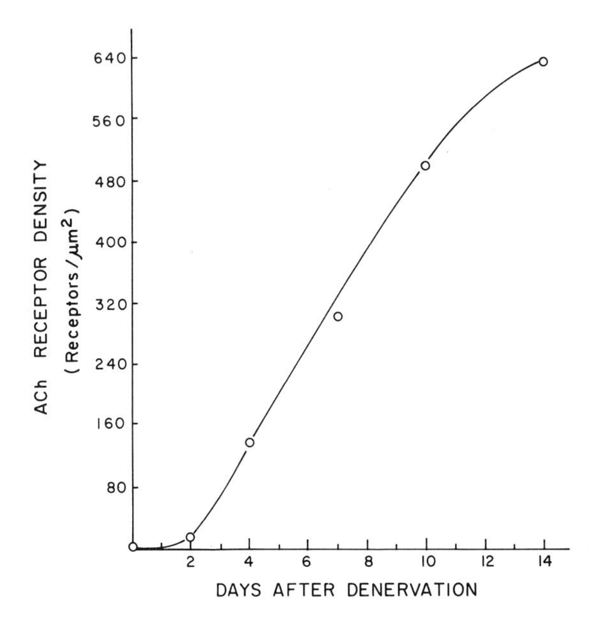

FIG. 2. Time course of increase in extrajunctional ACh receptors following denervation of rat hemidiaphragm.

ACh receptors are synthesized, processed, incorporated into plasma membranes, and even degraded are almost surely qualitatively the same. Indeed, our study of these processes so far has confirmed this supposition and has demonstrated a close quantitative similarity as well.

Early observations indicated that receptors accumulate on primary chick myotubes in culture because the rate at which new receptors are incorporated into the surface membrane is somewhat greater than the rate at which receptors are removed from the membrane. Since the rate for degradation is not a function of the age of the culture (*vide infra*), the difference between rates of incorporation and degradation of receptors is greatest during the earliest part of receptor accumulation and decreases progressively. By the time receptor accumulation has reached a plateau level (about 10 days in culture), the rates have become equal and turnover of receptors in the plateau region of the accumulation curve simply results in replacement of receptors removed from the membrane.

A MODEL FOR RECEPTOR BIOSYNTHESIS AND TURNOVER

Our study of receptor biosynthesis and turnover has led to a model which is illustrated diagrammatically in Fig. 3. Messenger RNA molecules direct the synthesis of nascent receptor polypeptide chains. The nascent chains are quickly assembled into receptors which are indistinguishable from func-

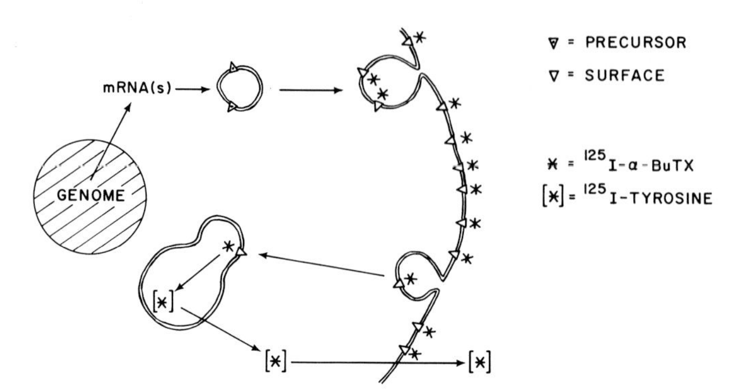

FIG. 3. Hypothetical "life cycle" of ACh receptors in cultured chick skeletal muscle. The figure depicts a cross section through a myotube with receptors symbolized as triangles in membrane profiles. Solid arrows indicate processes consistent with data on receptor metabolism. Symbols for precursor and surface receptors and for [^{125}I]α-bungarotoxin and [^{125}I]tyrosine are indicated on figure.

tional surface receptors in molecular size, buoyant density, electrophoretic mobility, and interaction with cholinergic agonists and antagonists. These "precursor" receptors then undergo a 2- to 3-hr transit before reaching the surface membrane. During this transit the "precursor" receptors are found on an internal membrane system from which they can be solubilized by nonionic detergents. Upon reaching the surface membrane, new receptors are either incorporated uniformly over the surface, or are rapidly distributed over the surface (if they are incorporated at discrete locations). There is an energy-dependent gate, blocked by 2,4-DNP, very close in the temporal sequence to the incorporation step. Once members of the surface membrane population, receptors have a half-life of 22 hr and are removed by a random process which does not discriminate between "new" and "old" receptors. There is an energy-dependent step close in the temporal sequence to the removal of receptors from the membrane. This energy dependence likely involves internalization of receptors by phagocytosis and formation of phagolysosomes (10), although the degradation step itself may be energy dependent (11). Subsequently, receptors are found in secondary lysosomes. If [^{125}I]monoiodo-α-bungarotoxin-ACh receptor complexes are internalized, the [^{125}I]monoiodo-α-bungarotoxin is degraded concomitantly with the receptor and [^{125}I]monoiodotyrosine appears in the medium.

The model presented above is derived primarily from evidence obtained from embryonic chick myotubes developing in tissue culture (12,13). The observations which have led to the model are the main subject of this chapter. Many of the experiments performed on chick myotubes are much more difficult or impossible to perform in an exploratory manner in other systems, for technical reasons. However, once a consistent observation has been made on chick cultures and one knows what to look for, other systems can be "spot-checked" to see if the mechanism involved in a given system resembles that found in the chick myogenic cultures. Several "spot-check" experiments have been performed on embryonic rat myotubes in culture and on several types of adult muscle following denervation and will be presented below.

Receptor Biosynthesis

One of the first predictions of the model illustrated in Fig. 3 is that new receptors which are incorporated into the surface membrane are synthesized *de novo* within myotubes. This can be definitively shown by labeling the receptor through its amino acid precursors. One faces two major technical problems in attempting to label the receptor with radioactive precursors. First, it is difficult to demonstrate that the receptor itself carries the radioactive tracer. One must first independently demonstrate complete purification of the receptor and show that the radioactive precursor is incorporated into the purified fraction. Complete purification of the ACh receptor has only been achieved from electric organs of fish, and the receptor represents about 1 part in 10^5 of the cell protein in our tissue culture system. Second, because the receptor is present in such small amounts in embryonic tissue, large quantities radioactive precursors of high specific activity must be employed. These technical difficulties have been partially overcome by Brockes and Hall (14), who obtained good evidence for the incorporation of [^{35}S]methionine into new ACh receptors appearing in denervated rat diaphragm. The diaphragms were incubated in small volumes so that sufficient concentrations of carrier-free [^{35}S]methionine could be employed. The receptor was purified about 500-fold by affinity chromatography and to a greater, but unmeasured, extent by sucrose gradient sedimentation.

Because receptors exist at even lower densities in chick myogenic cultures than in the denervated rat diaphragm, and since the method of Brockes and Hall (14) is not practical for routine experiments, we have chosen to label the receptor in another manner. The receptor has been labeled with amino acids which change a physical property of the receptor: the buoyant density (13). Since newly synthesized receptors of altered density can be detected as α-bungarotoxin–receptor complexes of altered density, no purification of the receptor is necessary. [^{2}H]- or [^{13}C]amino acids are incorporated into newly synthesized receptors by incubation of myotube cultures, under conditions which allow active receptor production, in medium containing these amino acids. After sufficient incubation of the myotubes with [^{2}H]- or [^{13}C]amino acids, receptors of increased buoyant density begin to appear on myotube surfaces. These newly synthesized ^{2}H or ^{13}C receptors are labeled with [^{125}I]monoiodo-α-bungarotoxin to form [^{125}I]monoiodo-α-bungarotoxin–^{2}H- or ^{13}C- receptor complexes.

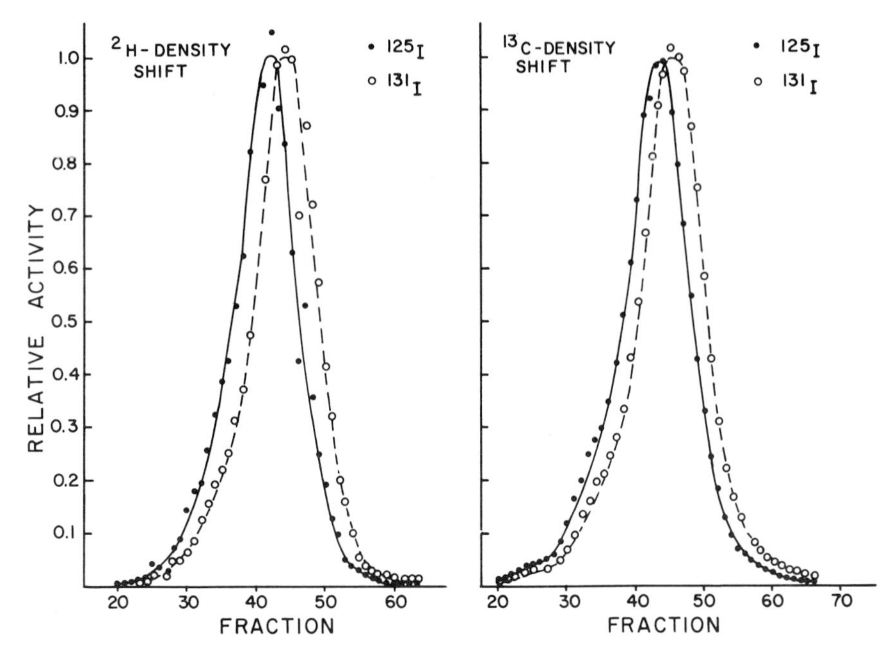

FIG. 4. Density shift of newly synthesized receptors. Several large 5-day-old cultures were incubated with 0.2 μg/ml unlabeled α-bungarotoxin overnight. After 16 hr incubation, the medium was switched to medium containing amino acids substituted with ^{2}H or ^{13}C (preparation of the medium is described below), but still containing 0.2 μg/ml unlabeled α-bungarotoxin. After 5 more hours of incubation the cultures were washed to remove unbound α-bungarotoxin and incubated in "heavy" medium for an additional 2 hr. During the final 30 min of incubation [^{125}I]α-bungarotoxin was added to the cultures such that the concentration was 0.2 μg/ml. The cultures were washed at 4°C, the cells scraped from the culture dishes, homogenized in 10 mM Tris, and centrifuged at 17,000 rpm for 45 min. The pellet was extracted in a small volume of 1% Triton-10 mM Tris. An aliquot of the extract was mixed with a marker of [^{131}I]α-bungarotoxin–receptor complexes, which were extracted from an independent set of cultures. The mixture was loaded in a thin band on a step metrizamide gradient containing 0.95 ml of each of five concentrations: 40%, 35%, 30%, 25%, 20% (w/w) of metrizamide. The solvent for the metrizamide was deuterium oxide. Centrifugation was carried out in a Beckman SW65 Rotor for 15 hr at 54,000 rpm at 4°C. Gradients were collected by puncturing the bottom of the tube. 100 four-drop fractions were collected from each gradient into scintillation vials. Radioactivity was monitored simultaneously on appropriate channels for ^{125}I and ^{131}I. After correction for cross-over between the channels, the data were plotted and smooth curves were drawn through each peak (peak fractions contained at least 800 cpm). Data were then normalized to the highest point on the smooth curve, and thus some points are greater than 1. Preparation of "heavy" medium: Heavy medium was our normal growth medium, except that 16 of the amino acids were substituted by an [^{2}H]- or [^{13}C]amino acid mixture, and serum normally added to the medium was exhaustively dialysed against Hank's. Amino acid mixtures were obtained from Merke, Sharpe and Dohme and were either 98% atom purity ^{2}H or 80% atom purity ^{13}C.

These complexes, extracted in nonionic detergents, are mixed with a marker of [^{131}I]monoiodo-α-bungarotoxin–^{1}H-, ^{12}C-receptor complexes. The latter are formed on surfaces of myotube cultures grown in medium containing only [^{1}H]-, [^{12}C]amino acids. The mixed extract is then centrifuged through a density gradient of metrizamide-D_2O. The experimental design is described in greater detail in the legend to Fig. 4. This figure illustrates that the newly synthesized receptors have been shifted to a higher density position in the gradient by incorporation of amino acids containing the stable isotopes.

Incorporation of Receptors into Plasma Membranes

According to our model of receptor metabolism (Fig. 3) newly synthesized receptors are incorporated into an intracellular membrane system which supplies the surface. Evidence for these "precursor" receptors comes first from experiments involving inhibitors of protein synthesis. One experiment, which employs puromycin to inhibit protein synthesis, is illustrated in Fig. 5. Protein synthesis is inhibited by 98% within 1 min of addition of puromycin at the start of the experiment. The block is maintained by the continued presence of puromycin. Incorporation of new receptors is interrupted about 2 hr following the addition of puromycin, suggesting that protein

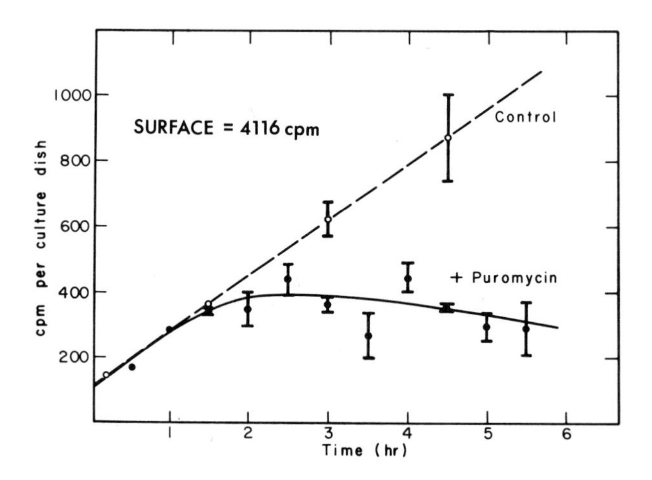

FIG. 5. Appearance of new ACh receptors following block of all old receptors with α-bungarotoxin. Chick muscle cultures were incubated in unlabeled α-bungarotoxin to saturate all receptors and the appearance of new ACh receptors was observed by incubation of sets of cultures in medium containing [^{125}I]α-bungarotoxin at various times subsequently (○). At time zero, 100 μg/ml puromycin was added to some cultures. In these puromycin-treated cultures, new receptors appear at normal rate for about 2 hr, after which there is no further incorporation of new receptors into plasma membranes (●). Total number of surface receptors on this culture set was determined, as described in Fig. 1, on four identical cultures which had not been pretreated with unlabeled α-bungarotoxin. (See ref. 12.)

synthesis is required for continued incorporation of new receptors. However, as the inhibition of new receptor incorporation takes between 2 and 3 hr to develop, it is suggested that there are presynthesized receptors which can supply the surface for a short time in the absence of new protein synthesis. This point is reiterated in Fig. 6. In this experiment protein synthesis has been interrupted by a puromycin "pulse" of 1 hr duration. Inhibition of incorporation of new receptors again develops 2–3 hr following the onset of inhibition of protein synthesis. Receptor incorporation in this case resumes, at the control rate, 2–3 hr following the removal of the block on protein synthesis. This experiment again suggests that there exist presynthesized receptors sufficient to supply the surface for 2–3 hr. In addition, it suggests that once new receptor synthesis is resumed these new receptors must undergo a 2–3 hr transit time before their exposure on the myotube surface.

Since the rate of incorporation of new receptors is about 5% per hour, the presynthesized receptors in transit should represent about 10–15% as many receptors as there are surface receptors. It is not required *a priori* that the "precursor" receptor equivalents be in a functional form recognizable as ACh receptors. For instance, the "precursor" receptors might exist as receptor subunits incapable of interaction with α-bungarotoxin. It has been found, however, that additional α-bungarotoxin binding sites, not available for interaction with extracellular α-bungarotoxin, are present in detergent

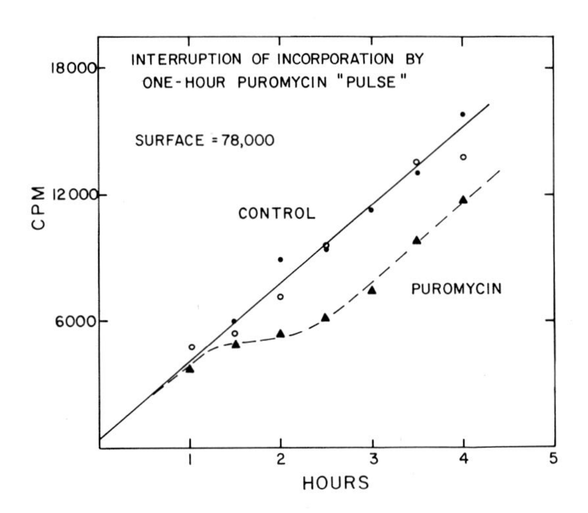

FIG. 6. Interruption of incorporation by puromycin. All sets of cultures (controls and puromycin) were incubated overnight with 0.2 μg/ml unlabeled α-bungarotoxin. During the final hour of incubation, one set of cultures (▲) was treated with 10 μg/ml puromycin. Another set (○) was preincubated with 10^{-4} DNP. Then all the puromycin, DNP, and unbound α-bungarotoxin was rinsed out. All the cultures were then returned to normal medium ($t=0$). The remainder of the experiment was carried out as normal incorporation measurement (see Fig. 5). DNP pretreatment has little effect on receptor incorporation as this rate is identical to control (●).

extracts of myotube cultures. Cultures first saturated with unlabeled α-bungarotoxin, are extracted with nonionic detergents. *In vitro* assays, with [^{125}I]monoiodo-α-bungarotoxin employed to probe for extra binding sites, reveal that there exist additional binding sites amounting to 10% as many sites as there are surface sites. These "precursor" receptors are indistinguishable in chromatographic, electrophoretic and sedimentation properties from functional surface receptors, and their α-bungarotoxin binding activity is antagonized by the cholinergic ligands *d*-tubocurarine and hexamethonium. As shown in Fig. 7, the number of additional binding sites found in the extracts rapidly decreases when the cultures are preincubated in puromycin. The kinetics of decrease match the kinetics of incorporation of new re-

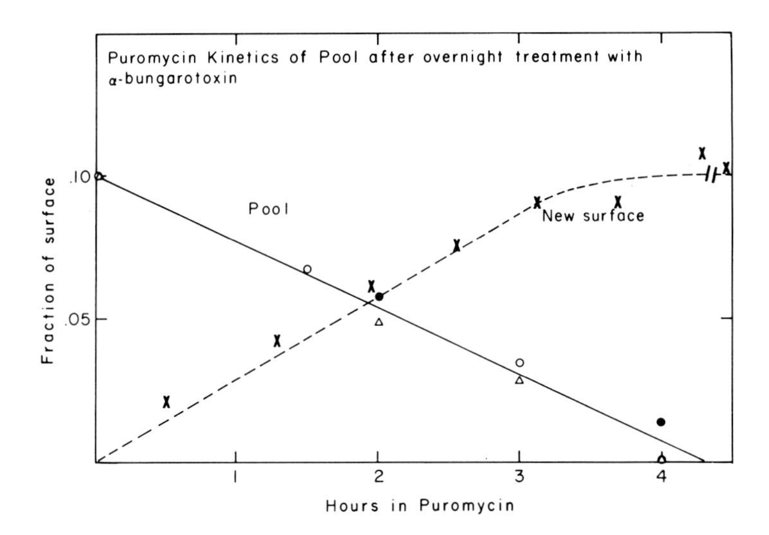

FIG. 7. Kinetics of precursor disappearance and surface receptor appearance. Decrease of "precursor" in presence of puromycin. All cultures were incubated overnight with 0.5–1.0 μg/ml unlabeled α-bungarotoxin. The next day, sets of 5–6 cultures were pretreated with 20 μg/ml puromycin for the indicated times. All cultures were then washed to remove unbound α-bungarotoxin. In one experiment (○), sets were immediately extracted in 1% Triton X-100. In the other two experiments (●), (△), the cells were scraped from the dish, homogenized by 25 strokes of a Dounce (pestle B) homogenizer (in 10 mM Tris at 0°C), and centrifuged at 50,000 × *g* × 30 min. The pellets were extracted with 1% Triton X-100, 10 mM Tris, pH 7.8. Each of the extracts was incubated with 0.01 μg/ml [^{125}I]α-bungarotoxin for 1 hr at room temperature. The complexes formed were freed of unbound α-bungarotoxin by Bio-Gel P-60 chromatography. The fractions in the excluded peak were pooled and an aliquot was taken for analysis by velocity sedimentation on sucrose gradients. In all cases the area under the 10 S peak in the gradient was measured. For each experiment, all values were normalized such that the control value equals 0.10 of the surface. Incorporation curve: Incorporation was measured as described in Fig. 5, except that cells were first saturated with unlabeled α-bungarotoxin for about 15 hr instead of 30 min. The actual plateau level reached in this experiment was about 8% of the total surface, but it has been normalized to 0.10 for comparison with the pool decrease. The long time point after the indicated break is at 6 hr 20 min (**X**). (See ref. 12.)

ceptors, suggesting that when receptors leave the "precursor" fraction they are incorporated into the surface.

The kinetics of decrease of the "precursor" fraction, illustrated in Fig. 7, appear to be linear from the time of puromycin addition: there is no lag between the onset of inhibition of protein synthesis by puromycin ($t = 0$) and the onset of the decrease in "precursor" receptors. This suggests that there is not a significant compartment of nonfunctional receptor equivalents which supplies the "precursor" fraction. In other words, it appears that receptor nascent chains are quickly assembled into functional receptor "precursors" which resemble surface receptors at least by the crude physical and biochemical criteria that have been employed.

The functional receptors present in the "precursor" fraction probably exist as membrane-bound molecules. To demonstrate this, myotubes were homogenized and a crude subcellular fractionation was carried out. All of the precursor α-bungarotoxin binding activity is sedimentable at $27{,}000 \times g$ for 1 hr. Several other enzymatic activities, including glycosyl transferases (15), hormone receptors (16), and 5′-nucleotidase (17), are common to both surface and internal membrane fractions, although a precursor–product relationship has not been established.

In contrast to the block in receptor incorporation by puromycin or cycloheximide, which takes several hours to develop, 2,4-dinitrophenol blocks receptor incorporation from the outset (data not shown). This suggests that there is an energy-dependent step very close to the time at which receptors are incorporated into the surface membrane.

Receptor Degradation

Once receptors have reached the surface membrane, they have a half-life in the surface of about 22 hr. The stability of receptors in the surface membrane can be measured by saturating the surface receptors with [^{125}I]monoiodo-α-bungarotoxin and then monitoring the release of radioactivity into the medium. Such an experiment is shown in Fig. 8. Over 90% of the radioactivity originally associated with surface receptors ([^{125}I]monoiodo-α-bungarotoxin–receptor complexes) at the start of the experiment is released into the medium with first-order kinetics. This suggests that both new and old receptors are degraded at random. The radioactivity released into the medium is the proteolytic degradation product of [^{125}I]monoiodo-α-bungarotoxin, [^{125}I]monoiodotyrosine. Figure 9 shows that during the first 8 hr after the initial formation of [^{125}I]monoiodo-α-bungarotoxin–receptor complexes on the myotube surface, a small amount of undegraded α-bungarotoxin is released. In all subsequent time periods, however, only degraded α-bungarotoxin is released by the cells. The identification of the released radioactivity as [^{125}I]monoiodotyrosine is shown in Fig. 10. The degradation product of [^{125}I]monoiodo-α-bungarotoxin chromatographs in

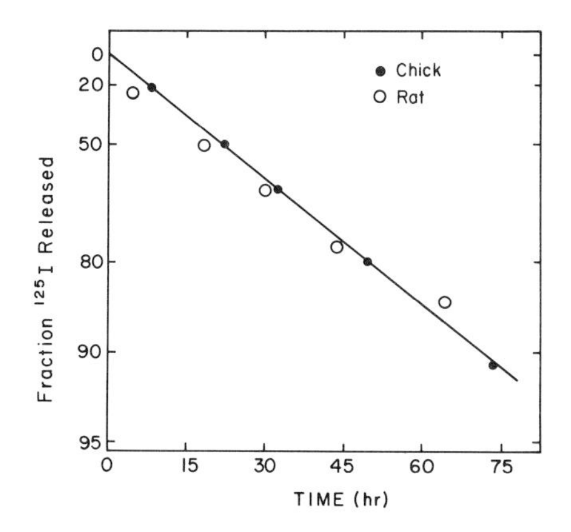

FIG. 8. First-order kinetics of ACh receptor degradation. Eight 6-day-old cultures grown on 35-mm dishes, were labeled with [^{125}I]α-bungarotoxin, washed, and returned to 2 ml of culture medium. All operations were carried out sterilely. At each of the indicated times, the entire 2 ml of medium was removed from each dish and was replaced by 2 ml of fresh medium. After the medium was centrifuged to remove cellular debris, 1 ml of the supernatant was counted and the other 1 ml was reserved for the experiment described in Fig. 9. After the medium had been removed for the last time point, the myotubes were extracted with 1% Triton to determine the number of remaining [^{125}I]α-bungarotoxin–ACh–receptor complexes. Rat, (○); chick, (●). Standard deviations for chick data are between $\pm$0.01 and $\pm$0.006. (See ref. 12.)

the previously established position of monoiodotyrosine. As shown in Fig. 8, similar results have been found for degradation of receptors in primary rat muscle cultures.

It was conceivable that the [^{125}I]monoiodo-α-bungarotoxin initially bound to surface receptors might be removed from the binding site and independently degraded, resulting in reactivation of the α-bungarotoxin binding site. To eliminate this possibility, it was necessary to show that the release of radioactivity, previously bound to surface receptors, quantitatively parallels the loss of α-bungarotoxin binding sites from the myotube surface. As has been shown above (see Fig. 5), following a 3-hr pretreatment of myotubes with puromycin, incorporation of new receptors into the myotube surface is completely inhibited. Figure 11 shows that the release of radioactivity to the medium is only slightly inhibited by puromycin and continues independent of incorporation for at least 8 hr. Thus, under these conditions, from 3 to 8 hr in puromycin, the rates of release of radioactivity previously bound to surface receptors and loss of original surface receptors from myotubes which were not pretreated with α-bungarotoxin can be compared. Figure 12 shows the results of such an experiment. The rate of loss of receptors and the rate of appearance of radioactivity in the medium, representing degraded [^{125}I]monoiodo-α-bungarotoxin–receptor complexes, are identical within experimental error. This suggests that both free receptors

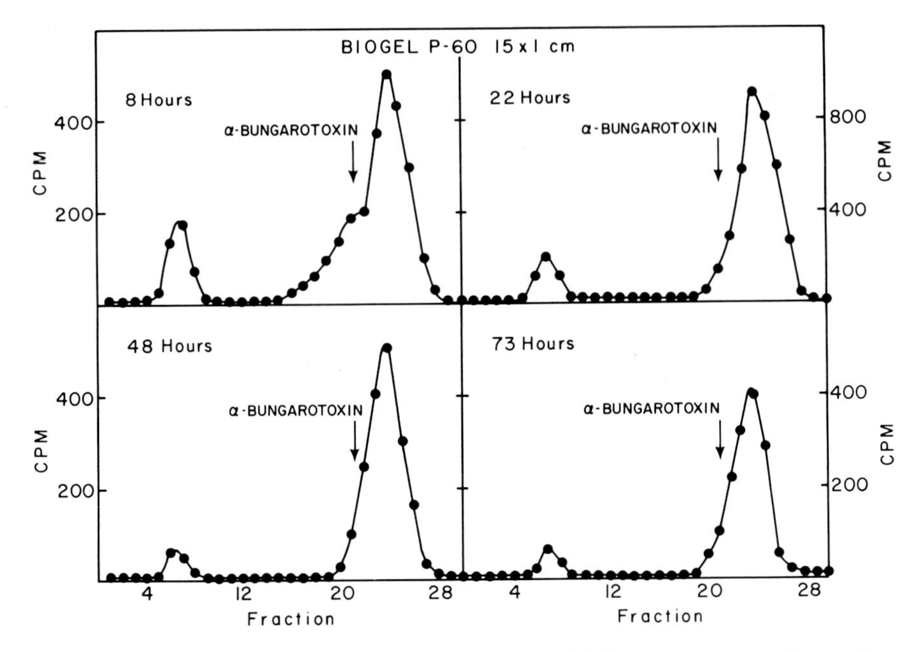

FIG. 9. Bio-Gel P-60 chromatography of the degraded [^{125}I]α-bungarotoxin. F12 medium was collected during each of the intervals 0–8 hr, 8–22 hr, 22–48 hr, 48–73 hr after chick muscle ACh receptors were labeled with [^{125}I]α-bungarotoxin. Each 1-ml sample was layered on a 15 × 1 cm Bio-Gel P-60 column. The elution buffer was 1% Triton X-100, 100 mM Tris-HCl pH 7.8. 0.5 ml fractions were collected at a flow rate of 6 ml/hr. Fractions were collected directly into scintillation vials and counted after the addition of 3.5 ml Triton-toluene flour. Recovery of radioactivity from such columns was always quantitative. (See ref. 12.)

and α-bungarotoxin–receptor complexes are degraded at similar rates and that the appearance of [^{125}I]monoiodotyrosine in the medium is a valid measure of receptor degradation. This experiment also suggests that the binding of α-bungarotoxin to surface receptors does not change the rate of degradation.

Preliminary characterization of the mechanism of degradation of surface receptors is consistent with a model (as in Fig. 3) in which receptors are internalized and then degraded. The initial kinetics of degradation, as shown in Fig. 8, are linear and extrapolate to 100% at the initial time point. This type of data alone suggests, that if intermediate steps exist in the degradation process they must be quite rapid and involve a very small fraction of the receptors. Accordingly, the initial part of the degradation experiment has been carried out in finer detail, measuring the [^{125}I]tyrosine released into the medium as a function of time after a very brief "pulse" incubation of myotubes with [^{125}I]α-bungarotoxin. The results are shown in Fig. 13. In the initial stages, the rate of release of [^{125}I]tyrosine into the medium from [^{125}I]monoiodo-α-bungarotoxin initially bound to surface receptors requires

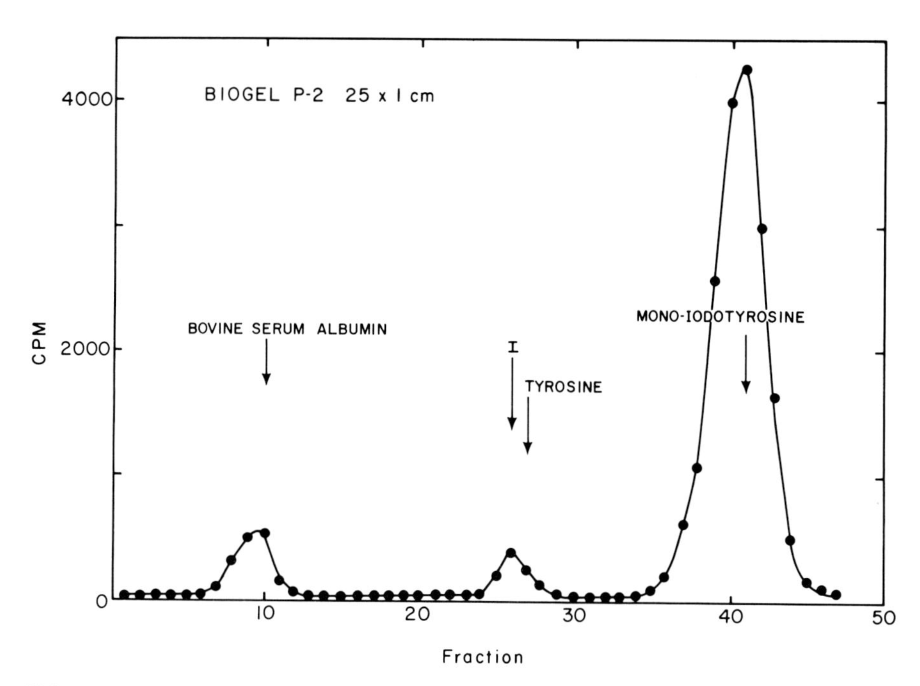

FIG. 10. Chromatography on Bio-Gel P-2 of radioactive material released from chick myotubes after [^{125}I]α-bungarotoxin was bound to ACh receptors. 15 chick muscle cultures were plated at 10^6 cells/100-mm dish and cultured in complete medium for 5 days. After incubation for 30 min with 0.1 μg [^{125}I]α-bungarotoxin/ml and extensive rinsing to remove unbound toxin, the cultures were incubated at 37°C for 16 hr in HEPES-buffered, balanced salt solution. The medium was removed, centrifuged at low speed to sediment cell debris, and the supernate was lyophilized, dissolved in 4 ml of distilled water, applied to a Bio-Gel P-2 column 1.0 × 25 cm and eluted with 0.05 M ammonium acetate, pH 5.0. The column was calibrated with markers as indicated and centrifuged α-bungarotoxin elutes in the void volume. One-ml fractions were collected. Recovery of radioactivity from such columns was between 65 and 70%. (See ref. 12.)

about 90 min to reach a maximum. During this time some of the [^{125}I]mono-iodo-α-bungarotoxin–receptor complexes, newly formed on the myotube surface, must move into a small compartment of receptors earmarked for degradation. Since the lag time, shown in Fig. 13, is equivalent to about one hour of degradation and since receptors are normally degraded at about 3% per hour, the compartment should contain about 3% as many receptors as there are surface receptors.

The effects of a number of experimental manipulations have suggested that receptors are degraded by a mechanism which involves internalization rather than one which occurs at the myotube surface or in the medium. As illustrated in Table 1 homogenates of myotubes labeled with [^{125}I]mono-iodo-α-bungarotoxin have little or no capacity to produce [^{125}I]mono-iodotyrosine. Degradation is inhibited by 2,4-dinitrophenol as shown in Fig. 13. As illustrated in Fig. 14, degradation of receptors is extremely tempera-

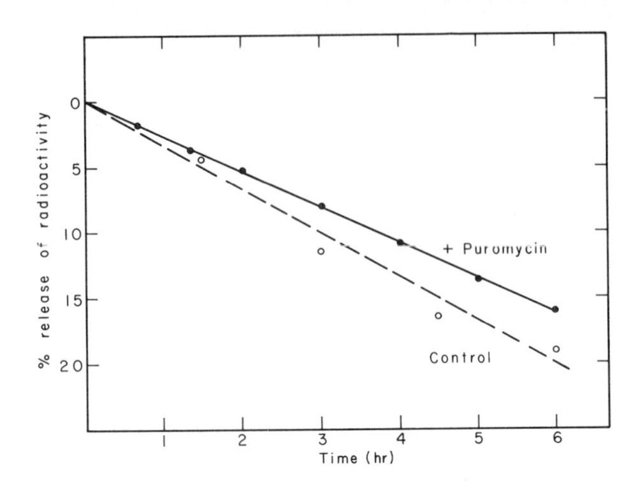

FIG. 11. Effect of puromycin on the release of ^{125}I from chick muscle cultures following binding of [^{125}I]α-bungarotoxin. Cultures were incubated in 0.1 μg/ml [^{125}I]α-bungarotoxin in medium at 37°C for 20 min. Then the cultures were washed extensively to remove unbound toxin and returned to 37°C with fresh medium $\pm$100 μg/ml puromycin. Sets of 4 cultures were then used at various times to measure bound and free ^{125}I. These data were obtained together with the data in Fig. 6 from a large set of chick muscle cultures. Note that, while puromycin immediately blocks protein synthesis (see text) and stops new receptor incorporation after about 2 hr (Fig. 6), it only slightly decreases the rate of toxin degradation (release of ^{125}I into the medium). Since this release rate is apparently equivalent to the rate of receptor degradation (see text), these data indicate a dissociation between the processes of receptor synthesis and incorporation into plasma membranes and receptor degradation. (See ref. 31.)

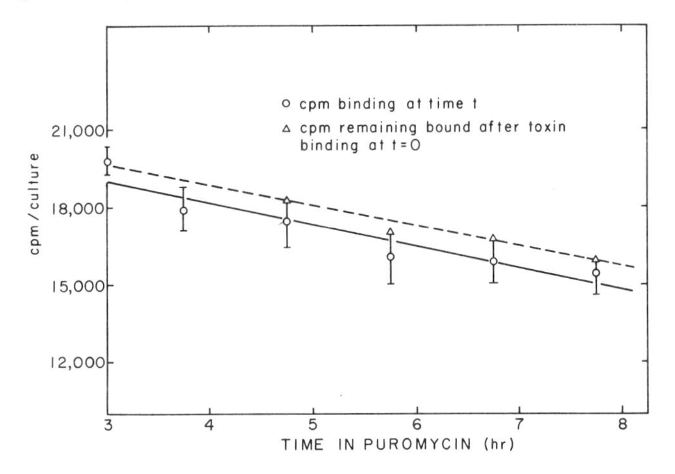

FIG. 12. Equivalence of α-bungarotoxin degradation and loss of receptors. Two sets (30 cultures and 24 cultures) of identical 6-day-old cultures, grown on 35-mm dishes, were pretreated with 20 μg/ml puromycin for 3 hr. One set of 30 cultures (△) was saturated with [^{125}I]α-bungarotoxin for the last 30 min of the 3-hr pretreatment. All of the cultures were then washed in tricine-buffered Hanks-BSA medium. The cultures were then returned to the incubator in F12 medium containing 20 μg/ml puromycin, except for six labeled cultures which were immediately extracted at the 3-hr time point. Before each of the indicated time points, six cultures from the set which had not been initially saturated with [^{125}I]α-bungarotoxin (○) were labeled for 30 min. At each of the time points, these six cultures and six from the set which had been initially labeled were washed and then extracted and counted. The lines drawn were fit to the data by least-squares linear regression: the error bars indicate standard deviation from the mean. (Error bars are ommitted from the degradation curve, for ease of illustration.) (See ref. 12.)

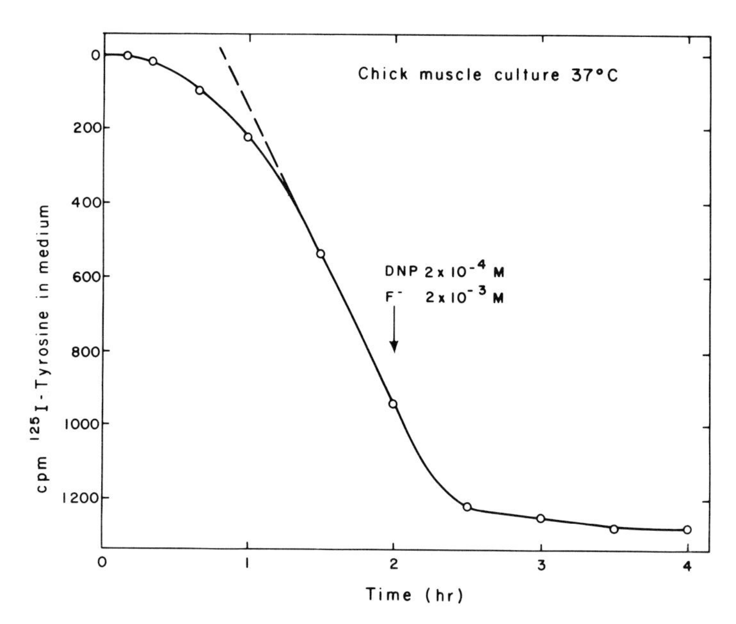

FIG. 13. Time course of [^{125}I]tyrosine release into the medium after brief exposure of myotubes to [^{125}I]α-bungarotoxin. Large plates (100 mm diam) of 5-day chick muscle cultures were transferred to HEPES-buffered medium and incubated for several hours in a 37°C room. Then ($t = 0$) warm medium containing 0.5 μg/ml [^{125}I]α-bungarotoxin was added to the cultures. After 5 min, it was removed by six (30 sec) rinses with warm medium. The medium was then replaced at the indicated times and the used medium (3 ml/dish) was centrifuged briefly to remove cell debris and then carrier l-tyrosine was added and an aliquot was fractionated on Bio-Gel P-2 and the fractions were assayed for radioactivity by scintillation spectrometry. The radioactivity in the l-tyrosine peak was calculated and the cumulative total [^{125}I]tyrosine released was plotted as a function of time. After the maximal rate of [^{125}I]tyrosine release was obtained (as judged from other experiment of this type), medium containing 2,4-dinitrophenol and fluoride replaced the normal medium. Time zero is the time of initial contact of [^{125}I]α-bungarotoxin and myotubes. (See ref. 13.)

ture-dependent, having a Q_{10} equal to 8. Several protease inhibitors that have been tested (12) have little effect on the degradation rate.

The strongest evidence supporting an internalization mechanism has been derived from electron microscope autoradiographs of myotube thin sections (18). In such micrographs of myotubes with receptors labeled by [^{125}I]-monoiodo-α-bungarotoxin under conditions of active receptor degradation (37°C in normal medium) about 3.5% of the grains were located over structures identified as secondary lysosomes. Some examples are shown in Fig. 15. About 60% of all readily identifiable secondary lysosomes had grains associated with them. In micrographs of myotubes labeled under conditions which inhibit receptor degradation (4°C) only about 0.4% of the grains were associated with secondary lysosomes and only about 4% of the identifiable secondary lysosomes were labeled.

The effects of many inhibitors of various cellular functions on degradation are presented in Table 2. Most remarkable, perhaps, is that these compounds

TABLE 1. *Degradation of α-bungarotoxin–receptor complexes in cell homogenates and isolated membranes*

Method of preparation	Incubation conditions	Iodo-tyrosine produced (%)	[I]tyrosine production rate constant (hr^{-1})
1. [a] Homogenization in 300 mM sucrose, 5 mM KCN, pH 7.4			
a. Supernate from 1,000 × *g*, 10-min centrifugation	300 mM sucrose, 5 mM KCN, pH 7.4, 37°C 24 hr	0	0
b. Same as 1a, except particulates were washed three times in F11[b] by centrifugation at 50,000 × *g*, 15 min	F11[b] plus 5 mM ATP	<0.6	<0.0008
2. [c] Nitrogen cavitation, 800 psi in 250 mM sucrose, 2 mM $MgSO_4$, 5 mM Tris, pH 7.4, 2 mM NaN_2; pellet from 20,000 × *g* 15 min centrifugation after addition of 1 mM EDTA	F11		
	37°C, 4½ hr	0	0
	37°C, 16 hr	<1	<0.0005
3. [a] Homogenization in F11 (Dounce homogenizer, "A" pestle)	37°C, 7 hr	1.4	0.002

In every case the entire incubation medium was made 1% in Triton X-100 and analyzed for iodotyrosine by Bio-Gel P-60 chromatography. Long columns were used to achieve complete separation of [^{125}I]α-bungarotoxin and [^{125}I]tyrosine. Percent I-tyrosine was calculated as the proportion of radioactivity eluting in the included fractions. Control rate of degradation is about 0.03 hr^{-1}.

[a] Before homogenization, chick muscle cultures were incubated with [^{125}I]α-bungarotoxin and washed to remove unbound [^{125}I]α-bungarotoxin.

[b] F11 is HEPES-buffered Ham's F12 with 2% embryo extract, 15% horse serum.

[c] The vesicles produced by nitrogen cavitation were labeled with [^{125}I]α-bungarotoxin in suspension, and unbound α-bungarotoxin was removed by repeated centrifugation.

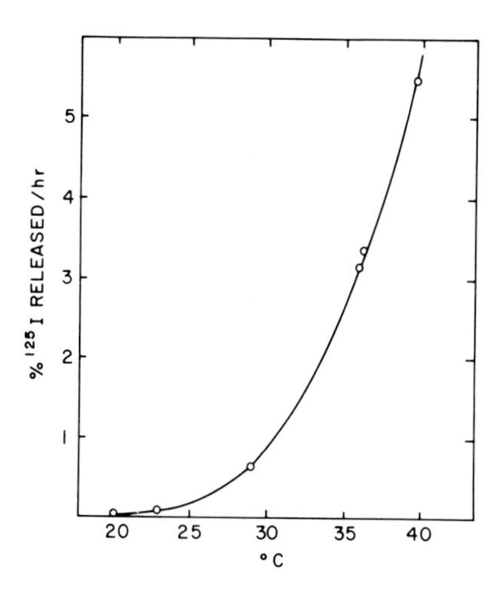

FIG. 14. Temperature dependence of ^{125}I release after binding of [^{125}I]α-bungarotoxin to cultures of chick muscle. Incubation time was 6 hr at indicated temperatures. The medium was then removed from each culture, centrifuged to remove any cell debris, and an aliquot counted by scintillation spectrometry. The bound ^{125}I was also determined for each culture after solubilization in 1% Triton X-100. Each point represents the average for 4 cultures.

all seem to have little effect on degradation. As shown in Table 3, the age of the culture has little effect on the rate of degradation of receptors. The accumulation of receptors on chick myotubes (Fig. 1) is complete by about 9 days in culture and there is no net increase at later times. Since degradation is always 3% per hour, it is inferred that the rate of incorporation of new receptors slows down in older cultures.

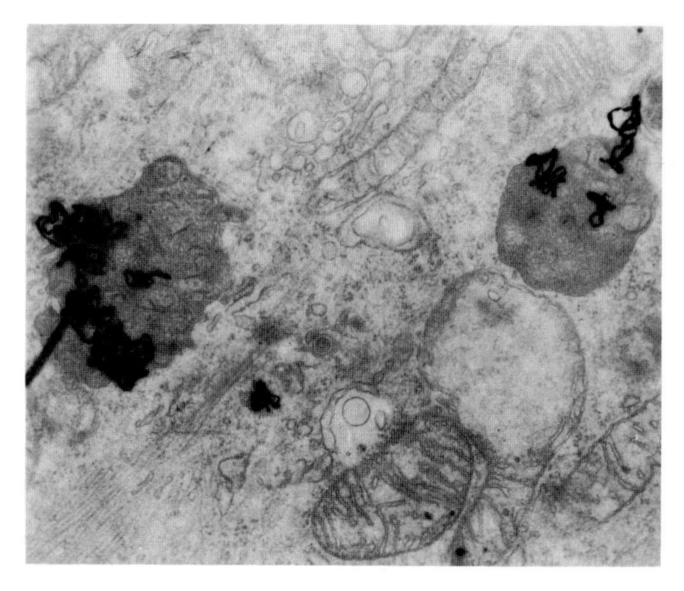

FIG. 15. Electron microscope autoradiograph of chick myotube after saturation (6 hr at 37°C) of ACh receptors with [^{125}I]α-bungarotoxin (0.1 μg/ml), showing grains over electron-dense organelles resembling secondary lysosomes. (See ref. 18.)

TABLE 2. *Effects of miscellaneous agents on ACh receptor degradation*

Additive	Medium[a]	Percent of control rate (mean ±SEM)	n^b
Cycloheximide 100 μg/ml	F11	80 ± 4	18
Puromycin 50 or 100 μg/ml	UnF11	83 ± 2	24
p-Fluorophenylalanine 0.1–1.0 mg/ml	UnF11	101 ± 4	12
Colchicine 100 μg/ml	F11 and UnF11	73 ± 2	20
Cytochalasin B 2.5 μg/ml in 1% DMSO	UnF11	85 ± 1	6
Choline 5×10^{-3} M	UnF11	119 ± 13	4
d-Tubocurarine	UnF11	103 ± 1	8
Triton X-100 0.01% (w/v)	F11	104 ± 6	4

[a] F11 is Ham's F12 with 2% embryo extract and 15% horse serum, buffered with 18 mM Na-HEPES at pH 7.4. UnF11 is F11 without serum or embryo extract.
[b] Number of cultures tested.

TABLE 3. *Rate of ACh receptor degradation as a function of culture age*

Age of culture (days)	Rate of degradation (mean ± SEM) (hr^{-1})	Half-life (hr)	n^a
3	0.039 ± 0.002	16	4
4	0.023 ± 0.001	28	5
5	0.023 ± 0.001	28	4
6	0.035 ± 0.001	18	10
7	0.034 ± 0.001	18	7
8	0.032 ± 0.001	19	4
9	0.028 ± 0.001	22	4
10	0.030 ± 0.001	21	4
13	0.023 ± 0.001	28	4
22	0.026 ± 0.002	24	4

Chick muscle cultures were plated at 1×10^5 cells/35-mm petri dish and cultured in F12 at 36°C. Degradation was measured as the fraction of radioactivity released into the medium per hour after [^{125}I]α-bungarotoxin was bound to ACh receptors in chick muscle cultures. Degradation was measured over a 6- or 8-hr period at 37°C in F11 medium (see footnote to Table 2). Half-lives were determined graphically by extrapolation. Mean half-life for all data was 22 hr.
[a] Number of cultures tested.

Internal Consistency of Accumulation, Incorporation, and Degradation Rates

So far, a reasonable, quantitative relationship between accumulation of receptors on myotube surfaces, incorporation of new receptors into the surface, and removal and degradation of receptors has been implied: the

TABLE 4. *Comparison of rate of accumulation, incorporation, and degradation*

Age (days)	Change in total receptor number per hour (%)			
	Incorporation	Degradation	Accumulation	I − D[a]
3–5	5.1 ± 0.9 (3.6–6.6)	2.8 ± 0.2 (2.3–3.9)	2.1 ± 0.2 (1.72–2.65)	2.3
5–7	4.3 ± 0.2 (3.8–5.0)	3.1 ± 0.1 (2.3–3.5)	1.12 ± 0.1 (0.94–1.26)	1.2

Each value listed is the mean and standard error of at least three such measurements from independently initiated culture sets. The lowest and highest values are listed in parentheses. Rates of accumulation were calculated for 4- and 6-day cultures from three curves such as that shown in Fig. 1. Examples of incorporation and degradation measurements are in Figs. 5, 6, 11, and 12.

[a] Incorporation rate minus degradation rate.

rate of accumulation must be equal to the difference between the rates of incorporation and degradation. The experimentally observed rate of accumulation can be compared with the rate calculated as the difference between experimentally observed rates of incorporation and degradation. Table 4 summarizes observations of accumulation, incorporation, and degradation rates made over a 2-year period. Agreement is found between the calculated and directly observed rate of accumulation. These rates can also be compared in another way, as a function of temperature. Figure 16 illustrates the temperature dependence of both incorporation and degradation. In Figure 17 a difference curve, calculated from the data in Fig. 16, is superimposed on the directly observed measurements of accumulation.

RECEPTOR METABOLISM IN ADULT SKELETAL MUSCLE

Some of the types of experiments performed *in vitro* on myotubes can be easily adapted for denervated adult muscles and two examples will be discussed. We have used the density label technique to demonstrate that extrajunctional receptors appearing on the surface of mouse extensor digitorum longus muscle following denervation are the result of *de novo* receptor synthesis (13). The details of the experimental procedure are similar to those for the *in vitro* experiment (Fig. 4) and are given in detail in the legend to Fig. 18. This experiment was performed 3 days following denervation and similar results were obtained from muscles 6 days after denervation.

Because of the diffusion barrier presented by the architecture of most adult muscles and the problem of nonspecific binding of [^{125}I]monoiodo-α-bungarotoxin to adult muscle tissue, degradation experiments, such as those presented in Figs. 8 and 12, are difficult to perform in adult muscle. Using a continuous perfusion apparatus for organ culture of adult muscle, the half-

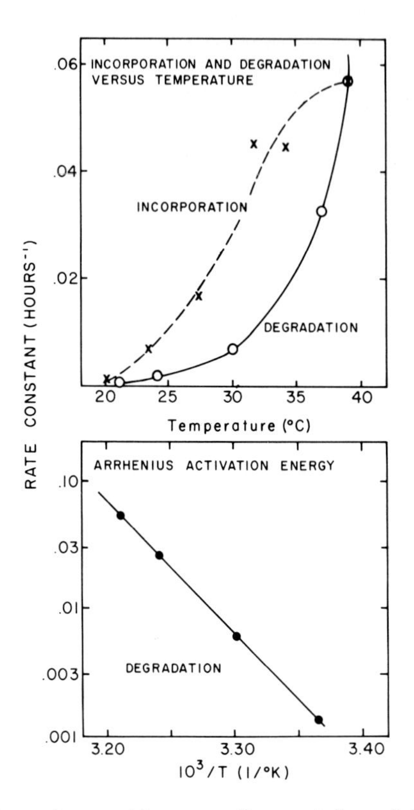

FIG. 16. Temperature dependence of incorporation and degradation and accumulation. Degradation curve: A large set of cultures were saturated with [^{125}I]α-bungarotoxin and unbound [^{125}I]α-bungarotoxin was removed. Five cultures were then incubated in 2 ml of medium at each of the indicated temperatures for 6 hr. Cultures were then assayed and the rate constant for degradation at each temperature was calculated. Cultures were maintained at these given temperatures by floating the culture dishes in water baths. Temperatures reported are those of the water bath, but agreed well with measurements made by a probe immersed directly in the culture medium (O). Incorporation curve: A large set of cultures was saturated with unlabeled α-bungarotoxin; the unbound α-bungarotoxin was removed. Four cultures were then incubated at each of the indicated temperatures in 2 ml of medium for 8 hr. All of the cultures were then equilibrated at 22°C and saturated with [^{125}I]α-bungarotoxin for 45 min, and the incorporation (cpm per hour) was divided by the total number of surface sites (cpm). Ordinate in this case is fraction surface sites per hour (**X**). Since accumulation was found to be zero at about 39–40°C, data on incorporation were normalized so that the value at 39°C equaled that of degradation, in order that incorporation and degradation curves determined on separate culture sets at different times could be compared. (See ref. 12.)

time for receptor degradation in adult denervated muscle has been found to be similar to that in chick myotubes (18). An incubation chamber for the mouse extensor digitorum longus muscle, continuously perfused with oxygenated medium, is placed in the well of a gamma counter. Denervated adult muscles, prelabeled with [^{125}I]monoiodo-α-bungarotoxin, are incu-

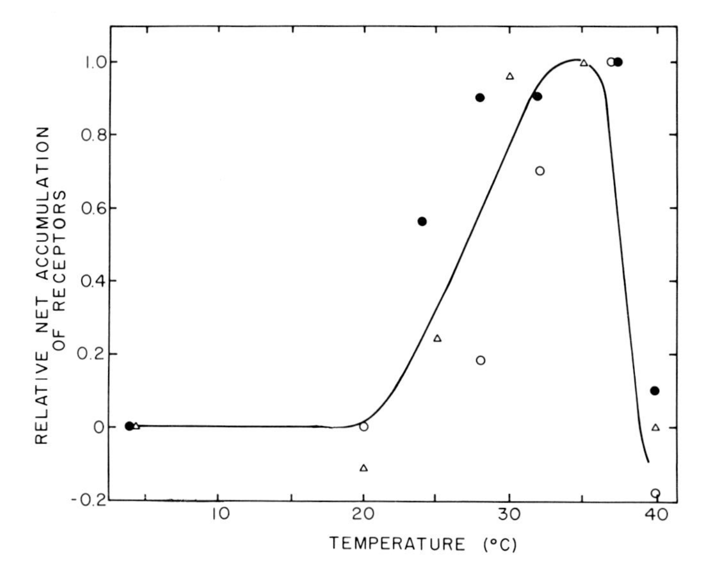

FIG. 17. ACh receptor accumulation in chick muscle cultures at different temperatures. Chick muscle cultures plated at 1×10^5 cells/35-mm dish were grown for 5 days in culture medium at 36°C. Then groups of 5 cultures were incubated at each of the indicated temperatures for 6 hr by floating the culture dishes in covered waterbaths. Cultures were subsequently incubated at 20°C with [^{125}I]α-bungarotoxin for 45 min, unbound toxin was removed, and the toxin–receptor complexes were extracted and counted by scintillation spectrometry. Data from three experiments are depicted with different symbols. Data from each experiment were normalized so that the largest accumulation in each experiment was equal to 1.0, and accumulation at the lowest temperature was equated to zero. The solid line is the difference curve between incorporation and degradation curves in Fig. 16. (See ref. 12.)

bated in the chamber and the amount of radioactivity remaining associated with the muscle is continuously monitored. The results of such an experiment are shown in Fig. 19. The initial rapid decline represents the removal of free α-bungarotoxin, and the slow decline represents receptor degradation. After corrections are made for nonspecific binding, the half-time for degradation can be estimated at 18–30 hr in agreement with a 22-hr half-time measured *in vitro*. A half-time of 19 hr has been reported by Huang and Chang (19) for clearance of [^{125}I]monoiodo-α-bungarotoxin from rat diaphragm *in vivo* after labeling of extrajunctional ACh receptors by intraperitoneal injection of [^{125}I]monoiodo-α-bungarotoxin.

FUTURE DIRECTIONS

So far we have only discussed some aspects of receptor biosynthesis and turnover in basically quiescent myotubes and muscle fibers deprived of innervation. It is clear from a vast body of evidence (20–22) that innervation and muscle activity profoundly affect the metabolism of ACh receptors. In

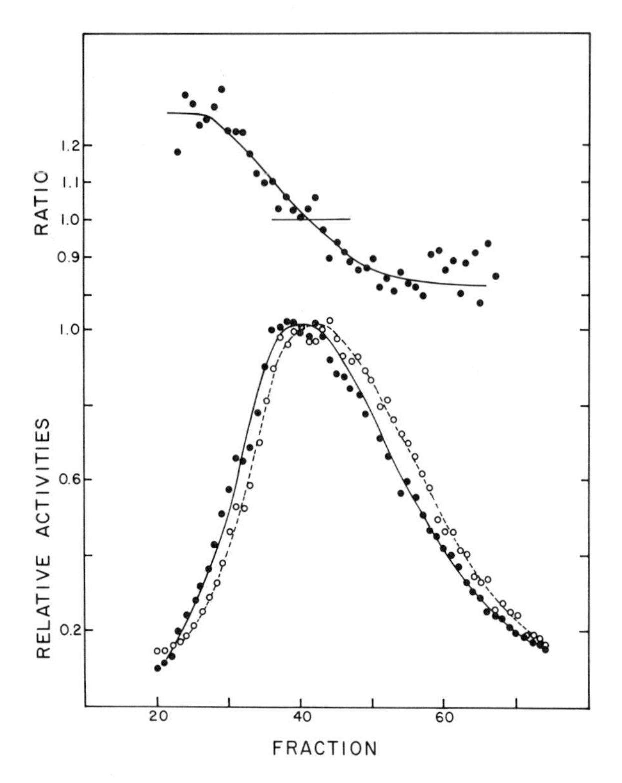

FIG. 18. Equilibrium density gradient sedimentation of 6-day denervated mouse extensor digitorum longus (EDL) muscle ACh receptors. [^{131}I]α-bungarotoxin–receptor complexes from 6-day denervated muscle (○). [^{125}I]α-bungarotoxin–receptor complexes appearing during 15 hr in organ culture in medium containing ^{2}H amino acids (●). α-Bungarotoxin–receptor complexes extracted from the muscles by homogenization in 1% Triton, 10 mM Tris buffer, pH 7.8 were prepurified by sucrose gradient velocity sedimentation. The ratio of ^{125}I/^{131}I values is shown above for peak fractions. (See ref. 13.)

active, innervated adult skeletal muscle extrajunctional ACh receptors are absent or sparse (6,23), whereas junctional ACh receptors are clustered at a density of about 20,000 sites/μm^2 at the postsynaptic membrane (24). Studies on the dissociation of radioactivity after labeled α-bungarotoxin has bound to junctional receptors (19,25,26) suggest a much slower turnover rate for these receptors compared with extrajunctional ones, although it is unlikely that disappearance of label strictly parallels destruction of ACh receptors in this case, for the estimated disappearance rate is similar to the rate of α-bungarotoxin–receptor dissociation (12). Determination of a reliable estimate for the turnover rate of synaptic ACh receptors remains a challenging task.

The disappearance of extrajunctional receptors following innervation of skeletal muscle can be mimicked by electrical stimulation of denervated

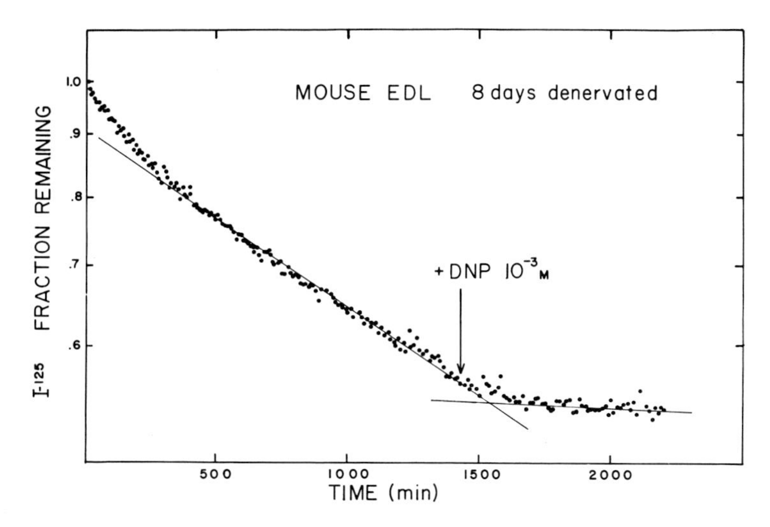

FIG. 19. Effect of 2,4-dinitrophenol on the release of radioactivity from 8-day denervated mouse EDL, following brief incubation with $[^{125}I]\alpha$-bungarotoxin. Measurements begin after the washout of most of the unbound α-bungarotoxin. (See ref. 18.)

muscle (27–29) or tissue cultured muscle (30) in the absence of innervation. The effects of muscle activity on ACh receptor metabolism are being vigorously investigated in several laboratories. Preliminary observations (J. Powell, *personal communication*) indicate that muscle activity in cultured mouse muscle results in a decline in ACh receptors with no statistically significant change in the rate of ACh receptor degradation. This suggests that some aspect of muscle activity has a controlling effect on receptor biosynthesis or incorporation into plasma membranes. The nature of this influence will be known in fair detail within the next few years.

REFERENCES

1. Sytkowski, A. J., Vogel, Z., and Nirenberg, M. W. (1973): Development of acetylcholine receptor clusters on cultured muscle cells. *Proc. Natl. Acad. Sci. USA,* 70:270–274.
2. Buckingham, M. E., Caput, D., Cohen, A., Whalen, R. G., and Gros, F. (1974): The synthesis and stability of cytoplasmic messenger RNA during myoblast differentiation in culture. *Proc. Natl. Acad. Sci. USA,* 71:1466–1470.
3. Patrick, J., Heinemann, S. F., Lindstrom, J., Schubert, D., and Steinbach, J. H. (1972): Appearance of acetylcholine receptors during differentiation of a myogenic cell line. *Proc. Natl. Acad. Sci. USA,* 69:2762–2766.
4. Fambrough, D. M. (1974): Acetylcholine receptors. Revised estimates of extrajunctional receptor density in denervated rat diaphragm. *J. Gen. Physiol.,* 64:468–472.
5. Berg, D. K., Kelly, R. B., Sargent, P. B., Williamson, P., and Hall, Z. (1972): Binding of α-bungarotoxin to acetylcholine receptors in mammalian muscle. *Proc. Natl. Acad. Sci., USA,* 69:147–151.
6. Hartzell, H. C., and Fambrough, D. M. (1972): Acetylcholine receptors: Distribution and extrajunctional density in rat diaphragm after denervation correlated with acetylcholine sensitivity. *J. Gen. Physiol.,* 60:248–262.

7. Miledi, R., and Potter, L. T. (1971): Acetylcholine receptors in muscle fibers. *Nature,* 233:599–603.
8. Libelius, R. (1974): Binding of ^{3}H-labelled cobra neurotoxin to cholinergic receptors in fast and slow mammalian muscles. *J. Neural Transm.,* 35:137–149.
9. Hartzell, H. C., and Fambrough, D. M. (1973): Acetylcholine receptor production and incorporation into membranes of developing muscle fibers. *Dev. Biol.,* 30:153–165.
10. Cohn, Z., and Stenham, R. (1975): 29th Annual Meeting of Society for General Physiologists (*Abstracts*).
11. Goldberg, A. L., Howell, E. M., Li, J. B., Martel, S. B., and Prouty, W. F. (1974): Physiological significance of protein degradation in animal and bacterial cells. *Fed. Proc.,* 33:1112–1119.
12. Devreotes, P. N., and Fambrough, D. M. (1975): Acetylcholine receptor turnover in membranes of developing muscle fibers. *J. Cell Biol.,* 63:335–358.
13. Devreotes, P. N., and Fambrough, D. M. (1976): Synthesis of acetylcholine receptors by cultured chick myotubes and denervated mouse extensor digitorum longus muscles. *Proc. Natl. Acad. Sci. USA,* 73:161–164.
14. Brockes, J., and Hall, Z. (1975): Synthesis of acetylcholine receptor by denervated rat diaphragm muscle. *Proc. Natl. Acad. Sci. USA,* 72:1368–1372.
15. Shur, B., and Roth, S. (1975): Cell surface glycosyl transferases. *Biophys. Biochem. Acta,* Reviews on Biomembranes, 415:473–512.
16. Bergeron, J. J., and Posner, B. I. (1975): Polypeptide hormone receptors in the golgi apparatus of rat hepatocytes. 15th Annual Meeting of American Society for Cell Biology (*Abstracts*).
17. Little, J. S., and Widnell, C. C. (1975): Evidence for the translocation of 5′-nucleotidase across hepatic membranes *in vivo. Proc. Natl. Acad. Sci. USA,* 72:4013–4017.
18. Devreotes, P. N., and Fambrough, D. M. (1976): Turnover of acetylcholine receptors in skeletal muscle. In *Cold Spring Harbor Symp. Qual. Biol.,* XL.
19. Chang, C. C., and Huang, M. C. (1975): Turnover of junctional and extrajunctional acetylcholine receptors of the rat diaphragm. *Nature,* 253:643–644.
20. Guth, L. (1968): "Trophic" influences of nerve on muscle. *Physiol. Rev.,* 48:645–687.
21. Harris, A. J. (1974): Inductive functions of the nervous system. *Annu. Rev. Physiol.,* 36:251–305.
22. Drachman, D. B. (editor) (1974): Trophic functions of the neuron. *Ann. NY Acad. Sci.,* 228.
23. Miledi, R., Stefani, E., and Zelena, J. (1968): Neural control of acetylcholine-sensitivity in rat muscle fibers. *Nature,* 220:497–498.
24. Fertuck, H. C., and Salpeter, M. M. (1974): Localization of acetylcholine receptors by ^{125}I-labeled α-bungarotoxin binding at mouse motor end-plates. *Proc. Natl. Acad. Sci. USA,* 71:1376–1378.
25. Berg, D. K., and Hall, Z. (1974): Fate of α-bungarotoxin bound to acetylcholine receptors of normal and denervated muscle. *Science,* 184:473–475.
26. Fertuck, H. C., Woodward, W., and Salpeter, M. M. (1975): *In vivo* recovery of muscle contraction after α-bungarotoxin binding. *J. Cell Biol.,* 66:209–213.
27. Lømo, T., and Rosenthal, J. (1972): Control of acetylcholine sensitivity by muscle activity in the rat. *J. Physiol.* (*Lond.*), 221:493–513.
28. Drachman, D. B., and Witzke, F. (1972): Trophic regulation of acetylcholine sensitivity of muscle: Effect of electrical stimulation. *Science,* 176:514–515.
29. Lømo, T. (1974): Neurotrophic control of colchicine effects on muscle. *Nature,* 249:473–474.
30. Cohen, S. A., and Fischbach, G. D. (1973): Regulation of muscle acetylcholine sensitivity by muscle activity in cell culture. *Science,* 181:76–78.
31. Fambrough, D. M., and Devreotes, P. N. (1974): Synthesis and degradation of acetylcholine receptors in cultured chick skeletal muscle. In: *Exploratory Concepts in Muscular Dystrophy,* Vol. II, edited by A. H. Milhorat.

Biogenesis and Turnover of Membrane Macromolecules,
edited by John S. Cook. Raven Press, New York 1976.

Induction and Inhibition of Friend Leukemic Cell Differentiation: The Role of Membrane-Active Compounds

Alan Bernstein, Alastair S. Boyd, Valerie Crichley, and Valerie Lamb

The Ontario Cancer Institute, and the Department of Medical Biophysics, University of Toronto, 500 Sherbourne Street, Toronto, Ontario M4X 1K9, Canada

Cell differentiation is characterized by a progressive restriction in the variety of proteins synthesized by the differentiating cell, and in the preferential synthesis of a small number of proteins unique to that cell lineage. For example, erythropoiesis in the mouse is initiated by the commitment of a pluripotent stem cell (1) to differentiate along the erythroid pathway. This commitment precludes the ability of these cells to mature into lymphocytes, granulocytes, or macrophages, and hence to synthesize the proteins associated with these alternate pathways of differentiation. After extensive proliferation, these committed stem cells mature into anucleated reticulocytes, with a protein-synthesizing capacity which is largely restricted to the making of adult hemoglobin (2).

The genetic and molecular events that control and accompany erythroid differentiation are largely unknown. Changes in the pattern of transcription and translation during erythropoiesis have been described (2–4), and these changes must play a central role in the control of differentiation. Other cellular structures may also be important during differentiation. In particular, the cell surface, which is now thought to play a determinant role in the contrasting behavior of normal and transformed cells (5–7), may also be involved in the regulation of gene expression during differentiation. Consistent with this hypothesis are the following observations. First, changes in the protein and glycoprotein composition of the cell surface (primary structure) occur during the differentiation of a variety of cell types (8–11). Second, changes in membrane fluidity (secondary structure) influence lymphocyte activation by plant lectins (12), and are correlated with the capacity of myeloid leukemic cell lines to differentiate (13). Third, mutations at a locus which has a profound effect on early embryonic development of the mouse (the T-Brachyury locus) have recently been shown to be associated with a specific antigen, the T antigen, localized at the cell surface (14,15).

One system of differentiation which has recently received a great deal of attention is that developed by Friend et al. (16). Friend cells are permanent mouse cell lines which can be induced to differentiate and produce hemoglobin upon stimulation with dimethylsulfoxide (DMSO).

Recently, we have found that local anesthetic amines inhibit this process of differentiation (17). In this chapter, we shall describe some properties of this inhibition phenomenon, and the information presently available obtained by ourselves, as well as others, on the properties of compounds that induce Friend cells to differentiate. Based on these observations, a model for the initiation of differentiation in Friend cells will be presented, suggesting that a crucial early step in this process involves the interaction of the inducing agents with the cell surface, resulting in changes in the overall fluidity of the plasma membrane.

FRIEND LEUKEMIC CELLS

Friend cells are *in vitro* cell lines that originated from the spleens of mice that had been infected with Friend leukemia virus (16). While the target cell for malignant transformation by Friend virus has not been identified, Friend cell lines clearly originate from cells with erythroid potentialities, since they retain the capacity to synthesize hemoglobin in small amounts. This "endogenous" synthesis can be stimulated 5- to 40-fold, 3 to 5 days after the addition of DMSO (16). Induction of hemoglobin synthesis in Friend cells by DMSO *in vitro* is accompanied by a number of additional phenotypic changes that also occur during normal erythropoietin-dependent erythroid differentiation *in vivo*. These are listed in Table 1.

A number of laboratories have now found that other compounds, in addition to DMSO, induce hemoglobin synthesis in Friend leukemic cells (28–33). Although it remains to be shown, it appears likely that cellular changes identical to those in Table 1 occur after induction of Friend cells with the other chemical inducers.

These observations immediately raise the question whether the wide array of Friend cell inducers act by a common mechanism, or, alternatively, whether there are several inductive pathways all leading to Friend cell

TABLE 1. *Cellular changes associated with the addition of DMSO to Friend leukemic cells*

Heme synthesis increases (16–18)
Iron transport increases (16)
Globin messenger RNA levels increase (19–21, 23)
Globin chain synthesis increases (22)
Erythroid antigen appears at the cell surface (24)
Friend virus release increases (25)
Alterations in purine metabolism (26)
Cell volume decreases (27)
Proliferative capacity markedly decreases (17,28)

References in parentheses.

differentiation. An understanding of this process could have wide implications both for normal erythropoiesis *in vivo,* and for other pathways of differentiation in mammalian cells.

INHIBITORS OF FRIEND CELL DIFFERENTIATION

Starting with the hypothesis that the induction of Friend cell differentiation is mediated by some common mechanism involving the cell surface, we reasoned that other known membrane-active compounds might also influence the induction of hemoglobin synthesis in these cells. The local anesthetic amines were chosen because they are all of low molecular weight, and have been shown to have a number of effects on cells and cell membranes, including protection of erythrocytes from osmotic hemolysis (34), increasing the susceptibility of normal mouse 3T3 cells to agglutination by concanavalin A (35,36), inhibition of surface capping of immunoglobulin molecules on B lymphocytes (37), and preferential lysis of Rous sarcoma virus-transformed chick embryo fibroblasts (38). Physical measurements of membrane fluidity have shown that local anesthetics increase membrane fluidity, in both artificial lipid bilayers (34,39–41) and in normal mouse fibroblasts (35,36).

Local Anesthetics Inhibit DMSO-Induced Hemoglobin Synthesis

Addition of local anesthetics to Friend cells, in the presence of DMSO, results in a marked inhibition in the incorporation of iron-59 into heme (17; Fig. 1), the number of benzidine-positive (hemoglobin-containing) cells (17), and the amount of hemoglobin in these cells, as measured spectrophotometrically (data not shown). On a molar basis, tetracaine appears to be 20 to 30 times more effective in inhibiting differentiation than procaine (Fig. 1). Both their relative potencies, and the absolute concentrations at which they inhibit Friend cell differentiation, agree very well with the concentrations at which these two anesthetics induce anesthesia (34,39), and protect erythrocytes from osmotic hemolysis (34). This concentration appears to be governed largely by the concentration of the drug in the membrane, or the partition coefficient (P) of the anesthetic in membrane/water (34). To a first approximation, the partition coefficient in octanol/water is a good measure of the relative solubility of a drug in biological membranes. The log P values for procaine and tetracaine in octanol/water are 1.92 and 3.73, respectively (42). Thus, tetracaine would be expected to be a more potent membrane-active compound than procaine because of its more hydrophobic properties.

Local Anesthetics Inhibit DMSO-Induced Cell Death

While there are several explanations for the inhibition of hemoglobin synthesis by local anesthetics, the most obvious is simply that the anes-

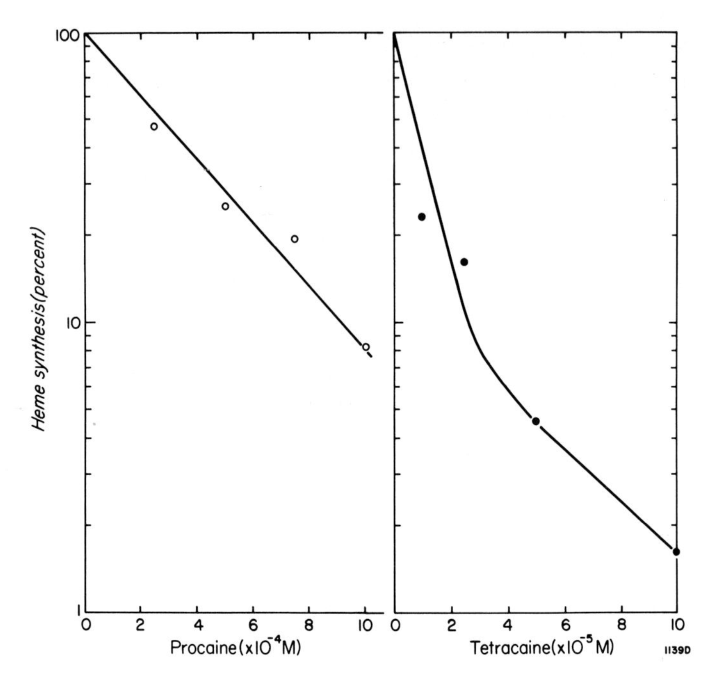

FIG. 1. Inhibition of DMSO-induced heme synthesis by the anesthetics, procaine and tetracaine. Friend cell clone M2 was induced with DMSO (1% v/v) and varying concentrations of anesthetic, in the presence of 0.125 μCi iron-59/ml. On day 5, the amount of iron in heme was determined as described previously (17).

thetics are toxic to Friend cells at the concentrations used in Fig. 1. For this reason, we have carefully examined the effects of procaine on Friend cell viability by measuring their growth rate in suspension culture, and plating efficiency in a semisolid medium of methylcellulose, after 5 or 6 days growth in suspension in the presence of anesthetic or DMSO, or both. At the concentrations used to inhibit differentiation, no toxic effects were observed (17). In fact, procaine appears to prevent partially the drop in cell viability that occurs 4 to 5 days after the addition of DMSO. This protective effect of procaine was determined both by measurements of plating efficiency in methylcellulose (17), and growth rate, expressed as the time required for the population to double in number in suspension culture (the "doubling time"; Fig. 2). It can be seen from Fig. 2 that cells grown in the presence of DMSO (1%) gradually stop dividing: by day 6 their doubling time is essentially infinite. By increasing the concentration of procaine in the medium, this increase in doubling time can be partially blocked (Fig. 2). Thus, procaine inhibits *both* the induction of hemoglobin synthesis, and the decrease in proliferative capacity that accompanies DMSO-induced differentiation.

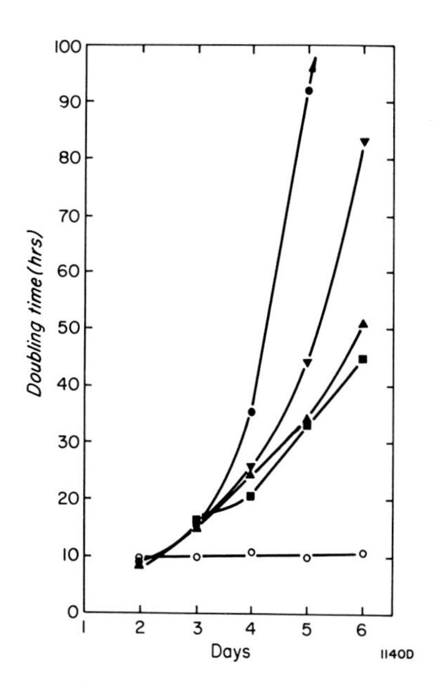

FIG. 2. Inhibition of DMSO-induced cell death by procaine HCl. On day 0, Friend cell, clone 745a, was subcultured at 10^5 cells/ml. Cultures were maintained in exponential phase ($2-5 \times 10^5$ cells/ml) by feeding every day. Cell counts were determined daily on a Coulter counter. (○) control; (●) 1% DMSO; (▼) 1% DMSO + 2.5×10^{-4} M procaine; (▲) 1% DMSO + 5×10^{-4} M procaine; (■) 1% DMSO + 7.5×10^{-4} M procaine.

Local Anesthetics Inhibit DMSO-Induced Cell Volume Changes

Differentiation of Friend cells *in vitro,* as well as normal erythroid differentiation *in vivo,* is also accompanied by a gradual decrease in cell volume to about one-half that of uninduced cells (27). This volume change appears to be an integral part of erythroid differentiation, and it was of interest, therefore, to see if local anesthetics could also block this parameter of differentiation. It can be seen from Fig. 3 that uninduced Friend cells have roughly constant volumes between days 3 and 5 (provided the cells are kept growing exponentially), whereas cells treated with DMSO become progressively smaller. On the other hand, cultures treated with both DMSO and procaine (10^{-3} M) show only a slight shift to smaller size by day 5. Because local anesthetics increase the surface area of human erythrocytes (34), the inhibition of cell shrinkage by procaine might be due directly to insertion of procaine into the lipid bilayer, rather than to any effect on differentiation per se. To test this possibility, induced cells that had been treated with DMSO alone for 5 days, and had the volume spectrum shown in Fig. 3, were allowed to interact with procaine (10^{-3} M) for 3 hr. After that time, volume spectra were again determined. No change in volume distribution was observed between the DMSO-treated cells and the same cells incubated with procaine for 3 hr, indicating that procaine itself has no direct measurable effect on Friend cell volume.

These volume spectra can be fitted to a Gaussian distribution by an itera-

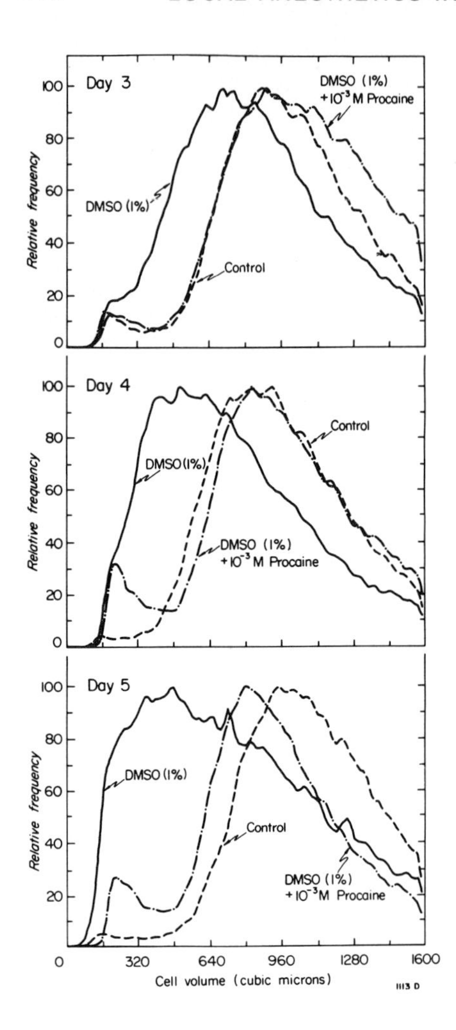

FIG. 3. The effect of DMSO and procaine HCl on Friend cell volume. Friend cell, clone 745a, was grown in medium containing 10% fetal calf serum, and DMSO or DMSO and procaine, as shown. Control groups were untreated. Cells were kept in the exponential phase of growth by feeding the cells everyday to maintain the cell density $2-4 \times 10^5$/ml. On days 3, 4, and 5, volume spectra were determined with a Coulter counter connected to a pulse height analyzer (43). The system was calibrated with latex beads, which had been independently sized on a STAPUT apparatus (27).

tive least-squares procedure (43), and the modal cell volume of different populations can be obtained in this manner. It can be seen that increasing concentrations of procaine are increasingly effective in blocking the decrease in modal cell volume associated with Friend cell differentiation (Fig. 4).

LOCAL ANESTHETICS INHIBIT OTHER FRIEND CELL INDUCERS

The experiments described so far do not provide any direct information on the mechanism of induction by DMSO, nor on the mechanism of inhibition by local anesthetics. However, if there is a common mechanism of induction, local anesthetics should inhibit differentiation by all the Friend cell inducers. Conversely, if the induction of differentiation by certain compounds was anesthetic-resistant, this would be strong evidence that there

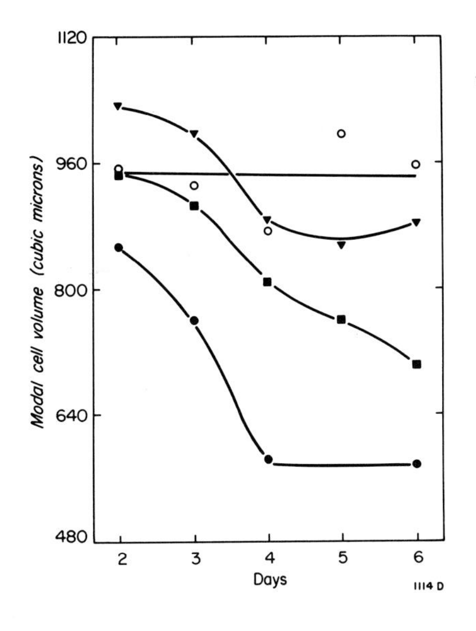

FIG. 4. Modal cell volumes of Friend cells at various times after the addition of DMSO and procaine. On day 0, Friend cell clone 745a, was subcultured to a density of 10^5 cells/ml and volume spectra, similar to those shown in Fig. 3, were determined every day. These volume spectra were then analyzed for their modal cell volume, as described previously (43). (○) control; (●) 1% DMSO; (■) 1% DMSO + 2.5×10^{-4} M procaine HCl; (▼) 1% DMSO + 10^{-3} M procaine HCl.

is more than one mechanism for the induction of erythroid differentiation in Friend cells. For this reason, we examined the ability of procaine to inhibit Friend cell differentiation by a number of different chemical inducers with widely different structures. As may be seen in Fig. 5, procaine effectively inhibited differentiation by all of the inducers that were examined. Furthermore, the extent of inhibition (the slope of the curve in Fig. 5) by procaine did not vary significantly with inducer (Fig. 5), and appears to be largely independent of inducer concentration (data not shown). It is of interest as well that DMSO-induced differentiation of two independent Friend cell clones, 745a and M2, both showed the same degree of sensitivity to procaine.

These observations suggest that the local anesthetics act at some step common to all of the inducers. The anesthetics may block interaction of the inducers with a common cellular "receptor," they may act directly at a common origin of induction, or they may block some subsequent step further along the inductive pathway that is shared by all the inducers.

INDUCERS OF FRIEND CELL DIFFERENTIATION

The fact that a wide spectrum of very different chemical compounds are effective inducers, plus the observation presented above that local anesthetics act as antagonists of all these inducers, suggested to us that there might be some common physicochemical properties which were shared by all of these active agents. It had previously been noted by others that Friend cell inducers all have a relatively low molecular weight (ranging between

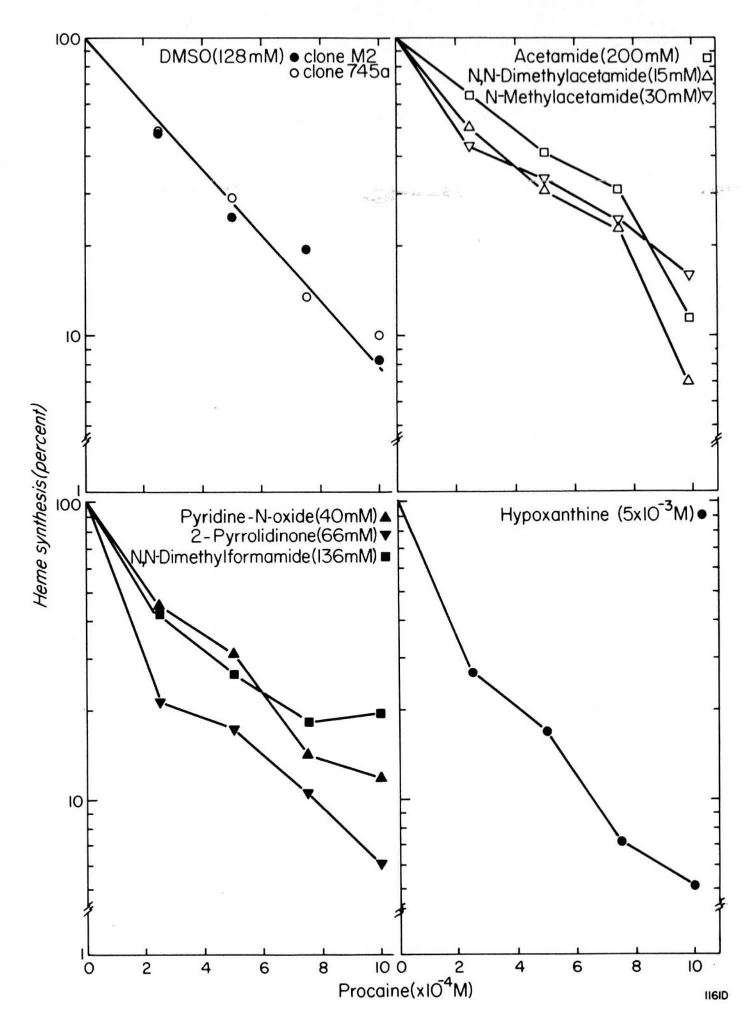

FIG. 5. Inhibition of heme synthesis by various Friend cell inducers. Except where indicated, Friend cell, clone M2, was used throughout. The accumulation of iron into heme on day 5 was measured as described previously (17).

~60 and 150 daltons), they are highly polar as determined by their high dipole moment (30) and they are all capable of acting as Lewis bases, by providing electron pairs for hydrogen bonding (33). This latter property is shared by most cryoprotective agents, and Preisler et al. (33) have noted that most cryoprotective agents will induce Friend cells.

We have also noted that all of the inducers, with the exception of butyric acid, do share one other common property, their partition coefficient (P) in octanol/water is less than 0.0 (log P between −0.5 and −2.0) (Fig. 6).

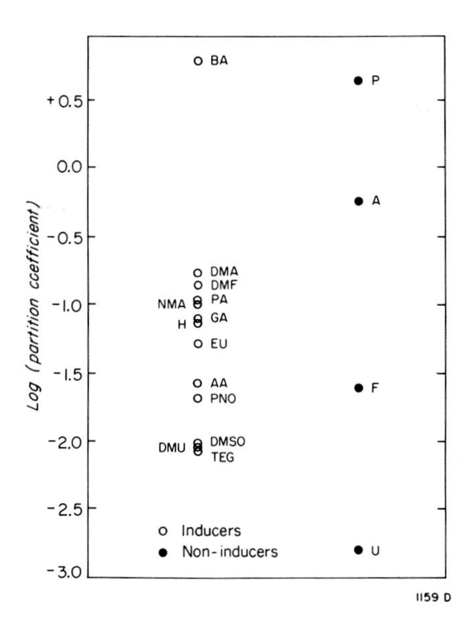

FIG. 6. Log (partition coefficients) in octanol/water (42) of Friend cell inducers (○), and several tested noninducers (●). Abbreviations used, and inducing concentrations, or concentrations that were found not to induce: BA, butyric acid ($1–2 \times 10^{-3}$ M); DMA, *N,N*-dimethylacetamide ($8–35 \times 10^{-3}$ M); DMF, *N,N*-dimethylformamide ($25–200 \times 10^{-3}$ M); PA, propionamide ($30–120 \times 10^{-3}$ M); NMA, *N*-methylacetamide ($20–50 \times 10^{-3}$ M); GA, glycolic acid ($10–30 \times 10^{-3}$ M); H, hypoxanthine (5×10^{-3} M); EU, ethylurea ($25–100 \times 10^{-3}$ M); AA, acetamide ($100–300 \times 10^{-3}$ M); PNO, pyridine-*N*-oxide ($10–130 \times 10^{-3}$ M); DMSO, dimethylsulfoxide ($70–350 \times 10^{-3}$ M); DMU, dimethylurea (9.6×10^{-3} M); TEG, triethyleneglycol ($100–300 \times 10^{-3}$ M); P, pyridine ($3–250 \times 10^{-3}$ M); A, acetone ($10–100 \times 10^{-3}$ M); F, formamide (0.23 M); U, urea ($15–280 \times 10^{-3}$ M).

Conversely, compounds that do not induce Friend cells are either very hydrophilic (log *P* less than –2.0) or very hydrophobic (log *P* greater than –0.5) (Fig. 6), with the possible exception of formamide (*vide infra*).

To test this correlation further, several additional compounds were selected on the basis of their low molecular weight, and partition coefficient, and examined for their ability to induce Friend cells. Several were too toxic, one was a potent Friend cell inducer (ethylurea; Fig. 6) and the last, glycolic acid (Fig. 6), did not, by itself, induce Friend cells. However, cells treated with both DMSO and glycolic acid synthesized reproducibly 2 to 3 times more heme than cells treated with DMSO alone (Table 2). It will be interesting to see if other compounds with the "correct" partition coefficient induce Friend cells, or, like glycolic acid, act synergistically with other known inducers.

Although the correlation between partition coefficient and inducibility is strong, we cannot say that a log *P* between –0.5 and –2.0 is a necessary and sufficient condition for inducibility of Friend cells. There are at least five expected exceptions to this rule. (*i*) Compounds that are too toxic, and yet still have the correct partition coefficient, will obviously not induce. (*ii*) Only those compounds whose molecular weight lies in a narrow range between ~60 and 150 daltons appear to be capable of inducing Friend cells. The failure of formamide to induce (Fig. 6) could be due to its low molecular weight (45.04 daltons) or to the fact that only one concentration of formamide was tested for inducibility (31). (*iii*) Compounds that are natural metabolites, such as the amino acid leucine (log $P = -1.71$) might not induce because they quickly enter metabolic pools. Metabolism would deplete the

TABLE 2. *Synergistic effect of glycolic acid on DMSO-induced Friend cell differentiation*

	Day 4	Day 6
Untreated control	5[a] ± 0.006[b]	5 ± 0.009
Glycolic acid		
15 mM	—	4 ± 0.009
25 mM	—	4 ± 0.009
DMSO (1% v/v)	100 ± 27	100 ± 3
DMSO (1%) plus glycolic acid (15 mM)	148 ± 15	163 ± 12
DMSO (1%) plus glycolic acid (25 mM)	190 ± 10	314 ± 4

The incorporation of iron-59 into heme was determined as described previously (17). Friend cell clone M2 was used in these experiments.

[a] Heme synthesis is expressed as percent of DMSO control for each day. For day 4, 100% equals 2,653 cpm per 10^6 cells. For day 6, 100% equals 4,977 cpm per 10^6 cells.

[b] Average ± SEM.

supply available for induction, and also possibly alter the partition coefficient in the membrane of the potential inducer. It is interesting to note, therefore, that hypoxanthine, which can be metabolized to inosine monophosphate by the enzyme hypoxanthine phosphoribosyl transferase (HPRT) and then used in nucleic acid synthesis, will also induce Friend cells (28; Fig. 5). Incorporation of hypoxanthine into DNA does not appear to be necessary for induction, since Friend cell mutants, selected for resistance to the hypoxanthine analogue, 6-thioguanine, are still inducible by hypoxanthine, even though they lack the activity for HPRT and thus cannot incorporate hypoxanthine into DNA (28). (*iv*) Compounds such as butyric acid, with a partition coefficient outside the predicted range, might induce either because they are metabolized in the cell to yield a product with a suitable log *P*, or because their presence in the cell alters the composition of the cellular structure involved in differentiation. For example, incubation of HeLa cells with the fatty acid butyric acid increases the amount of one glycosphingolipid, sialyllactosylceramide, and induces the synthesis of the corresponding enzyme activity, CMP-sialic acid: lactosylceramide sialyl transferase, within 7–20 hr (44). The ability of butyric acid to induce Friend cells might be due, therefore, not to its physicochemical properties directly (i.e., log *P* value), but to its biochemical effects on cell surface composition. This argument does not, of course, distinguish between common or distinct induction mechanisms for butyric acid and the other inducers in Fig. 6. (*v*) The natural inducers of normal erythropoiesis, such as erythropoietin and the β-androgen metabolites, are undoubtedly much different in molecular weight, and possibly partition coefficient, than the inducers listed in Table 1. The action of these natural inducers is probably mediated via specific cellular receptors. Thus, although their chemical properties are different, they may or may not act by the same underlying mechanism as the small molecular weight inducers of Friend leukemic cells.

With the exception discussed above, the inducers shown in Fig. 6 share the two common properties of low molecular weight, and a log partition coefficient in octanol/water between −0.5 and −2.0. This observation does not, by itself, provide information on the site of action of the different inducers. It is consistent with the hypothesis that there is a common mechanism of induction for the various compounds listed in Fig. 6. Nor does the range in partition coefficients explain by itself the wide range in molar concentrations necessary for optimum induction (see legend to Fig. 6). For the acetamide family of inducers (AA, NMA, and DMA), there does appear to be a good linear relationship between log P and log $(1/C)$, where C is the optimum dose for induction (data not shown). This is in agreement with the Meyer–Overton rule for anesthesia (34), and appears to be a general empirical observation for a wide variety of membrane-active compounds (45).

A MODEL FOR FRIEND CELL DIFFERENTIATION

From the above discussion, it seems clear that the inducers of Friend cell differentiation share two important properties, their low molecular weight, and their relative solubility in octanol/water (except butyric acid, which is known to affect membrane structure in other cells). In addition, it is important to point out that the local anesthetic amines, which inhibit Friend cell differentiation, have profound effects on cell membranes, and in particular, increase the fluidity of both artificial lipid bilayers (34,40) and biological membranes (36).

On the basis of these properties of inducers and inhibitors, we wish to propose a model for the modulation of gene expression in Friend cells. The central thesis of the model is that changes in membrane structure, and in particular, in membrane fluidity, can alter the pattern of gene expression in Friend cells, and possibly in other differentiating systems. The properties of low molecular weight and relative solubility in octanol/water are both thermodynamically and physically compatible with the ability of these compounds to insert into the lipid bilayer. Thus, combined with the known action of anesthetics on membrane fluidity mentioned earlier, the model is consistent with present information on the induction and inhibition of Friend cell differentiation. It is, of course, important to recognize that the mechanism by which epigenetic events, such as cellular differentiation, might be coupled to alterations in membrane structure are completely unknown. The model, therefore, will not deal with this problem of information transfer.

An important feature of this model is that it makes certain predictions which are testable. First, it predicts that other compounds with physicochemical properties similar to the inducers described above should also stimulate Friend cell differentiation. Second, it should be possible to manipulate membrane structure, and in particular membrane fluidity, by means other than the addition of local anesthetics, and these manipulations should affect the process of differentiation. For example, membrane fluidity can be altered by manipulating membrane cholesterol levels (46,47), or by altering the fatty acid composition of membrane phospholipids (48,50). Certain membrane mutations in mammalian cells may also affect membrane fluidity (51). Third, the model predicts that Friend cell inducers should decrease membrane fluidity in both artificial lipid bilayers and in whole cells. Fluidity changes can be monitored in several ways: (i) the relative mobility of fluorescent compounds in Friend cell membranes can be measured using the technique of fluorescence polarization (47); (ii) the relative mobility in cells of nitroxide-labeled probes, such as stearic acid analogues, can be determined by measuring their paramagnetic resonance spectra (Weber et al., *this volume;* also ref. 52); (iii) the effects of inducers and inhibitors of Friend cell differentiation on the fluidity of artificial lipid bilayers can be determined by measuring, with a differential scanning calorimeter, the temperatures at which phase transitions in the bilayer occur (40). Using the latter technique, Lyman et al. (53) recently observed that incorporation of Friend cell inducers into artificial lipid bilayers does result in a decrease in the fluidity of the bilayer. Furthermore, the addition of local anesthetics, similar to the ones used in this study, can reverse this decrease in fluidity caused by Friend cell inducers. Fourth, the model suggests that Friend cell inducers need not cross the membrane to induce differentiation. It should be noted, however, that this is not a necessary prediction of the model since the inducers may also interact with some intracellular target in addition to their primary interaction at the cell surface.

It is recognized that other models could be envisaged to account for the information presently available on Friend cell differentiation. However, it seems clear that any such model will have to accommodate what seems to be a primary role of the membrane in this process. Hopefully the experiments outlined above will help to elucidate this problem.

ACKNOWLEDGMENTS

The authors wish to thank J. Gusella, D. Housman, F. Loritz, and H. Preisler for communicating information prior to publication. We gratefully acknowledge the assistance of R. G. Miller in analyzing the volume spectra and the continued interest shown by A. Axelrad during the course of this work. We also thank L. Siminovitch and G. Price for their help in the preparation of this manuscript. The work described in this chapter was supported by grants from the Medical Research Council of Canada, and the National Cancer Institute of Canada.

REFERENCES

1. McCulloch, E. A., Gregory, C. J., and Till, J. E. (1973): Cellular communication early in haemopoietic differentiation. In: *Ciba Foundation Symposium on Haemopoietic Stem Cells,* pp. 183–199. Elsevier, Amsterdam.
2. Benz, E. J., and Forget, B. G. (1974): The biosynthesis of hemoglobin. *Semin. Hematol.,* 11:463–523.
3. Marks, P. A., and Rifkind, R. A. (1972): Protein synthesis: Its control in erythropoiesis. *Science,* 175:955–961.
4. Glass, J. G., Lavidor, L. M., and Robinson, S. H. (1975): Studies of murine erythroid cell development: Synthesis of heme and hemoglobin. *J. Cell Biol.,* 65:298–308.
5. Tooze, J. (editor) (1973): *The Molecular Biology of Tumor Viruses.* Cold Spring Harbor Laboratory, New York.
6. Brady, R. O., and Fishman, P. H. (1974): Biosynthesis of glycolipids in virus transformed cells. *Biochim. Biophys. Acta,* 355:121–148.
7. Hynes, R. O. (1974): Role of surface alterations in cell transformation: The importance of proteases and surface proteins. *Cell,* 1:147–156.
8. Weiser, M. M. (1973): Intestinal epithelial cell surface membrane glycoprotein synthesis. I. An indicator of cellular differentiation. *J. Biol. Chem.,* 248:2536–2541.
9. Isselbacher, K. J. (1975): The intestinal cell surface: properties of normal, undifferentiated, and malignant cells. In: *The Harvey Lectures,* Series 69, pp. 197–222. Academic Press, New York.
10. Akeson, R., and Herschman, H. R. (1974): Neural antigens of morphologically differentiated neuroblastoma cells. *Nature,* 249:620–623.
11. Akeson, R., and Herschman, H. R. (1974): Modulation of cell-surface antigens of a murine neuroblastoma. *Proc. Natl. Acad. Sci. USA,* 71:187–191.
12. Alderson, J. C. E., and Green, C. (1975): Enrichment of lymphocytes with cholesterol and its effect on lymphocyte activation. *FEBS Lett.,* 52:208–211.
13. Sachs, L., Inbar, M., and Shinitsky, M. (1974): Mobility of lectin sites on the surface membrane and the control of cell growth and differentiation. In: *Control of Proliferation in Animal Cells,* pp. 283–296, edited by B. Clarkson and R. Baserga. Cold Spring Harbor Laboratory, New York.
14. Bennett, D., Goldberg, E., Dunn, L. C., and Boyse, E. A. (1972): Serological detection of a cell-surface antigent specified by the T (Brachyury) mutant gene in the house mouse. *Proc. Natl. Acad. Sci. USA,* 69:2076–2080.
15. Artzt, K., Bennett, D., and Jacob, F. (1974): Primitive teratocarcinoma cells express a differentiation antigen specified by a gene at the T-locus in the mouse. *Proc. Natl. Acad. Sci. USA,* 71:811–814.
16. Friend, C., Scher, W., Holland, J. G., and Sato, T. (1971): Hemoglobin synthesis in murine virus-induced leukemic cells *in vitro:* Stimulation of erythroid differentiation by dimethylsulfoxide. *Proc. Natl. Acad. Sci. USA,* 68:378–382.
17. Bernstein, A. (1976): Local anesthetics inhibit the induction of hemoglobin synthesis in Friend erythroleukemic cells. (*Manuscript in preparation.*)
18. Ebert, P. S., and Ikawa, Y. (1974): Induction of delta-aminolevulinic acid synthetase during erythroid differentiation of cultured leukemia cells. *Proc. Soc. Exp. Biol. Med.,* 146: 601–604.
19. Ross, J., Ikawa, Y., and Leder, P. (1972): Globin messenger-RNA induction during erythroid differentiation of cultured leukemia cells. *Proc. Natl. Acad. Sci. USA,* 69:3620–3623.
20. Ross, J., Gielen, J., Packman, S., Ikawa, Y., and Leder, P. (1974): Globin gene expression in cultured erythroleukemic cells. *J. Mol. Biol.,* 87:697–714.
21. Preisler, H. D., Housman, D., Scher, W., and Friend, C. (1973): Effects of 5-bromo-2′-deoxyuridine on production of globin messenger RNA in dimethylsulfoxide-stimulated Friend leukemia cells. *Proc. Natl. Acad. Sci. USA,* 70:2956–2959.
22. Ostertag, W., Crozier, T., Kluge, N., Melderis, H., and Dube, S. (1973): Action of 5-bromodeoxyuridine on the induction of haemoglobin synthesis in mouse leukemia cells resistant to 5-BUdR. *Nature* [*New Biol.*], 243:203–205.

23. Gilmour, R. S., Harrison, P. R., Windass, J. D., Affara, N. A., and Paul, J. (1974): Globin messenger RNA synthesis and processing during haemoglobin induction in Friend cells. I. Evidence for transcriptional control in clone M_2. *Cell Differ.*, 3:9–22.
24. Ikawa, Y., Ross, J., Leder, O., Gielen, J., Packman, S., Ebert, P., Hayashi, K., and Sugano, H. (1973): Erythrodifferentiation of cultured Friend leukemia cells. In: *Differentiation and Control of Malignancy of Tumor Cells*, pp. 515–546, edited by W. Nakamura, T. Ono, T. Sugimura, and H. Sugano. University Park Press, Baltimore, Md.
25. Dube, S. K., Pragnell, I. B., Kluge, N., Gaedicke, G., Steinheider, G., and Ostertag, W. (1975): Induction of endogenous and of spleen focus-forming viruses during dimethylsulfoxide-induced differentiation of mouse erythroleukemia cells transformed by spleen focus-forming virus. *Proc. Natl. Acad. Sci. USA*, 72:1863–1867.
26. Reem, G. H., and Friend, C. (1975): Purine metabolism in murine virus-induced erythroleukemic cells during differentiation *in vitro*. *Proc. Natl. Acad. Sci. USA*, 72:1630–1634.
27. Loritz, F., Bernstein, A., and Miller, R. G. (1976): (*Submitted for publication*.)
28. Gusella, J., and Housman, D. (1976): Cell, *in press*.
29. Scher, W., Preisler, H. D., and Friend, C. (1973): Hemoglobin synthesis in murine virus-induced leukemic cells *in vitro*. III. Effects of 5-bromo-2′-deoxyuridine, dimethylformamide and dimethylsulfoxide. *J. Cell. Physiol.*, 81:63–70.
30. Tanaka, M., Levy, J., Terada, M., Breslow, R., Rifkind, R. A., and Marks, P. A. (1975): Induction of erythroid differentiation in murine virus-infected erythroleukemia cells by highly polar compounds. *Proc. Natl. Acad. Sci. USA*, 72:1003–1006.
31. Preisler, H. D., and Lyman, G. (1975): Differentiation of erythroleukemia cells *in vitro:* Properties of chemical inducers. *Cell Differ.*, 4:179–185.
32. Leder, A., and Leder, P. (1975): Butyric acid, a potent inducer of erythroid differentiation in cultured erythroleukemic cells. *Cell,* 5:319–322.
33. Preisler, H. D., Christoff, G., and Taylor, E. (1975): Personal communication.
34. Seeman, P. (1972): The membrane action of anesthetics and tranquilizers. *Pharmacol. Rev.*, 24:583–655.
35. Poste, G., Papahadjopoulos, D., Jacobson, K., and Vail, W. J. (1975): Local anesthetics increase susceptibility of untransformed cells to agglutination by concanavalin A. *Nature,* 253:552–554.
36. Poste, G., Papahadjopoulos, D., Jacobson, K., and Vail, W. J. (1975): Effects of local anesthetics on membrane properties. II. Enhancement of the susceptibility of mammalian cells to agglutination by plant lectins. *Biochim. Biophys. Acta,* 394:520–539.
37. Ryan, G. B., Unanue, E. R., and Karnovsky, M. J. (1974): Inhibition of surface capping of macromolecules by local anesthetics and tranquillisers. *Nature,* 250:56–57.
38. Rifkin, D. B., and Reich, E. (1971): Selective lysis of cells transformed by Rous sarcoma virus. *Virology,* 45:172–181.
39. Papahadjopoulos, D. (1972): Studies on the mechanism of action of local anesthetics with phospholipid model membranes. *Biochim. Biophys. Acta,* 265:169–186.
40. Papahadjopoulos, D., Jacobson, K., Poste, G., and Shepherd, G. (1975): Effects of local anesthetics on membrane properties. I. Changes in the fluidity of phospholipid bilayers. *Biochim. Biophys. Acta,* 394:504–519.
41. Jain, M. K., Wu, N. Y., and Wray, L. V. (1975): Drug-induced phase change in bilayers as possible mode of action of membrane expanding drugs. *Nature,* 255:494–495.
42. Leo, A., Hansch, C., and Elkins, D. (1971): Partition coefficients and their uses. *Chem. Rev.,* 71:525–615.
43. Miller, R. G., Wuest, L. J., and Cowan, D. H. (1972): Volume analysis of human red cells. I. The general procedure. *Ser. Haematol.,* 5:105–127.
44. Simmons, J. L., Fishman, P. H., Freese, E., and Brady, R. O. (1975): Morphological alterations and ganglioside sialyltransferase activity induced by small fatty acids in HeLa cells. *J. Cell Biol.,* 66:414–424.
45. Hansch, C., and Glave, W. R. (1970): Structure–activity relationships in membrane-perturbing agents hemolytic, narcotic, and antibacterial compounds. *Mol. Pharmacol.,* 7: 337–354.
46. Oldfield, E., and Chapman, D. (1972): Dynamics of lipids in membranes: heterogeneity and the role of cholesterol. *FEBS Lett.,* 23:285–297.
47. Shinitzky, M., and Inbar, M. (1974): Difference in microviscosity induced by different

cholesterol levels in the surface membrane lipid layer of normal lymphocytes and malignant lymphoma cells. *J. Mol. Biol.*, 85:603–615.
48. Wisnieski, B. J., Williams, R. E., and Fox, C. F. (1973): Manipulation of fatty acid composition in animal cells grown in culture. *Proc. Natl. Acad. Sci. USA*, 70:3669–3673.
49. Williams, R. E., Wisnieski, B. J., Rittenhouse, H. G., and Fox, C. F. (1974): Utilization of fatty acid supplements by cultured animal cells. *Biochemistry*, 13:1969–1977.
50. Ferguson, K. A., Glaser, M., Bayer, W. H., and Vagelos, P. R. (1975): Alteration of fatty acid composition of LM cells by supplementation and temperature. *Biochemistry*, 14: 146–151.
51. Ling, V. (1975): Drug resistance and membrane alteration in mammalian cells. *Can. J. Genet. Cytol.*, 17:503–515.
52. Gaffney, B. J. (1975): Fatty acid chain flexibility in the membranes of normal and transformed fibroblasts. *Proc. Natl. Acad. Sci. USA*, 72:664–668.
53. Lyman, G., Preisler, H., and Papahadjopoulos, D. (1975): Personal communication.

Note added in proof: We have recently observed that ouabain is a potent inducer of hemoglobin synthesis in ouabain-resistant Friend cells (A. Bernstein, M. Hunt, V. Crichley, and T. Mak, *manuscript in preparation*). Given the strong binding specificity of ouabain for the plasma membrane enzyme Na/K-ATPase (see chapters by Cook et al. and Baker, *this volume*, for discussions on ouabain and the properties of ouabain-resistant cells), this observation provides additional support for the hypothesis that the induction of erythroid differentiation in Friend cells results from the interaction of inducing compounds with the cell surface.

Biogenesis and Turnover of Membrane Macromolecules,
edited by John S. Cook. Raven Press, New York 1976.

Effect of Methylprednisolone on Renal Na-K-ATPase Deficiency in the Postobstructive Kidney

D. R. Wilson, W. H. Knox, J. Sax, and A. K. Sen

Departments of Medicine and Pharmacology, Faculty of Medicine, University of Toronto, Toronto, Ontario M5S 1A8, Canada

Renal Na-K-ATPase activity has been shown to be related to the renal tubular reabsorption of sodium (1,2) and water (1–3). An increase in filtered sodium load and tubular reabsorption of sodium are associated with increased renal Na-K-ATPase activity following uninephrectomy [renal hypertrophy (1,4)] or methylprednisolone administration (1). Decreased filtered sodium load and tubular reabsorption of sodium have been associated with enzyme deficiency after adrenalectomy (5) and in experimental hypothyroidism (6).

Previous experiments from this laboratory have shown that, after relief of 24 hr bilateral ureteral obstruction in the rat, there is diminished renal tubular reabsorption of sodium and water which persists for at least 7 days (7). The postobstructive diuresis was associated with a decrease in renal Na-K-ATPase activity. It is the purpose of this chapter to review these observations and describe further experiments, including the effects of methylprednisolone administration on renal function and Na-K-ATPase activity in the postobstructive kidney.

METHODS

Male albino rats of the Wistar strain weighing 250–350 g were kept in metabolic cages and had free access to water and a standard rat pellet diet unless otherwise specified. Each experimental group included 5–8 animals. Methods used have been described previously (7).

Experimental Groups

Ureteral Ligation

Under light sodium pentobarbitol anesthesia (50 mg/kg body wt intraperitoneally), a small midline incision was made in the lower abdomen, the lower end of both ureters was identified, ureteral ligation was carried out, and the abdomen closed. The animals were returned to metabolic cages and allowed food but no water for the next 24 hr. The rats were then anesthetized again, the abdomen reopened, and the ligatures removed from both ureters. Thereafter, animals had free access to food and water until the evening before the experiment when food was withdrawn. Clearance studies and enzyme assay were usually carried out 24 hr after relief of obstruction, (24 hr postobstruction) or as indicated at time intervals of 2 hr, 3 days, and 7 days after relief of obstruction.

Methylprednisolone Experiments

This group of animals was treated in identical fashion to the 24-hr postobstruction group, except they received methylprednisolone (2.5 mg IM) daily for 4 doses. The 4 doses of hormone were given for 2 days before the ureteral ligation, on the day of ligation and on the day of relief of ligation.

Sham-Operated

These animals were treated in identical fashion to the ureteral ligation experiments except that no ligature was placed about the ureters.

Renal Function Studies

On the day of the experiment animals were anesthetized with inactin (80 mg/kg body wt intraperitoneally), placed on a heated animal board, and a tracheostomy was performed. Appropriate polyethylene tubing was inserted in the jugular vein for intravenous infusions, in the femoral artery to measure blood pressure, using a mercury manometer, and in the bladder for urine collections. Clearance studies were carried out using [^{3}H]inulin, with three urine collections being obtained in each animal. Arterial blood and urine were analyzed as required for blood urea nitrogen, hematocrit, radioactive inulin, sodium, and osmolality by methods previously described (7).

Preparation of Kidney Homogenate and Assay of Enzyme Activity

In clearance experiments the kidneys were removed, immediately chilled in ice cold 0.9% saline, and sliced separating the cortex and outer red medulla as previously described (7). Two kidneys were combined for analysis. Microsomal Na-K-ATPase was isolated according to the method of Post and Sen (8) using heparin treatment to separate the heavy microsomal fraction. Sodium- and potassium-dependent ATPase activity was determined as previously described and protein was measured by the method of Lowry et al. (9).

RESULTS

Changes in Renal Function in the Postobstructive Kidney

Figure 1 illustrates the changes in blood urea nitrogen level, which rose to 170 ± 18 mg% (±1 SEM) 24 hr after relief of 24 hr bilateral ureteral ligation, and then subsequently decreased to normal levels by 7 days. Glomerular filtration rate also showed a gradual increase from very low values when studied 2 hr, 24 hr, 3 days, and 7 days after relief of obstruction. Urine osmolality, which is an index of urine-concentrating ability, was markedly decreased in the postobstructive kidney and improved as shown. Urine flow rate and sodium excretion rate were markedly increased, by a factor of ~10 at the peak of the postobstructive diuresis, and remained increased 7 days after relief of obstruction. The tubular reabsorption of sodium was markedly decreased in the postobstructive kidney and gradually improved with time (Fig. 2).

Changes in Renal Na-K-ATPase in the Postobstructive Kidney

As shown in Fig. 3 there was a marked decrease in enzyme activity in the whole homogenate and heavy microsomal fraction of both the cortex and outer medulla 24 hr after relief of obstruction. Decreased activity persisted 3 days after relief of obstruction and remained low in the outer medulla of postobstructive kidneys at 7 days. In contrast to this finding, there was no change in magnesium ATPase activity in the postobstructive kidney (Fig. 4).

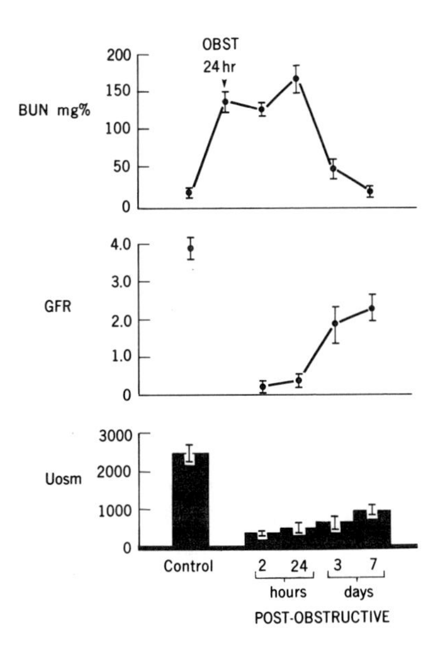

FIG. 1. *Upper,* Changes in blood urea nitrogen level (BUN, mg%) 24 hr after bilateral ureteral ligation and at intervals of 2 hr, 24 hr, 3 days, and 7 days after relief of obstruction. *Middle,* Changes in glomerular filtration rate (GFR, ml/min-kg body wt). *Lower,* Changes in urine osmolality (Uosm, mOsm/kg H_2O).

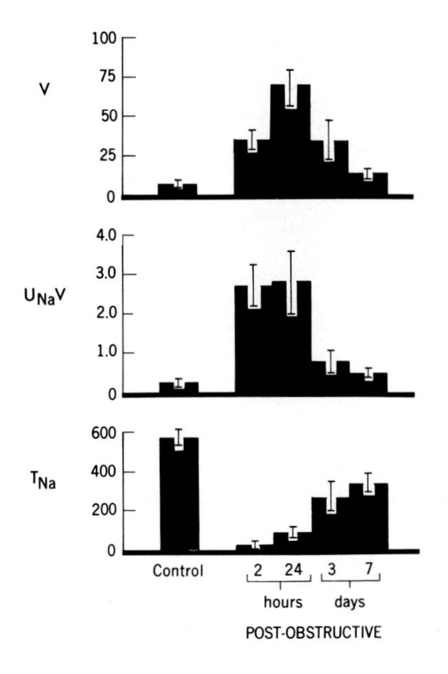

FIG. 2. *Upper,* Changes in urine flow rate (V = ml/min-kg body wt) after relief of obstruction. *Middle,* Changes in sodium excretion ($U_{Na}V$ = μeq/min-kg body wt). *Lower,* Changes in net tubular reabsorption of sodium (T_{Na} = μeq/min-kg body wt).

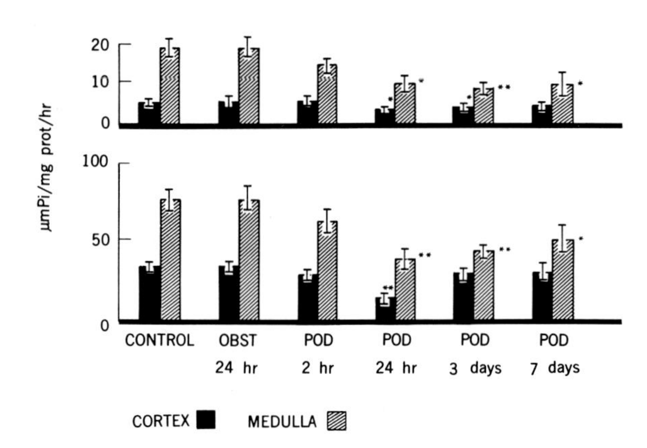

FIG. 3. Renal Na-K-ATPase activity after 24 hr bilateral ureteral ligation and at intervals of 2 hr, 24 hr, 3 days, and 7 days after relief of obstruction are shown for the renal cortex and outer medullary zone as measured in the whole homogenate and heavy microsomal fractions. * = $p < 0.05$; ** = $p < 0.01$ compared to control values. *Upper,* whole homogenate; *lower,* heavy microsomes.

Effect of Uremia on Na-K-ATPase Activity in Other Organs

Rats undergoing postobstructive diuresis, in which decreased renal Na-K-ATPase activity was demonstrated, had elevated blood urea nitrogen levels and were sick for 48 hr prior to the enzyme determinations. Uremia or renal failure has been shown to inhibit Na-K-ATPase activity in several other tissues (10–13). It was therefore possible that our findings of renal Na-K-

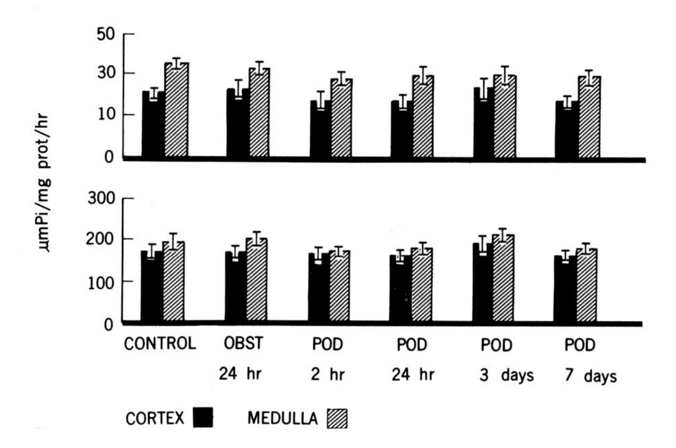

FIG. 4. Changes in Mg^{2+}-ATPase activity before and after relief of bilateral ureteral obstruction in the whole homogenate and heavy microsomal fractions. There were no significant changes. *Upper,* whole homogenate; *lower,* heavy microsomes.

TABLE 1. *Na-K-ATPase activity in brain and liver of rats after relief or ureteral ligation ("uremic")*

	Control	Postobst. 24 hr	*N*
Brain (whole homogenate)			
Na-K-ATPase[a]	19.6 ± 0.5	18.4 ± 1.3	7 NS[b]
Mg^{2+}-ATPase	15.4 ± 0.8	16.7 ± 0.9	7 NS
Liver (heavy microsomes)			
Na-K-ATPase	6.5 ± 0.8	6.3 ± 0.9	6 NS
Mg^{2+}-ATPase	28.4 ± 1.3	32.2 ± 1.6	6 NS

Enzyme activity was measured 24 hr postobstruction.
[a] µmoles P_i/mg protein-hr.
[b] NS = not significant, $p > 0.05$.

ATPase deficiency in the postobstructive kidney were secondary to non-specific effects of uremia or renal failure.

Accordingly, enzyme activity was determined in the liver and brain of rats 24 hr after relief of 24 hr bilateral ureteral obstruction, in which the mean blood urea nitrogen level was 170 ± 18 mg%. As shown in Table 1, there was no significant difference between Na-K-ATPase activity in the brain or liver of sham-operated controls and animals undergoing post-obstructive diuresis.

Effects of Methylprednisolone on Renal Function and Na-K-ATPase Activity

Another series of animals received 4 injections of methylprednisolone before and during the period of ureteral obstruction. As shown in Table 2,

TABLE 2. *Effects of methylprednisolone on renal function*

	Control ($N = 7$)	Postobst. 24 hr ($N = 8$)	Methylpred. + postobst. 24 hr ($N = 8$)	*p* values: Control	*p* values: Postobst.
GFR (ml/min-kg body wt)	3.9 ± 0.3	0.67 ± 0.15	2.54 ± 0.4	<0.05	<0.01
V (μl/min-kg body wt)	7.3 ± 1.1	69.3 ± 12.2	43.6 ± 6.9	<0.01	NS
V/GFR (×100)	0.19 ± 0.03	13.2 ± 2.2	2.4 ± 0.7	<0.01	<0.01
$U_{Na}V$ (μeq/min-kg body wt)	0.27 ± 0.06	2.77 ± 0.77	1.48 ± 0.24	<0.01	NS
%FENa	0.05 ± 0.01	4.2 ± 1.1	0.51 ± 0.12	<0.01	<0.01
T_{Na} (μeq/min-kg body wt)	572 ± 43	94 ± 23	369 ± 58	<0.05	<0.01
U_{osm} (mOsm/kg H_2O)	2466 ± 221	558 ± 64	870 ± 102	<0.01	<0.05

Abbreviations: GFR = glomerular filtration rate; V = urine flow rate; V/GFR (×100) = fractional excretion of water; $U_{Na}V$ = sodium excretion rate; %FENa = fractional excretion of sodium; T_{Na} = rate of tubular reabsorption of sodium; Uosm = urine osmolality.
All values are expressed as mean ±1 SEM.

there was considerable increase in glomerular filtration rate, in the tubular reabsorption of sodium, in urine osmolality, and in other parameters of renal function in these animals, when compared to untreated rats at 24 hr after relief of obstruction. However, renal function was not returned to normal and increased sodium and water excretion continued when compared to sham-operated controls.

In contrast, renal Na-K-ATPase activity was markedly increased in the methylprednisolone-treated postobstructive animals and was similar to untreated control rats (Table 3).

TABLE 3. *Effects of methylprednisolone on renal Na-K-ATPase activity (heavy microsomes, μMP_i/mg protein-hr)*

	Control ($N = 7$)	Postobst. 24 hr ($N = 8$)	Methylpred. +Postobst. 24 hr ($N = 8$)
Cortex	33.5 ± 2.4	12.1 ± 1.9	34.6 ± 4.7*
Outer medulla	75.5 ± 6.6	38.2 ± 5.0	80.9 ± 7.1*

*$p < 0.01$ compared to postobstruction 24 hr; all values are mean ±1 SEM.

DISCUSSION

Renal Na-K-ATPase deficiency is present in the postobstructive kidney, particularly in the outer medullary region, and this change correlates with decreased tubular reabsorption of sodium and water (Figs. 2 and 3). Renal Mg-ATPase and 5′-nucleotidase were not affected in the postobstructive kidney, indicating that the abnormality was enzyme-specific. Renal failure itself, presumably due to retained toxic products, has been shown to cause decreased Na-K-ATPase activity in red blood cells of humans (10,11), and in brain and intestine of rats (12,13). Animals undergoing postobstructive diuresis in the present experiments did not have severe renal failure, but did demonstrate considerable elevation of blood urea nitrogen levels. However, it was found that there was no inhibition of Na-K-ATPase activity in the liver or brain of such rats, indicating that the decrease in renal Na-K-ATPase activity was organ-specific.

Decreased renal Na-K-ATPase activity in the postobstructive kidney may be secondary to a decrease in the filtered load of sodium, which is the major determinant of tubular sodium reabsorption, or may be a direct effect of increased intrarenal pressure on tubular function. Previous experiments have shown that methylprednisolone administration increases glomerular filtration rate, tubular reabsorption of sodium, and renal Na-K-ATPase activity (1,14), while studies of renal hypertrophy following unilateral nephrectomy have yielded similar results (1,4). Aldosterone and deoxycorticosterone acetate (DOCA) administration also increase renal Na-K-ATPase activity (14,15). In the present experiments, methylprednisolone administration before and during the period of ureteral obstruction resulted in the maintenance of normal renal Na-K-ATPase activity in the postobstructive kidney (Table 3). Filtered sodium load (as indicated by glomerular filtration rate) and tubular reabsorption of sodium were improved, but not returned to normal levels (Table 2). Further studies involving simultaneous methylprednisolone-treated normal rats, as well as postobstructive animals, are in progress. The present experiments allow two preliminary conclusions. First, glucocorticoids may have a direct effect on renal Na-K-ATPase activity which is not dependent on an increase in filtered load of sodium. Dissociation between renal Na-K-ATPase activity and tubular reabsorption of sodium has also been recently described by Fisher, Welt, and Hayslett (16). Second, renal Na-K-ATPase deficiency is probably partially responsible for the impaired sodium and water reabsorption observed in the postobstructive kidney in view of the marked improvement noted with restoration of enzyme levels to the normal range.

REFERENCES

1. Katz, A. I., and Epstein, F. H. (1967): The role of sodium–potassium-activated adenosine triphosphatase in the reabsorption of sodium by the kidney. *J. Clin. Invest.*, 46:1999–2011.

2. Martinez-Maldonado, M., Allen, J. C., Inagaki, C., Tsaparase, N., and Schwartz, A. (1972): Renal sodium-potassium-activated adenosine triphosphatase and sodium reabsorption. *J. Clin. Invest.*, 51:2544–2551.
3. Martinez-Maldonado, M., Allen, J. C., Eknoyan, G., Suki, W., and Schwartz, A. (1969): Renal concentrating mechanism: Possible role for sodium–potassium-activated adenosine triphosphatase. *Science*, 165:807–808.
4. Schmidt, U., and Dubach, U. C. (1974): Induction of Na-K-ATPase in the proximal and distal convolution of the rat nephron after uninephrectomy. *Pfluegers Arch.*, 346:39–48.
5. Hendler, E. D., Torretti, J., Kupor, L., and Epstein, F. H. (1972): Effects of adrenalectomy and hormone replacement on Na-K-ATPase in renal tissue. *Am. J. Physiol.*, 222:754–760.
6. Katz, A. I., and Lindheimer, M. D. (1973): Renal sodium- and potassium-activated adenosine triphosphatase and sodium reabsorption in the hypothyroid rat. *J. Clin. Invest.*, 52:796–804.
7. Wilson, D. R., Knox, W., Hall, E., and Sen, A. K. (1974): Renal sodium- and potassium-activated adenosine triphosphatase deficiency during post-obstructive diuresis in the rat. *Can. J. Physiol. Pharmacol.*, 52:105–113.
8. Post, R. L., and Sen, A. K. (1967): *Methods Enzymol.*, edited by S. P. Colowick and N. O. Kaplan, p. 762. Academic Press, New York.
9. Lowry, O. H., Rosebrough, N. J., Farr, A. L., and Randall, R. J. (1951): Protein measurement with the Folin phenol reagent. *J. Biol. Chem.*, 193:265–275.
10. Cole, C. H., Balfe, J. W., and Welt, L. G. (1968): Induction of a ouabain-sensitive ATPase defect by uremic plasma. *Trans. Assoc. Am. Phys.*, 81:213–220.
11. Cole, C. H. (1973): Decreased ouabain-sensitive adenosine triphosphatase activity in the erythrocyte membrane of patients with chronic renal disease. *Clin. Sci. Mol. Med.*, 45: 775–784.
12. Minkoff, L., Gaertner, G., Darab, M., Mercier, C., and Levin, M. L. (1972): Inhibition of brain sodium-potassium ATPase in uremic rats. *J. Lab. Clin. Med.*, 80:71–78.
13. Kramer, H. J., Backer, A., and Kruck, F. (1974): Inhibition of intestinal (Na^+-K^+)-ATPase in experimental uremia. *Clin. Chim. Acta*, 50:13–18.
14. Charney, A. N., Silva, P., Besarab, A., and Epstein, F. H. (1974): Separate effects of aldosterone, DOCA, and methylprednisolone on renal Na-K-ATPase. *Am. J. Physiol.*, 227:345–350.
15. Schmidt, U., Schmid, J., Schmid, H., and Dubach, U. C. (1975): Sodium- and potassium-activated ATPase. A possible target of aldosterone. *J. Clin. Invest.*, 55:655–660.
16. Fisher, K., Welt, L. G., and Hayslett, J. P. (1974): Dissociation of Na-K-ATPase from tubular reabsorption of sodium. (abstract). *Clin. Res.*, 22:526A.

Biogenesis and Turnover of Membrane Macromolecules,
edited by John S. Cook. Raven Press, New York 1976.

Thyroid Hormone: Thermogenesis and the Biosynthesis of Na^+ Pumps

I. S. Edelman

Cardiovascular Research Institute; and the Departments of Medicine and Biochemistry and Biophysics of the University of California School of Medicine, San Francisco, California 94143

Magnus-Levy (1) discovered in 1895 that hyperthyroid patients had high rates of oxygen consumption (Q_{O_2}) and that feeding desiccated thyroid to normal man increased the basal metabolic rate. In the intervening decades, abundant evidence has been adduced indicating that thyroid hormone is a dominant regulator of heat production in mammals and that this effect is preserved in isolated tissues (2).

In the early 1950s thyroid thermogenesis was attributed to uncoupling of mitochondrial oxidative phosphorylation (3–5). The validity of the uncoupling hypothesis, however, has been questioned for several reasons: The Q_{O_2} of mitochondria after treatment with thyroxine was not increased (5). The physiological effects (e.g., on the circulation or on protein synthesis) of thyroid hormone were not elicited by a well-defined uncoupler, dinitrophenol (6). Mitochondria from liver or skeletal muscle of thyroid-treated rats had normal P/O and respiratory control ratios (7).

MITOCHONDRIAL SITE OF ACTION

In rats, thyroid hormone increased the size, number, and cristae of skeletal muscle mitochondria, as well as the respiratory capacity and rate of protein synthesis of liver mitochondria (7–9). Evidence of enhanced functional capacity of mitochondria was also adduced recently and these changes have been implicated in the thermogenic response. Thyroid hormone increased the activity ratios of hexokinase/citrate synthase and glycerolphosphate dehydrogenase/triosephosphate dehydrogenase, and transport of reducing equivalents from glycerol to O_2 (10,11). Direct changes in mitochondrial performance have also been described, in that thyroid hormone augmented "carrier-mediated" transport of ADP into liver mitochondria (12). If, however, coupling of ATP synthesis to respiration remains intact, the steady-state, sustained thermogenic response cannot be attributed simply to an increase in mitochondrial oxidative capacity, in that the local concentrations of ADP and P_i (or the $ADP \cdot P_i/ATP$ ratio) will determine the Q_{O_2}. This inference raises the question of whether energy utilization mediates the action of thyroid hormone on cellular respiration.

SODIUM-DEPENDENT RESPIRATION AND THYROID THERMOGENESIS

The hypothesis that increased energy expenditure for transmembrane active Na^+ transport mediates a significant fraction of the thermogenic response has been examined in a variety of tissues (13). In these studies, two techniques were used to quantify the rate of energy utilization in support of active Na^+ transport: Inhibition of the Na^+ pump with ouabain, or removal of Na^+ from the bathing media. The use of ouabain to estimate sodium-dependent respiration [$Qo_2(t)$] is illustrated in Table 1. Thus, under resting conditions, active Na^+ transport appears to use a high proportion of the total energy supply, i.e., from 16 to 40% in tissues of the euthyroid rat. The contribution of augmented Na^+-linked energy expenditure to thyroid thermogenesis was evaluated by this technique (± ouabain) in rat liver, kidney, and skeletal muscle (14,15). Thyroidectomized or euthyroid rats were injected with triiodothyronine (T_3), 50 μg/100 g body wt on alternate days for a total of three injections. The Qo_2 of liver slices from thyroidectomized rats increased 104%, and from euthyroid rats 63%, after administration of T_3. In these slices, $Qo_2(t)$ increased 330% in the thyroidectomized group and 123% in the euthyroid group. The increase in $Qo_2(t)$ accounted for 90–100% of the increase in Qo_2 in the transitions from hypothyroid → euthyroid → hyperthyroid levels. Similar results were obtained in diaphragm and kidney. In the rat diaphragm, the increase in $Qo_2(t)$ accounted for 45% of the increase in Qo_2 in the transition from hypothyroid → euthyroid, and for 90% of the increase in the transition from euthyroid → hyperthyroid. In the rat kidney, the increase in $Qo_2(t)$ accounted for 29% of the increase in

TABLE 1. *Comparison of respiratory indices of various tissues of the rat*

	Qo_2	Qo'_2[a]	$Qo_2(t)$[b]	$Qo_2(t)/Qo_2$
Brain cortex (slices)	10.3	6.2	4.1	0.40
Kidney cortex (slices)	26.2	16.9	9.3	0.35
Liver (slices)	8.2	5.1	3.1	0.38
Diaphragm	7.7	6.5	1.2	0.16

Respiration is in microliters O_2 per milligram dry wt of tissue per hour.

[a] Qo_2' denotes respiration independent of Na^+ transport (e.g., in presence of ouabain).

[b] $Qo_2(t)$ denotes Na^+ transport-dependent respiration.

From F. Ismail-Beigi, Ph.D. dissertation. University of California, San Francisco, 1972.

Qo_2 in thyroidectomized rats given T_3 and for 46% of the increase in euthyroid rats. These results implied that the energy demands of the Na^+ pump mediated a major fraction of the metabolic response to thyroid hormone, presumably as a result of enhanced hydrolysis of ATP in support of active extrusion of Na^+ and coupled uptake of K^+. To analyze the basis for thyroidal stimulation of energy turnover coupled to Na^+ transport, a conventional model was assumed.

POSSIBLE MECHANISMS IN THYROIDAL ACTIVATION OF Na^+ TRANSPORT

A model of the Na^+ pump–mitochondrial coupled system is shown in Fig. 1. If thyroid hormone increased passive permeability to Na^+ and K^+ (site 1 in Fig. 1), the resultant rise in intracellular Na^+ and fall in intracellular K^+ concentrations could stimulate pump activity and ATP hydrolysis. The liberated ADP + P_i would then stimulate coupled oxidative phosphorylation. Alternatively, thyroid hormone may shift the efficiency of the Na^+ pump to a lower level, obligating increased ATP hydrolysis per equivalent of Na^+ transported (site 2 in Fig. 1). A third possibility would be enhanced mitochondrial synthesis of ATP resulting in a rise in the driving force (ATP/ADP · P_i) of the Na^+ pump (site 3 in Fig. 1). A fourth possibility is that Na^+ transport is increased either by direct activation of preexisting Na^+ pumps or by increasing the number of such pumps (site 4 in Fig. 1).

Thyroidal modulation of the pathways labeled 1 and 2 in Fig. 1 predict that the hormone would increase the ratio of intracellular Na^+ (Na^+_i) to

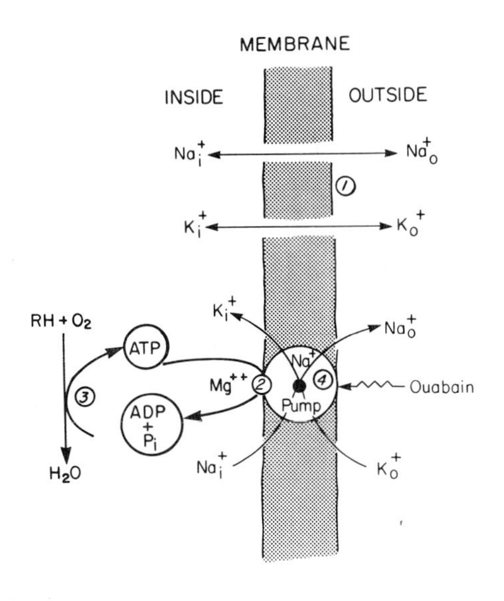

FIG. 1. Sites of regulation of Na^+ transport across cell membranes. Site 1 indicates passive Na^+ and K^+ permeability channels. Site 2 is the coupling of ATP hydrolysis to active Na^+ transport. Site 3 is mitochondrial regeneration of ATP (coupled oxidative phosphorylation). Site 4 is the Na^+ pump. [Reprinted from I. S. Edelman (23).]

intracellular K^+ (K^+_i) concentrations. In contrast, activation at site 3 or 4 should lower the Na^+_i/K^+_i ratio. These measurements were made in liver slices after incubation *in vitro* and in diaphragm and heart *in vivo* (16). Administration of T_3 to hypothyroid and euthyroid rats lowered the Na^+_i/K^+_i ratio by 20–40% without affecting plasma Na^+ and K^+ concentrations. These results implied that either mechanism 3 or mechanism 4 dominated the response to the hormone. Measurements of cellular ATP/ADP ratios failed to provide support for an action on mitochondrial augmentation of the ATP/ADP ratio (i.e., mechanism 3) (17). The plausibility of thyroidal activation of the Na^+ pump was evaluated by measuring the effect of the hormone on the transport enzyme, $Na^+ + K^+$-activated adenosine triphosphatase (NaK-ATPase). Liver homogenates from T_3-treated, thyroidectomized rats had 54% higher NaK-ATPase activity than the paired control, and from T_3-treated euthyroid rats, 81% higher activity (15). In kidney cortex homogenates, T_3 given to thyroidectomized rats elicited a 69% increase, and to euthyroid rats a 21% increase. The significance of these findings was reinforced by the results obtained on the cerebral cortex, which showed no changes in Qo_2, $Qo_2(t)$ or NaK-ATPase activity. The selectivity of the thyroid-induced changes in NaK-ATPase activity was evaluated by measuring Mg-ATPase, 5′-nucleotidase and adenyl cyclase activities. In plasma-membrane enriched fractions, T_3 had lesser or insignificant effects on Mg-ATPase and 5′-nucleotidase activities (15) and in liver homogenates, adenyl cyclase activities (basal, epinephrine-stimulated, and F1-stimulated) were the same when prepared from hypothyroid, euthyroid, or hyperthyroid rats (18). These results supported the inference that thyroid hormone selectively enhanced NaK-ATPase activity, and that this reflected activation of the Na^+ pump. These inferences also predicted that thyroid hormone should increase the rate of efflux of Na^+ from target tissues. This prediction was evaluated with liver slices prepared from rats given prior injections of T_3 or the diluent (16). Administration of T_3 to thyroidectomized rats resulted in a 105% increase in the Na^+ efflux rate constant (k_2) and in euthyroid rats elicited a 30% increase in k_2, indicating that thyroid status modulates Na^+ efflux as predicted.

A variety of indices were used to evaluate both the role of augmented energy expenditure in thyroid thermogenesis and the action of the hormone on the Na^+ transport apparatus, including Qo_2, $Qo_2(t)$, tissue K^+ concentration, Na^+ efflux (k_2), and NaK-ATPase activity. The effects of thyroid status on these indices in liver are summarized in Table 2. On a qualitative basis, it is evident that all of these indices of Na^+ and K^+ transport yield a positive correlation with thyroid status. The magnitude of the changes in K^+ content, however, were modest in comparison with the changes in $Qo_2(t)$. The relatively small change in K^+_i could be indicative of effects on the passive permeability of the liver cell membrane to Na^+ and K^+, masked by the greater effects on the Na^+ pump (compare $\Delta Qo_2(t)$ and ΔNaK-ATPase).

TABLE 2. *Summary of the effects of thyroid hormone on various indices of Na$^+$ transport in rat liver*

Thyroid status	Δk_2 (%)	Δ "Initial" tissue K$^+$ (%)	ΔNaK-ATPase (%)	$\Delta Qo_2(t)$ (%)
Thyroidectomized ± T_3	+105	+47	+54	+330
Euthyroid ± T_3	+30	+24	+81	+123

These figures were compiled from various sources and represent the percent increase in each parameter following administration of T_3 (50 μg/100 g body wt × 3) to either thyroidectomized or euthyroid rats. Computed from refs. 15 and 16.

QUANTITATIVE RELATIONSHIPS BETWEEN Qo_2, $Qo_2(t)$, AND NaK-ATPase

A crucial issue in judging the role of Na$^+$ transport in thyroid thermogenesis concerns the validity of the estimates of $Qo_2(t)$. The use of ouabain for this purpose rests on two assumptions: (i) selective inhibition of the Na$^+$ pump, and (ii) the assumption that ouabain-dependent changes in intracellular Na$^+$ and K$^+$ concentrations do not alter oxidative metabolism. Ouabain does not appear to have a toxic effect on mitochondrial oxidative activity, in that no effect on respiration of liver slices was detected when ouabain was added to Na$^+$-free media (14). These results support the first assumption. The second assumption was evaluated in a recent series of experiments on rat skeletal muscle (19). Estimates of $Qo_2(t)$ were obtained with both ouabain and with Na$^+$-free, K$^+$-enriched media. As shown in Table 3, the derived values for $Qo_2(t)$ were about the same by both techniques. In the presence of ouabain, cellular K$^+$ concentrations were reduced to ~15 meq/liter and cellular Na$^+$ concentrations were increased to ~155 meq/liter. Conversely, incubation in choline- or sucrose-Ringers supplemented with 40 meq/liter K$^+$, lowered cell Na$^+$ to less than 5 meq/liter, and maintained cell K$^+$ at ~125 meq/liter. The similarity in the estimates of $Qo_2(t)$ in these solutions implies that the respiratory rates were independent

TABLE 3. *Effect of T_3 on respiration of rat diaphragm*

Medium	meq/liters		$\Delta Qo_2(t)/\Delta Qo_2$	
	Na$^+$	K$^+$	Thyroidectomized	Euthyroid
−K$^+$ + Ouabain	140	0	0.39	0.91
Choline-Ringers	0	20	0.55	0.77
Choline-Ringers	0	40	0.52	0.91
Sucrose-Ringers	0	40	0.42	0.77
Mean			0.47	0.84

Compiled from the results of Asano et al. (19). See text.

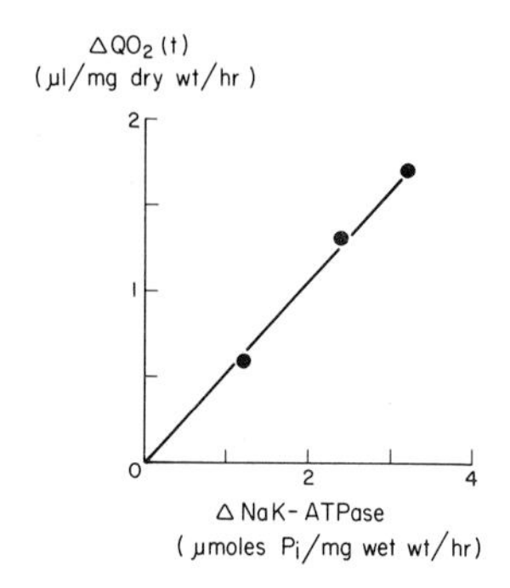

FIG. 2. The relationship between the change in $Qo_2(t)$ [i.e., $\Delta Qo_2(t)$] and the change in NaK-ATPase activity (i.e., ΔNaK-ATPase) in response to various doses of T_3. Pairs of thyroidectomized rats were injected with either the diluent or 10, 50, or 250 μg of T_3/100 g body wt, and skeletal muscle was assayed 48 hr after the injection. [These results were taken from Asano et al. (19).]

of intracellular ion composition during the time required to measure Qo_2 and $Qo_2(t)$, i.e., 60–90 min.

The role of augmentation of Na^+ transport in thyroid thermogenesis was also evaluated by comparing the time courses of the changes in respiration with those in NaK-ATPase activity. After injection of T_3 into either thyroidectomized or euthyroid rats, hepatic Qo_2 and $Qo_2(t)$ began to increase at 12 hr, reached a peak at 48 hr, and returned to control levels over the next 4 days (20). Analysis of the changes in $Qo_2(t)$ and NaK-ATPase activity revealed time courses that were similar to those of Qo_2 with peak increases at 48 hr and a return to control levels 6 days after single injections of T_3. When prolonged elevations in hepatic Qo_2 were produced by repeated injections of T_3, Qo_2, $Qo_2(t)$, and NaK-ATPase activity increased almost in parallel to plateau levels at 48 hr (20).

Support for the Na^+ transport hypothesis was also obtained in studies on dose responses in rat skeletal muscle (19). Single injections of T_3, varying from 10 to 250 μg/100 g body wt, were given to thyroidectomized rats and skeletal muscle was assayed 48 hr later. The magnitude of the absolute increases in Qo_2, $Qo_2(t)$, and NaK-ATPase were in fixed proportions (2 : 1 : 1). The results in Fig. 2 illustrate the linear correlation of $\Delta Qo_2(t)$ vs ΔNaK-ATPase activity and that the regression line intersects the origin. These findings imply that thyroidal stimulation of Na^+-dependent respiration is a consequence of augmentation of Na^+ pump activity.

MECHANISM OF THYROIDAL ENHANCEMENT OF NaK-ATPase ACTIVITY

The conventional assay for NaK-ATPase activity estimates the maximal rate of hydrolysis, presumably equivalent to the maximum transport capacity of the Na^+ pump. *In vivo,* active Na^+ transport is probably regulated at a level well below this maximum capacity. Thus, an important question is to determine whether thyroid hormone modifies the kinetic properties of the transport enzyme as well as the maximal rate (V_{max}). Two series of experiments were done to explore this question. In the first series, thyroidectomized rats were given a single dose of T_3 (250 μg/100 g body wt), and skeletal muscle was assayed 48 hr later (19). In these experiments, T_3 had no effect on the K_m for ATP, but increased V_{max} by 40%. In the second series, three doses of T_3 (50 μg/100 g body wt) were injected into thyroidectomized rats on alternate days, and the kidney cortex was assayed 48 hr after the third injection (21). The results indicated that T_3 had no effect on the K_ms for Na^+, K^+ or ATP, but increased the V_{max}s by 79, 46, and 134% for Na^+, K^+, and ATP, respectively. The increase in enzyme activity with invariant K_ms implies that thyroid hormone increases the number of pump sites. This inference was tested in a further set of experiments with $AT^{32}P(\gamma)$, which, in the presence of Na^+ and Mg^{2+}, forms a phosphorylated intermediate at the catalytic site of the NaK-ATPase (21). Thyroidectomized rats were injected with T_3 (50 μg/100 g body wt) at 48-hr intervals for a total of three doses. The renal membrane fractions were then assayed for NaK-ATPase activity and for formation of the phosphorylated intermediate (Table 4). Renal cortical NaK-ATPase activity was increased 70% and ^{32}P incorporation was increased 79% by administration of T_3. Regression analysis of the dependence of NaK-ATPase activity on the number of enzyme units (^{32}P content) revealed a 1 : 1 correspondence, indicating that the entire increase in enzyme activity is attributable to a corresponding increase in the number of pump sites.

The finding of hormonal augmentation of the number of NaK-ATPase units reinforces the possibility of effects on the synthesis of new pumps. This inference is in accord with the anabolic action, including enhanced RNA and protein synthesis, of thyroid hormone (8). To evaluate the effect of T_3 on the synthesis of a major component of the Na^+ pump, the differential incorporation of [^3H]methionine and [^{35}S]methionine into the phosphorylated intermediate of the NaK-ATPase system was studied (22). Thyroidectomized rats were injected with either the diluent or T_3 (50 μg/100 g

TABLE 4. *Effect of triiodothyronine (T_3) on renal cortical NaK-ATPase activity and Na^+/Mg^{2+}-dependent ^{32}P incorporation*

	NaK-ATPase (μmoles P_i/mg prot./hr)	Phosphorylated intermediate (pmoles P/mg prot.)
$-T_3$ ($N = 5$)	14.8 ± 1.3	36.6 ± 4.8
$+T_3$ ($N = 6$)	25.2 ± 1.6	65.4 ± 4.0
p	<0.01	<0.01
% Difference	70%	79%

Mean ± SEM. Thyroidectomized rats were injected with 50 μg T_3/100 g body wt q. 48 hr × 3.
From C. S. Lo et al. (21).

body wt) and 20 hr later, the rats were infused with either [^{3}H]methionine or [^{35}S]methionine (and in repeat series the isotopes were reversed) for 1 hr. The kidneys were removed either 8 hr or 20 hr after terminating the infusions of labeled amino acids. The kidneys of each pair of rats (T_3- and diluent-treated) were pooled and the renal cortical membrane fractions were assayed for NaK-ATPase activity, and aliquots of these fractions were processed via a procedure that yields eightfold purification of the enzyme. The partially purified enzyme fractions were extracted with sodium dodecyl sulfate (SDS) and analyzed by SDS gel electrophoresis. In control experiments, diluent–diluent pairs and T_3-treated–T_3-treated pairs were also analyzed by these procedures. The ^{32}P-labeled protein band gave isotope ratios of about unity in the two control sets of experiments and enhanced incorporation of methionine in the T_3/diluent pairs. At 8 hr after labeling, the T_3/control ratio was 1.44, and at 20 hr, 1.61. These results indicate that T_3 augmented amino acid incorporation to the same extent as the increase in NaK-ATPase activity. The selectivity of this effect was indicated by the finding that two other protein bands, adjacent to the ^{32}P-labeled band, showed no change in the ratio of labeled amino acids incorporated after treatment with T_3. In additional experiments, estimates were made of the rate of degradation of the ^{32}P-labeled peptide. The turnover rate of the ^{32}P-labeled peptide was 15% per day and was independent of thyroid status. These findings indicate that thyroid hormone induces the synthesis of the major subunit of the NaK-ATPase. The resulting increase in the number of Na^+ pumps per cell accounts, at least in part, for the increase in Na^+-dependent respiration.

Thyroid thermogenesis appears to result from a complex series of events, including an expanded mitochondrial capacity for oxidative phosphorylation, an increase in energy expenditure in active Na^+ transport as a result of the biosynthesis of new pumps, and probably a simultaneous increase in Na^+ (perhaps K^+ as well) permeability and other thermogenic pathways that have not yet been identified.

REFERENCES

1. Magnus-Levy, A. (1895): Uber den respiratorischen Gaswechsel unter dem Einfluss der Thyreoiden sowie unter verschiedenen pathologischen Zustanden. *Berlin. Klin. Wochenschrift,* 32:650–652.
2. Barker, S. B. (1964): Physiological activity of thyroid hormones and analogues. In: *The Thyroid Gland,* edited by R. Pitt-Rivers and W. R. Trotter, Vol. I, pp. 199–236. Butterworths, London.
3. Martius, C., and Hess, B. (1951): The mode of action of thyroxine. *Arch. Biochem. Biophys.,* 33:486–487.
4. Lardy, H., and Feldott, G. (1951): Metabolic effects of thyroxine *in vitro. Ann. NY Acad. Sci.,* 54:636–648.
5. Hoch, F., and Lipmann, F. (1954): The uncoupling of respiration and phosphorylation by thyroid hormones. *Proc. Natl. Acad. Sci. USA,* 40:909–921.
6. Pitt-Rivers, R., and Tata, J. R. (1959): *The Thyroid Hormones.* Pergamon Press, London.

7. Tata, J. R., Ernster, L., and Lindberg, O. (1962): Control of basal metabolic rate by thyroid hormones and cellular function. *Nature,* 193:1058–1060.
8. Tata, J. R., Ernster, L., Lindberg, O., Arrhenius, E., Pederson, S., and Hedman, R. (1963): The action of thyroid hormones at the cell level. *Biochem. J.,* 86:408–428.
9. Gustafsson, R., Tata, J. R., Lindberg, O., and Ernster, L. (1965): The relationship between the structure and activity of rat skeletal muscle mitochondria after thyroidectomy and thyroid hormone treatment. *J. Cell Biol.,* 26:555–578.
10. Kubišta, V., Kubištova, J., and Pette, D. (1971): Thyroid hormone induced changes in the enzyme activity pattern of energy-supplying metabolism of fast (white), slow (red), and heart muscle of the rat. *Eur. J. Biochem.,* 18:553–559.
11. Werner, H. V., and Berry, M. N. (1974): Stimulatory effects of thyroxine administration on reducing-equivalent transfer from substrate to oxygen during hepatic metabolism of sorbitol and glycerol. *Eur. J. Biochem.,* 42:315–324.
12. Babior, B. M., Creagan, S., Ingbar, S. H., and Kipnes, R. S. (1973): Stimulation of mitochondrial adenosine diphosphate uptake by thyroid hormones. *Proc. Natl. Acad. Sci. USA,* 70:98–102.
13. Edelman, I. S., and Ismail-Beigi, F. (1974): Thyroid thermogenesis and active Na^+ transport. *Recent Prog. Horm. Res.,* 30:235–257.
14. Ismail-Beigi, F., and Edelman, I. S. (1970): The mechanism of thyroid calorigenesis: Role of active sodium transport. *Proc. Natl. Acad. Sci. USA,* 67:1071–1078.
15. Ismail-Beigi, F., and Edelman, I. S. (1971): The mechanism of the calorigenic action of thyroid hormone: Stimulation of $Na^+ + K^+$-activated adenosine triphosphatase activity. *J. Gen. Physiol.,* 57:710–722.
16. Ismail-Beigi, F., and Edelman, I. S. (1973): Effects of thyroid status on electrolyte distribution in rat tissues. *Am. J. Physiol.,* 225:1172–1177.
17. Ismail-Beigi, F., Salibian, A., Kirsten, E., and Edelman, I. S. (1973): Effects of thyroid hormone on adenine nucleotide content of rat liver. *Proc. Soc. Exp. Biol. Med.,* 144:471–474.
18. Jones, J. K., Ismail-Beigi, F., and Edelman, I. S. (1972): Rat liver adenyl cyclase activity in various thyroid states. *J. Clin. Invest.,* 51:2498–2501.
19. Asano, Y., Liberman, U. A., and Edelman, I. S. (1976): Thyroid Thermogenesis: Relationships between Na^+-dependent respiration and $Na^+ + K^+$-adenosine triphosphatase activity in rat skeletal muscle. *J. Clin. Invest.,* 57:368.
20. Ismail-Beigi, F., and Edelman, I. S. (1974): Time-course of the effects of thyroid hormone on respiration and $Na^+ + K^+$-ATPase activity in rat liver. *Proc. Soc. Exp. Biol. Med.,* 146:983–988.
21. Lo, C. S., August, T., and Edelman, I. S.: Effects of triiodothyronine on $Na^+ + K^+$-activated adenosine triphosphatase in the rat kidney. (*In preparation.*)
22. Lo, C. S., and Edelman, I. S.: Effect of triiodothyronine on amino acid incorporation into $Na^+ + K^+$-ATPase of renal cortex. (*In preparation.*)
23. Edelman, I. S. (1974): Thyroid thermogenesis. *N. Engl. J. Med.,*290:1303–1308.

Biogenesis and Turnover of Membrane Macromolecules,
edited by John S. Cook. Raven Press, New York 1976.

Hormonal Regulation of Hormone Receptor Concentration: A Possible Mechanism for Altered Sensitivity to Hormones

Andrew H. Soll*

Veterans Administration Wadsworth Hospital Center, University of California at Los Angeles, School of Medicine, Los Angeles, California 90073, and Diabetes Branch, NIAMDD, National Institutes of Health, Bethesda, Maryland 21214

The notions that (*i*) a sustained absence of a hormone fosters increased sensitivity to that specific hormone, and (*ii*) a sustained elevation of a hormone promotes hormone resistance have predated the current explosion of work with cell-surface hormone receptors. It now appears that the ability of a hormone to regulate the concentration of its specific receptor provides a possible mechanism for these phenomena. The plasma membrane receptor, aside from being a necessary first step in hormone action and accounting for target cell specificity (1,2), also appears to affect cellular sensitivity to hormone. Furthermore, the concentration of receptors for at least some hormones is under feedback control to the extent that sustained—but not acute—changes in hormone concentration induce a change in receptor concentration, thus potentially modulating hormone action. This chapter will detail the alterations in hormone receptors that lead to variations in cellular sensitivity to hormone action and will consider the factors that regulate these changes in hormone receptors. Several examples of hormone receptor regulation are described, with emphasis upon the insulin receptor.

Regulation of the insulin receptor has been studied, using animal models of obesity, circulating mononuclear cells from obese human patients, and a lymphoblastoid cell line in continuous culture. Insulin binding has been determined in these studies by the use of a biologically active monoiodo-[^{125}I]insulin (3). Biological specificity of the binding interaction of [^{125}I] insulin is evidenced by a close correlation between the potencies of a group of insulin analogues to inhibit the binding of [^{125}I]insulin to liver membranes and to stimulate glucose oxidation in isolated fat cells (4). Binding to specific receptors and degradation of [^{125}I]insulin by liver membranes are clearly separate processes, as indicated by different specificities for analogues of insulin, affinities, and temperature and pH dependencies (5). In fact, [^{125}I] insulin bound to receptors is protected against degradation (5). Generation of binding-inhibition curves using various concentrations of unlabeled insulin allows assessment of the apparent affinity and capacity of the

* Present address: Department of Medicine, Division of Gastroenterology, UCLA Center for the Health Sciences, Los Angeles, California 90024.

receptor population (6). Scatchard analysis has been useful for quantitating these data. However, the recent demonstration of negative cooperativity in insulin receptor site–site interactions (7) has made interpretation of the traditional treatment of affinity constants difficult; nonetheless, the intercept on the abscissa remains an accurate estimate of total receptor concentration, and apparent affinities can be compared at similar fractional saturations of the receptor population. Such quantitative analyses are quite useful for comparing different receptor populations (8,9).

DECREASED INSULIN BINDING IN OBESITY

Studies with the ob/ob Mouse

In the obese hyperglycemic mouse (ob/ob) with its characteristic hyperglycemia, hyperinsulinemia, and extreme resistance to insulin (10), there is a decrease in insulin binding to liver membranes (11,12), isolated hepatocytes (12), fat cell membranes (13), thymocytes (14), and heart muscle membranes (15). This alteration in insulin binding appears to be specific for the insulin receptor rather than being a nonspecific effect from extreme obesity, i.e., the fully purified plasma membrane fraction of liver possesses normal ultrastructural morphology, normal activities of 5′-nucleotidase and adenyl cyclase, and normal protein subunit composition on disc gel electrophoresis (11,12). In addition, growth hormone, glucagon, and catecholamine binding to the liver membrane fraction of the ob/ob mouse is not significantly altered (11,12).

The insulin bound to the liver membrane fraction of the ob/ob mouse is only 25 to 35% of that observed in the thin controls (Fig. 1, Table 1), and this decrease in binding is fully accounted for by a decrease in receptor concentration, as reflected by the intercept on the abscissa of the Scatchard plot (Fig. 2)(8). In liver it is clear that this decrease in receptor concentration is not due simply to the distribution of a fixed number of receptors over the larger surface area of cells from the obese mice, because a similar relative decrease is evident, regardless of whether insulin binding to intact hepatocytes is expressed per unit of surface area, per milligram of membrane protein, or per cell (12).

Occupancy of available binding sites by the elevated plasma insulin in the ob/ob mouse does not account for the observed differences in the binding of insulin, for binding is not decreased in normal mice given very large doses of insulin just prior to killing (11,12). The receptors that remain in the ob/ob membranes are indistinguishable from normal by other criteria, including affinity for insulin, kinetics of association and dissociation, temperature dependence of binding, and the biological specificity of the binding reaction (8). The receptor in the liver membranes of the ob/ob mouse is also indistinguishable with respect to insulin receptor site–site interactions (8).

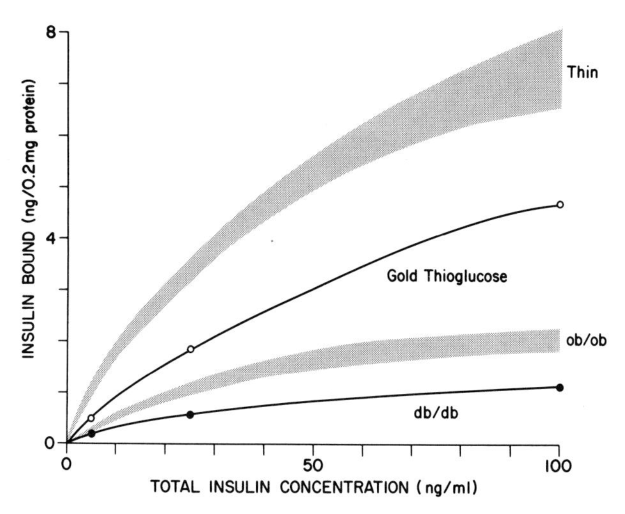

FIG. 1. Insulin binding to mouse liver membranes in ob/ob, db/db, and gold thioglucose obese mice. The indicated groups of mice were fed *ad lib.* until sacrifice. Liver membranes were prepared and incubated with[^{125}I]insulin at 0.1 ng/ml and unlabeled insulin over the range of 0 to 50,000 ng/ml. After 6 hr at 20°C, triplicate samples were centrifuged, and the radioactivity in the membrane pellet counted. Radioactivity that was bound to the membrane pellet in the absence of unlabeled insulin is referred to as "total binding"; radioactivity bound to the membrane in the presence of excess unlabeled insulin is referred to as "nonspecific binding." Specific [^{125}I]insulin binding is calculated as the difference between the total and nonspecific binding and has been normalized to a membrane concentration of 0.2 mg/ml. The specific insulin bound is plotted as a function of the total insulin concentration. Data for the thin mice are the mean ±SEM of triplicate determinations in 19 experiments on 10 different preparations of liver membranes and for the ob/ob mice from 15 experiments on 7 membrane preparations. Data for db/db (●) and gold thioglucose obese mice (○) are the mean ±SEM in three experiments. Data for the thin controls for the db/db and gold thioglucose mice fell within the range of the thin controls for the ob/ob mice, and therefore all of the thin controls were considered as one group. [From Soll et al. (16), used with permission.]

Neither degradation of insulin nor *in vitro* degradation of the receptor itself accounts for the differences observed in binding (8).

Insulin binding to liver membranes from known heterozygotes (ob/+) is normal (16), and this fact, in addition to the fact that the receptors that remain are indistinguishable from normal by many criteria, argues against the defect of the insulin receptor being a direct result of the ob gene mutation. It appears instead that the insulin receptor defect is a general phenomenon in obesity.

Similar findings of decreased insulin binding to liver membranes from ob/ob mice with only a slight decrease in glucagon binding have recently been confirmed by Chang et al. (17)(Table 1). These investigators also find a 35% decrease in lectin binding when expressed per milligram membrane protein. Assuming that lectin binding provides quantitative assessment of membrane glycoproteins, they conclude that there is a major alteration in

TABLE 1. *Insulin, glucagon, and lectin binding in ob/ob mice: Effects of starvation*

	Thin[a]		ob/ob[a]		ob/ob[b] / Thin	
	Fed	Fasted	Fed	Fasted	Fed (%)	Fasted (%)
Insulin	79	100	22	58	28	58
(fasted vs. fed)		(26% ↑)		(164% ↑)		
Glucagon	210	240	166	148	79	62
(fasted vs. fed)		(14% ↑)		(8% ↓)		
Wheat germ agglutinin	2×10^3	2.5×10^3	1.3×10^3	1.6×10^3	65	64
(fasted vs. fed)		(25% ↑)		(23% ↓)		

Specific ^{125}I-ligand binding to liver membrane fractions from ob/ob and normal mice fed and after 3 days of fasting. Data adapted from Chang et al. (17).

[a] Data expressed as cpm $\times 10^{-3}$/mg membrane protein. In parentheses, change of ligand binding with fasting expressed as percent of fed.

[b] ^{125}I-ligand binding for ob/ob mouse expressed as percent of thin control for both fed and fasted states, respectively.

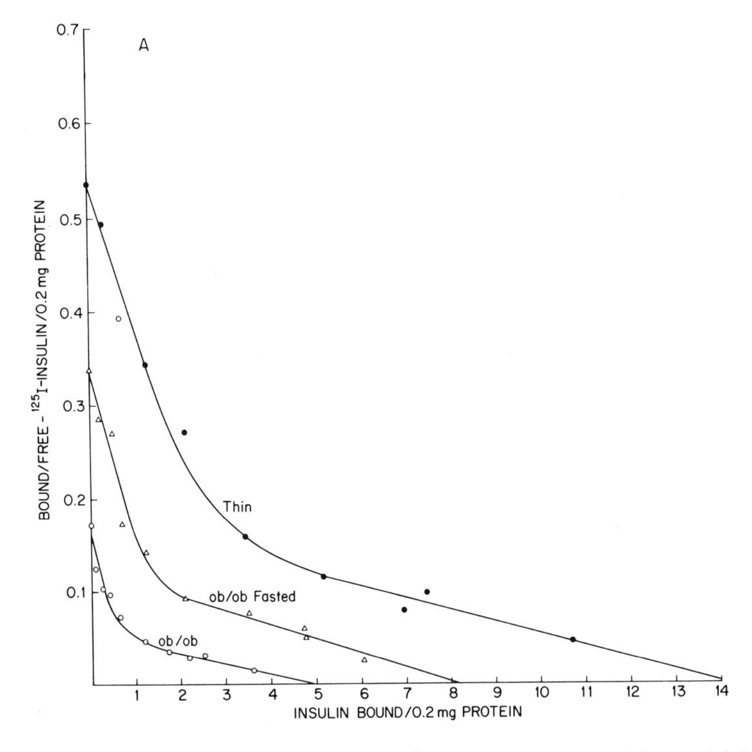

FIG. 2. Scatchard plot for the ob/ob liver membranes were prepared from ob/ob mice, their thin littermates, and ob/ob mice which had been fasted for 24 hr. Specific insulin binding was determined as in the legend to Fig. 1. The bound/free of [^{125}I]insulin is plotted as a function of the bound insulin, both axes normalized to a membrane protein concentration of 0.2 mg/ml. [From Soll et al. (16), used with permission.]

the composition of plasma membranes in ob/ob mice, with the decrease in insulin binding being a nonspecific effect of these alterations. In addition, they find a decrease in insulin and lectin binding to kidney cells isolated by collagenase and trypsin digestion. Although these findings are of great interest, several considerations hamper their interpretation. Lectins bind to membrane loci on intact cells and are useful membrane markers; however, with fractionated cells they lose their specificity for the surface membrane (18). Since the membrane fraction used by Chang et al. (17) is a relatively crude membrane fraction, lectin binding in their preparation might not be a quantitative marker for membrane glycoproteins. Further, lectin binding is critically dependent upon the terminal sugar residues of glycoprotein molecules. One cannot assume that the disordered carbohydrate metabolism in obese diabetic mice spares these carbohydrate components of membrane glycoproteins, and such a change in the membrane carbohydrate might alter lectin binding without reflecting a major change in membrane glycoprotein composition. Furthermore, the apparent preservation of membrane integrity, structure, and function in the ob/ob mice (11,12) would be difficult to envision in the face of 35% decrease in membrane glycoprotein. Finally, the isolated kidney cells were exposed to enzyme digestion and subsequent short-term cultures. The effects of such manipulations on the quantitative aspects of insulin and lectin binding have not been studied. Resolution of this controversy will await direct quantification of membrane glycoproteins. Regardless of these considerations, the observation stands that, when expressed per milligram of membrane protein, the binding of insulin is decreased, whereas glucagon binding is essentially unchanged. Arguing that there is a change in membrane composition affecting the membrane preparation in the ob/ob mouse such that membrane protein is rendered a misleading denominator for expressing the binding data, one could calculate the data using lectin binding as a denominator. Doing this, insulin binding would be somewhat decreased with glucagon binding increased: the result remains a decrease in the ratio of insulin receptors to glucagon receptors. The conclusion that these alterations in hormone receptors enter into the genesis of insulin resistance displayed by these obese mice is unavoidable.

DECREASED INSULIN BINDING IN OTHER ANIMAL MODELS OF OBESITY

Decreased insulin binding to liver membranes has also been described for two other forms of genetic obesity in mice. Insulin binding in mice homozygous for the db gene studied in their hyperinsulinemic phase is decreased to the same extent as that found in the ob/ob mice (Fig. 1)(16). One also finds decreased insulin binding (19) in NZO mice, where inheritance is polygenic rather than autosomal recessive as with ob and db mice. Similarly, with obesity induced in normal mice by the hypothalamic toxin, gold thioglucose, hyperinsulinism and a defect in insulin binding are found, with the magnitude of the decrease in binding and of the rise in serum insulin being proportional to weight gain (Fig. 3). Quantitative analyses indicate that in each of these instances the major factor accounting for the decrease in insulin binding is a decrease in the insulin receptor concentration with no change in apparent affinity (16).

A correlation between weight gain and an insulin binding defect in liver membranes is evident only in gold thioglucose obese mice (16). Insulin binding does not change as ob/ob mice gain weight between 6 and 26 weeks of age (16), probably because at 6 weeks of age ob/ob mice are already markedly obese and have the maximal insulin binding defect seen in such a setting. Normal mice were also studied between 6 and 26 weeks of age, during which time their body weight increased by 69%, while the insulin binding to liver membranes remained unchanged (16). With normal mice, the gain in body weight does not reflect true obesity with all of its concomitant metabolic changes, although there is certainly some increase in the proportion of body fat with aging. Goldfine et al. (20) also found no change in insulin binding to rat thymocytes with age, although a decrease in insulin stimulation of amino acid transport was noted. Apparently the decrease in insulin action seen with aging is not mediated by an alteration

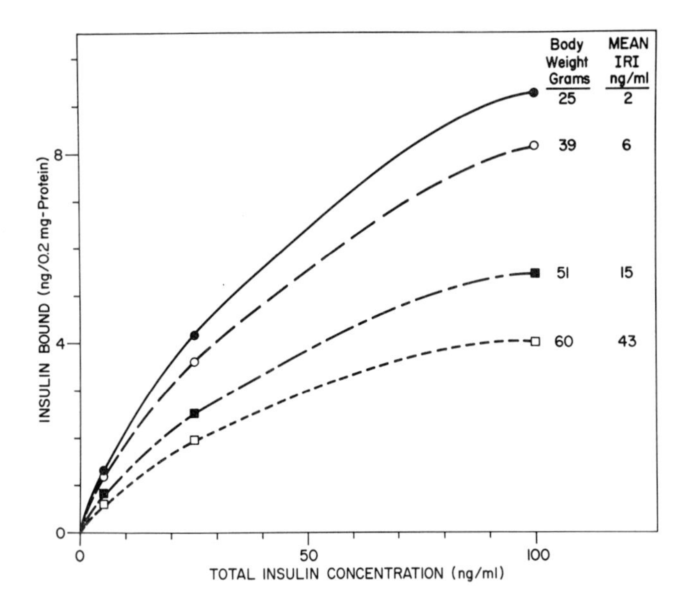

FIG. 3. The effect of body weight and immunoreactive insulin (IRI) on insulin binding in gold thioglucose obese mice. Normal C57B1/6J mice were treated with gold thioglucose (0.5 mg/g body wt) at 4 weeks of age. At 33 weeks of age mice were separated into groups with the mean body weight as indicated, and liver membranes prepared. Specific insulin binding was determined as indicated in the legend to Fig. 1. The data are the mean of triplicate determinations in 3 experiments. Immunoreactive insulin levels in the plasma were determined at the time of sacrifice. [From Soll et al. (16), used with permission.]

in the insulin receptor, but rather results from changes in other rate-limiting metabolic steps.

Livingston et al. (21) failed to find differences in insulin binding to adipocytes isolated from rats of increasing weight. However, in their study the heavier rats were older rats, rather than truly obese rats. These studies are thus consistent with the data cited above, which were obtained with liver membranes from normal mice, and indicate that increase in body weight and increase in adipocyte cell size with aging is not associated with any change in the insulin receptor. Any alteration in cell response to insulin in this instance is probably mediated by another mechanism.

THE EFFECTS OF DIETARY MANIPULATION ON INSULIN BINDING IN OBESE MICE

Studies utilizing diet restriction have elucidated a few points regarding the dynamics of the insulin receptor defect in obesity. When gold thioglucose treated obese mice are dieted to the weight of their thin controls, serum insulin falls to normal, and insulin binding returns to or to very nearly normal levels (Fig. 4)(16). In ob/ob mice dieted to the weight of their thin controls,

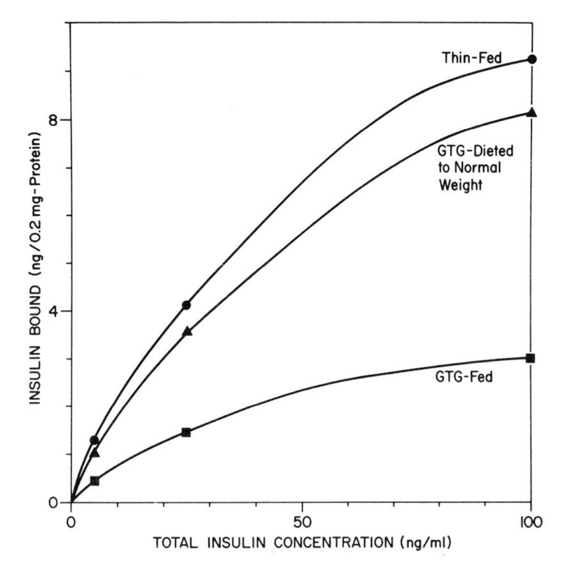

FIG. 4. Effect of food restriction on insulin binding in gold thioglucose induced obesity. Gold thioglucose obese mice from the group described in Fig. 3 were divided at an age of 33 weeks into two groups, each with a mean weight of 53 g. The mice designated GTG-fed had free access to food, and at the time of killing at 38 wk weighed 59 g and had a mean IRI of 36.6 ± 5.9 ng/ml. The group designated GTG-dieted to normal weight were restricted to 4 g of Purina Rat Chow daily and weighed 24 g at the time of killing, with IRI of 1.7 ± 0.1 ng/ml. The thin mice weighed 24 g and had IRI of 2.0 ± 1.2 ng/ml. Specific [^{125}I]insulin binding to liver membranes was determined as in Fig. 1. The data are the mean of triplicate determination in 3 experiments. [From Soll et al. (16), used with permission.]

serum insulin decreases, but remains somewhat above the normal range, and insulin binding increases from about 30 to 60% of normal (16). The failure of insulin binding in the dieted obese mice to return fully to normal is probably related to their persistent increase in the proportion of body fat as well as persistent hyperinsulinemia, although this point has not been carefully studied. Fasting ob/ob mice for 24 hr produces a small (less than 10%) loss in body weight, whereas the fall in serum insulin and rise in insulin binding are similar to those observed in the ob/ob mice dieted to normal weight (Fig. 5a)(16). The change in binding capacity in both these instances is again related to a change in receptor concentration (Fig. 2). Forgue et al. (15) have studied insulin binding using a heart muscle membrane preparation from ob/ob mice and have found that after 40 hr of fasting insulin binding was increased from 30 to 80% of that found with fed controls. Chang et al. (17) have studied insulin binding to liver membranes of ob/ob mice after 3 days of starvation (Table 1). They found a 165% increase in insulin binding which rose from 28 to 58% of the respective controls. Changes in glucagon and lectin binding with fasting were much less im-

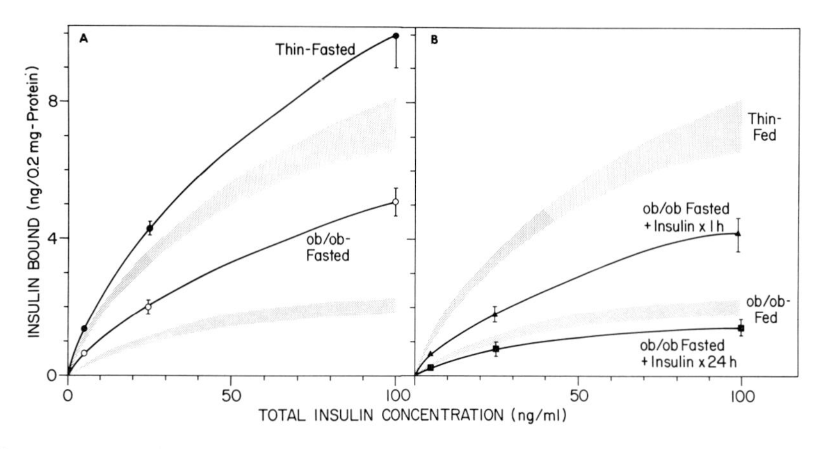

FIG. 5. The effect of fasting and the effect of acute and sustained hyperinsulinemia on insulin binding. **A:** ob/ob and thin mice were fasted for 24 hr. Specific [^{125}I]insulin binding was determined as in Fig. 1. The data are the mean of triplicate determinations in 4 experiments on 2 preparations of membranes from the thin fasted mice and the mean of 9 experiments on 4 preparations from the ob/ob fasted mice. The data for the thin fed and ob/ob fed mice that were age- and sex-matched with the fasted mice fell within the range of their respective groups in Fig. 1, and these latter data are therefore used for comparison. **B:** ob/ob mice fasted for 24 hr were given subcutaneous injections of the long-acting NPH insulin at the start of the fast and after 12 hr of fasting. In 3 groups of mice so treated with 6, 24, and 48 units of insulin, respectively, specific insulin binding (■) was in the range of the ob/ob fed mice. The data for these three groups were not statistically different, and, therefore, have been combined and expressed as the mean ± SEM for a total of 7 experiments. Insulin binding was also determined in a group of ob/ob mice fasted 24 hr and given 4 U regular insulin 1 hr before killing. These data are the mean ± SEM of triplicate determinations in 4 experiments. [From Soll et al. (16), used with permission.]

pressive. Although the statistical significance of these changes is not given, the data indicate that the major alteration with starvation in the ob/ob mouse is an increase in insulin binding rather than any change in glucagon or lectin binding. Therefore these data support the argument that the major abnormality in the liver membranes from fed obese mice is the decrease in insulin binding. Furthermore, the results obtained with dietary manipulation indicate that the decrease in insulin binding seen in obesity is directly conditioned by the fed obese state and is partially or completely reversible by food restriction.

THE CORRELATION BETWEEN SERUM INSULIN CONCENTRATION AND INSULIN BINDING IN OBESE MICE

In the studies cited thus far a trend is evident in that the defect in insulin binding is apparent when the serum insulin concentration is elevated, whereas dietary manipulations which lead to a fall in serum insulin cause a rise in insulin binding. When data from all of the groups of mice discussed thus far are examined collectively (Fig. 6), an inverse correlation is apparent between insulin binding and the serum insulin concentrations. With the exception of the gold thioglucose-treated obese mice, there was no overall

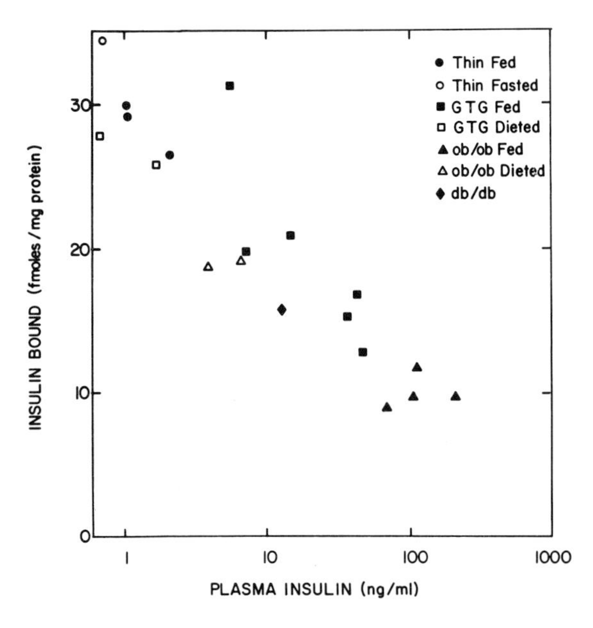

FIG. 6. The correlation of insulin binding and hyperinsulinemia. The amount of insulin bound (expressed as fmol/mg membrane protein) at an insulin concentration of 0.1 ng/ml (1.7×10^{-11} M) plotted for the indicated groups as a function of their mean plasma insulin concentrations at the time of sacrifice. [From Soll et al. (16), used with permission.]

correlation between insulin binding and body weight, degree of obesity, or plasma glucose. These data indicate that the hyperinsulinemia itself might be a major factor causing the decrease in the insulin receptor concentration. To test this hypothesis, ob/ob mice were fasted for 24 hr, but given exogenous insulin during the entire fast to prevent the usual fall in serum insulin (16). In these insulin-treated fasted mice, insulin binding did not increase, as would be expected with fasting alone (Fig. 5b). Another group of fasted ob/ob mice were given regular insulin 1 hr prior to the end of the fast, and despite the extreme acute hyperinsulinemia in these mice, insulin binding increased as expected for fasting alone. Furthermore, when ob/ob mice are treated with the beta cell toxin, streptozotocin, there is an improvement in insulin sensitivity and a fall in serum insulin (22); concomitantly, the insulin binding defect to liver membranes is largely repaired (23). These several experiments indicate that sustained—in contrast to acute—hyperinsulinemia is an important factor influencing the insulin receptor concentration in obesity.

STUDIES WITH THE INSULIN RECEPTOR IN HUMAN OBESITY

Human obesity is also characterized by hyperinsulinemia, glucose intolerance, and insulin resistance. Insulin resistance has been directly demonstrated by *in vivo* forearm perfusion studies (24) and with isolated adipocytes (25). Decreased insulin binding to fat cell membranes (26) and to peripheral mononuclear cells isolated by the Ficoll-Hypaque technique (27,28) has been demonstrated. In the latter studies (28), an increase in insulin binding following weight loss was demonstrated. More recently, Schwartz et al. (29,27) have shown that the cell responsible for the binding of insulin in the Ficoll-Hypaque preparation was the monocyte, which has a biologically specific surface membrane receptor for insulin with binding clearly independent of phagocytosis. In light of these observations, Bar et al. (30) have reexamined insulin binding in obese patients and have found that insulin binding to monocytes from obese human subjects is also decreased, with this decrease due to a lower receptor concentration rather than an alteration in affinity. Both short-term fasting and long-term dieting return insulin binding toward normal. In parallel to the animal studies cited previously, Bar et al. (30) find an inverse correlation between the insulin receptor concentration and the degree of hyperinsulinemia. There is considerable metabolic variation among obese patients: hyperinsulinemia and insulin resistance are not always demonstrated, especially when dietary composition is not carefully controlled (25). Bar et al. (30) find that their obese patients who are metabolically normal, with normal serum insulin values, have normal insulin binding to their peripheral monocytes.

Amatruda et al. (31), studying adipocytes isolated by a collagenase technique from human tissue obtained at operation, have found no difference in the number of insulin receptors per cell nor their affinity when comparing obese and normal human subjects. Several problems must be dealt with before the interpretation of this study is clear. Only pooled data for the obese patients is presented, and although hyperinsulinemia was found, there is no measure of the insulin resistance of individual patients at the time of operation. Because of the considerable individual variation in insulin resistance in obese subjects, significant deviations from normal might have been present in only a few of the patients in Amatruda's (31) study. Any precedent dietary change, even short periods of fasting, stress, or altered dietary composition, could alter insulin sensitivity (25) and insulin binding. If the cells were not actually insulin-resistant at the time of harvesting, then an insulin binding defect would not be expected. A minor problem is presented by the collagenase preparation of the isolated cells which might significantly alter the quantitative aspects of binding. Finally, the adipocytes from the obese patients have a mean cell diameter of 124 μm, compared to 77.6 μm for the normal weight subjects, and thus the surface area of the adipocytes from the obese patients is increased by 155%. Assuming that the number of receptors per cell is the same in the two groups, as Amatruda et al. suggest, then the insulin receptor concentration, i.e., the number of receptors per surface area, would be decreased to 40% of normal in the adipocytes from the obese patients. When the data are expressed in this manner, one finds a profound decrease in the insulin receptor concentration in adipose tissue from obese human patients, a defect of similar magnitude to that found in the ob/ob mouse. Thus, the presence of an insulin binding defect in these studies with adipose tissue depends upon the choice of the denominator for expressing receptor concentration. Whether a cell retains its insulin sensitivity with changes in cell size with a constant total number of receptors or with a constant receptor concentration is not known. This problem of choosing the biologically pertinent denominator for expressing receptor number is most troublesome with studies using adipocytes, because of their very great increase in size with obesity. Isolated hepatocytes from the ob/ob mouse are only slightly larger than those from the thin mice, and thus the relative insulin binding defect is evident regardless of how the data are expressed (12). This problem also does not interfere with the interpretation of data obtained with liver cell membranes, muscle, isolated thymocytes, or peripheral monocytes, and thus the weight of evidence available at present strongly supports the presence of a specific defect in insulin binding with obesity in both mice and men. This binding defect appears in all cases studied to date to be due to a decrease in receptor concentration rather than an alteration of receptor affinity, and it appears that sustained, but not acute, hyperinsulinemia is the factor most consistently associated with this change. It is likely that a decrease in receptor concentration accounts for the decrease in cellular sensitivity to insulin in obesity; yet direct studies correlating receptor concentration and cellular response to insulin have not yet been reported.

THE INSULIN RECEPTOR IN OTHER STATES OF ALTERED INSULIN SENSITIVITY

States of Decreased Insulin Sensitivity

Both glucocorticoid excess and growth hormone excess lead to hyperinsulinemia, glucose intolerance, and insulin resistance. The insulin receptor appears to be involved in the insulin resistance of glucocorticoid excess in that rats implanted with MtT tumors, which secrete growth hormone, ACTH, and prolactin, have decreased insulin binding to liver membranes, which returns to normal after adrenalectomy, despite the continued secretion of these three hormones by the tumor (32,33). In addition, rats chronically treated with ACTH or with glucocorticoids have decreased insulin binding to liver membranes (32) and to adipocytes and hepatocytes (34), whereas rats given exogenous growth hormone for 1 week show no change in insulin binding to liver membranes (32). In both the tumor-bearing and corticoid-treated rats, [^{125}I]glucagon binding to liver membranes remains unchanged (32), indicating that the decrease in insulin binding is a specific change. The differences between these two insulin-resistant states highlight the complexity of the role of the insulin receptor in states of altered insulin sensitivity. With glucocorticoid excess the rate-limiting step might be the initial binding of hormone to its cell surface receptor, whereas in growth hormone excess the resistance to hormone action appears to result from alterations of metabolic steps distal to the initial binding of hormone to its cell-surface receptor. It is of interest that the sustained hyperinsulinemia induced by growth hormone excess does not "down regulate" the insulin receptor concentration, as is apparently the case with obesity, glucocorticoid excess, and cultured lymphocytes exposed to insulin (*vide infra*). Perhaps the cellular resistance to insulin induced by growth hormone also affects the pathway by which insulin regulates its own receptor.

Recently attention has been directed to a unique form of insulin resistance occurring in nonobese females who also have the skin condition, acanthosis nigricans (35). In six patients studied to date, hyperinsulinemia and insulin resistance were found but not associated with any previously recognized causes of insulin resistance (36). Instead these patients all had a profound defect in the specific binding of insulin to their circulating monocytes, with the range from 5 to 30% of normal (36). Fasting resulted in lowered serum insulin levels, but did not change insulin binding in these six patients, as occurs with fasting in obese patients; this indicates that the insulin binding defect was primary and not directly conditioned by the hyperinsulinemia itself. In three of these patients a circulating factor has been found which directly interferes with the binding of insulin to human monocytes, cultured human lymphocytes, highly purified plasma membranes of rat liver, and fresh avian erythrocytes (37). The binding of growth hormone is not affected, indicating that the factor is specific for the insulin receptor. The factor appears to be an antibody directed at or near the surface receptor for insulin. A similar phenomenon has been described in patients with myasthenia gravis, who have been found to have high titers of antibodies that bind to membrane preparations containing receptors for acetylcholine (38).

An attenuated fall in plasma glucose following an intravenous injection of insulin has been observed in nonobese, nonketotic, adult diabetic patients with fasting hyperglycemia (39), indicating that these patients are insulin-resistant. Olefsky and Reaven (40) have found a 50% decrease in insulin binding to a Ficoll-Hypaque preparation of mononuclear cells from a similar group of diabetic patients who also displayed fasting hyperinsulinemia. Whether the insulin binding defect was primary or related to the observed hyperinsulinemia could not be ascertained. The monocyte fraction of the cell preparation was not quantitated (29), and therefore the results are open to some question.

States of Increased Insulin Sensitivity

Only preliminary data are available regarding the insulin receptor in states of increased insulin sensitivity. Hypophysectomized (32,41) and adrenalectomized (32) rats are hypoinsulinemic and show a small, but consistent, increase in insulin binding to liver membranes, consistent with their increase in insulin sensitivity and with the concept of a negative feedback regulation of the insulin receptor by insulin. In addition, thymocytes from hypophysectomized rats have increased insulin binding and increased sensitivity to insulin as assessed by insulin-stimulated amino acid transport (20). The effect of hypophysectomy on insulin binding might not be mediated by hypoinsulinemia alone in that rats implanted with an extrasellar pituitary gland capable of secreting only prolactin do not show a rise in insulin binding with hypophysectomy (39). This implication of prolactin as a regulator of the insulin receptor is in conflict with previously cited studies using rats implanted with MtT tumors wherein the persistent elevation of prolactin, ACTH, and growth hormone did not prevent the return of insulin binding to normal after adrenalectomy.

In obese mice and obese humans, fasting causes a decrease in hyperinsulinemia and insulin resistance and the insulin binding defect is repaired. Consistent with the concept of down regulation of the receptor by insulin, there is a fall in insulin concentrations and a small rise in insulin binding in fasted *normal* mice (Fig. 5a)(16). Chang et al. (17) have also noted a 25% rise in insulin binding in fasted normal mice (Table 1). However, even though a rise in the insulin receptor concentration in liver occurs, resistance to insulin action has been demonstrated *in vitro* (42). Assuming that these studies of insulin binding are correct, one must postulate that the insulin–receptor interaction does not in this setting exert rate-limiting control over cellular sensitivity to insulin. Some of these apparent inconsistencies might also be explained by different receptor dynamics in different tissues.

IN VITRO DEMONSTRATION OF A REGULATORY EFFECT OF INSULIN ON ITS OWN RECEPTOR

Using a human lymphocyte cell line in continuous culture, Gavin et al. (43), demonstrated that insulin can exert a regulatory effect on the concentration of its own receptor. In these experiments lymphocytes were preincubated in full growth media with various concentrations of insulin, followed by a thorough wash step, after which the insulin binding capacity was determined by incubation with [^{125} I]insulin. With 10^{-8} M insulin in the preincubation media, insulin binding was unchanged at 2 hr, decreased by 30% at 5 hr, and decreased by 55% at 16 hr (Fig. 7). The fact that insulin binding reached equilibrium by 30 min under these conditions reasonably excludes the possibility of occupancy of receptor sites by insulin as the

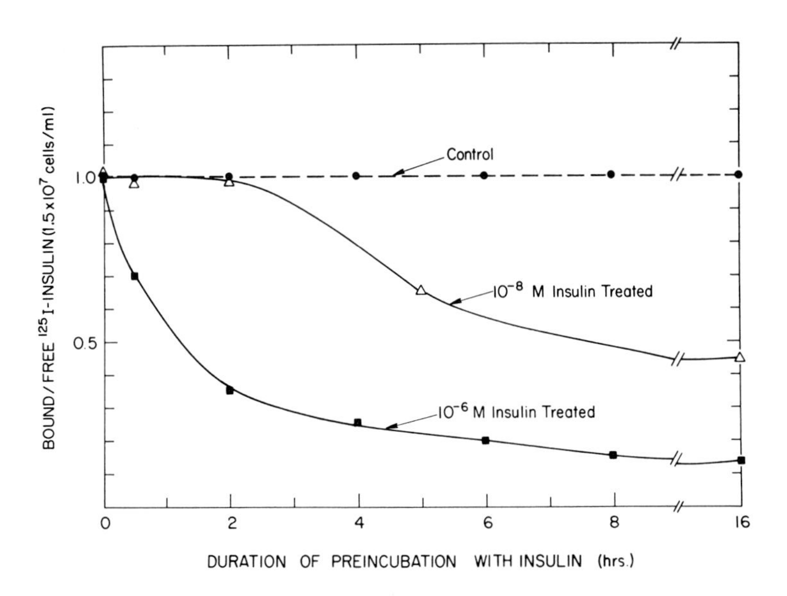

FIG. 7. Effect of duration of exposure to insulin on the binding of [^{125}I]insulin. Human lymphoblastoid cells from continuous culture (IM-9) were preincubated for up to 16 hr in Eagle's medium with the various concentrations of insulin. At the indicated times, cells were centrifuged, washed, resuspended, and exposed to [^{125}I]insulin (10^{-12} M) for 90 min at 15°C. The bound/free ratio of the [^{125}I]insulin is plotted as a function of the duration of preincubation. [From Gavin et al. (43), used with permission.]

mechanism for the apparent decrease in receptor concentration. With concentrations of insulin higher than 10^{-8} M, the down regulatory effect was more pronounced and evident at an earlier time (Fig. 7). The change in binding was due to a decrease in receptor concentration, rather than a change in affinity. When cells after 16 hr of preincubation with 10^{-8} M insulin were washed and incubated for another 16 hr in an insulin-free medium, there was full recovery of the insulin receptor population, and this recovery was inhibited by cycloheximide. This recovery was also inhibited by 10^{-9} M insulin, a concentration that was insufficient to induce the initial decrease. This latter observation might explain why ob/ob mice that were dieted to normal weight with decreased but persistent hyperinsulinemia failed to fully recover their insulin binding capacity. The site controlling receptor regulation appears to be indistinguishable from the binding site and the site for activating cellular metabolism in that among the insulin analogues tested, specificity for all three phenomena has been quite similar (44).

Cultured lymphocytes exposed only to cycloheximide, in concentrations that reversibly block protein synthesis but do not impair cell viability, show a small decrease in insulin receptor concentration (45)(Table 2). When cycloheximide and insulin are added together during preincubation, the down regulatory effect of insulin is partially attenuated (Table 2). The down regulatory response thus appears to have two components: (*i*) a cycloheximide-insensitive component that probably results from activation of a preexisting degrading mechanism (e.g., shedding of receptors, *in situ* en-

TABLE 2. *Cycloheximide effects on down regulation of receptors in lymphocytes (IM-9)*

	Insulin receptor (%)	hGH[a] receptor (%)
Control	100	100
CH (10^{-5} M)	88	80
Insulin (10^{-8} M)	50	100
+ CH	75	—
hGH (2×10^{-10} M)	100	40
+ CH	—	10

[a] Human growth hormone.

Cells were preincubated for 6 hr at 37°C in Eagle's medium with the indicated concentrations of hormones or cycloheximide (CH). After a thorough wash, specific binding was then determined by incubating the cells with [^{125}I]growth hormone or [^{125}I]insulin with and without excess concentrations of the respective unlabeled hormone. Specific binding has been expressed as percent of the control that was preincubated for 6 hr without any added agents. Data from Lesniak et al. (45).

zymic or nonenzymic degradation, or endocytosis) or from turning off of receptor synthesis; and (*ii*) a cycloheximide-sensitive component that probably represents synthesis of new components of the degrading system. The fact that cycloheximide attenuates receptor regulation by insulin hampers any quantitative correlation between a cycloheximide effect and receptor synthesis, since such an effect might be caused by a simultaneous synthesis turnoff of both receptor and degrading enzymes, thereby affecting the rate of synthesis and degradation.

OTHER EXAMPLES OF THE REGULATION OF A RECEPTOR BY ITS HORMONE (TABLES 3 AND 4)

Growth Hormone

In parallel with the studies by Gavin et al. (43), cultured human lymphocytes were exposed to growth hormone which induced a specific concentration and time-dependent fall in the growth hormone receptor concentration, with no alteration in receptor affinity (44,45). This down regulation of receptors was very sensitive to growth hormone and not simply due to occupancy of receptors by hormone, in that growth hormone at 0.1 nM caused a 50% decrease in the receptor concentration at 24 hr while it occupied only 10% of the receptor sites, and 4 nM caused an 80% decrease in receptor concentration while occupying only 50% of the receptors. Even with concentrations of growth hormone that fully saturate the receptors, the down regulatory effect did not exceed an 80% fall in receptor concentration. Down regulation of the receptor concentration was reversible when growth hormone was removed, with the recovery of receptors blocked by cycloheximide. Receptor regulation by growth hormone was specific in that the concentration of insulin receptors was constant during these various manipulations. Conversely, the number of growth hormone receptors was unchanged during down regulation of the insulin receptor by insulin (Table 2).

Cycloheximide alone caused a slightly greater fall in basal growth hormone receptor concentration than was observed with the insulin receptor, consistent with a somewhat more rapid rate of receptor turnover or with the possibility that turnover of the insulin receptor depends upon ongoing synthesis of the receptor and of degrading enzymes, both of which are inhibited by cycloheximide (Table 2). In contrast to its effect on receptor regulation by insulin, cycloheximide *enhanced* down regulation of the growth hormone receptor by growth hormone (Table 2). Receptor regulation by growth hormone thus appears to consist mainly of a cycloheximide-insensitive component (i.e., activation of a preexisting degrading mechanism) with the enhancement by cycloheximide probably due to inhibition of synthesis of new receptors. Thus down regulation by growth hormone consists primarily of activation of a preexisting degrading mechanism, while

TABLE 3. *Hormone receptor regulation*

Receptor	Animal/tissue	Regulating factor	Comments	Ref.
		I. Homotrophic Negative Receptor Regulation[a]		
Insulin	Cultured human lymphocytes	Insulin in media		Gavin et al. (43)
Insulin	Obese mice, liver membranes	Sustained hyperinsulinemia	Insulin resistance *in vivo* and *in vitro*	Soll et al. (16)
Insulin	Human peripheral monocytes	Sustained hyperinsulinemia	Insulin resistance	Bar et al. (30)
Growth hormone	Cultured human lymphocytes	Growth hormone in media		Lesniak et al. (45)
TRH	Clonal rat pituitary cells	TRH		Hinkle et al. (47)
Calcitonin	Rat kidney membranes	Parathyroidectomy ? calcitonin		Sraer et al. (55)
Beta-adrenergic	Frog red cell membranes	Beta-adrenergic agents, *in vivo*	↓ Adenylcyclase stimulation	Mukherjee et al. (46)
Progesterone	Guinea pig uterus	Progesterone	*In vivo* effect	Milgrom (49)
		II. Homotrophic positive receptor regulation.		
Prolactin	Rat liver	Hypophysectomy	↓ Receptor concentration	Posner et al. (41) Costlow et al. (48)
Prolactin	Rat liver	Extrasellar pituitary implant	↑ Receptor concentration	Posner et al. (41)
Prolactin	Rat liver	Prolactin injection	↑ Receptor concentration	Costlow et al. (48)

[a] The terms positive and negative homotrophic receptor regulation are used where there is a respective positive or negative correlation between the concentration of the receptor and the concentration of the hormone specific for that receptor. Down regulation refers to that instance of homotrophic negative receptor regulation where a sustained elevation in hormone produces a decrease in receptor concentration.

TABLE 4. *Heterotrophic receptor regulation*[a]

Receptor	Tissue	Regulating factor	Receptor regulation	Comments	Ref.
Estrogen	Female rat liver	Hypophysectomy	Repression		Costlow (48)
Estrogen	Female rat liver	Prolactin injection	Induction	Reverses effect of hypophysectomy	Costlow (48)
Estrogen	Murine uterus and mammary tissue	Prolactin	Induction	Effect in short-term culture	Leung et al. (53)
Estrogen	Murine uterus and mammary tissue	Progesterone	Repression of prolactin effect	Effect in short-term culture	Leung et al. (53)
Estrogen	Murine mammary tumor	Prolactin	Induction	Corresponds with tumor behavior	Sasaki et al. (54)
Estrogen	Murine mammary tumor	Progesterone	Repression of prolactin effect	Corresponds with tumor behavior	Sasaki et al. (54)
Estrogen	Human breast cancer	?	?	Usually predicts hormonal dependency	Jensen (50)
Progesterone	Guinea pig uterus	Estrogen	Induction		Milgrom (49)
MSH	Mouse melanoma cells	Cell cycle	Induction	↑ Receptor and response in G_2	Varga et al. (56)
ACTH	Rat adipocytes	Phenoxazones	Activation	↑ Sensitivity to ACTH	Lang et al. (58)
Acetylcholine	Rat diaphragm	Denervation ? factor	↑ In extra-junctional sites	↑ Sensitivity	Fambrough (59)

[a] Heterotrophic receptor regulation refers to receptor regulation by agents or hormones other than those specific for the receptor. Induction and repression have been used to indicate the direction of change in receptor concentration; however, it must not be assumed a priori that these terms imply only the turning on and off of receptor synthesis, as many other processes might affect receptor concentration under these circumstances.

down receptor regulation by insulin depends upon a similar process plus *de novo* protein synthesis. Furthermore, receptor synthesis appears to be enhanced during down regulation by growth hormone.

Catecholamines

Mukherjee et al. (46), utilizing [^{3}H](—)-alprenolol, a potent competitive beta-adrenergic antagonist, recently defined a stereospecific beta-adrenergic receptor site in a membrane fraction from frog erythrocytes. The physiological relevance of this binding site was corroborated by their finding of a direct correlation between the binding potency of 25 adrenergic agents and the ability of these agents to stimulate adenylate cyclase. Utilizing this system, frogs that were pretreated with beta-adrenergic catecholamines were found to have a 75% decrease in isoproterenol stimulation of adenylate cyclase activity, while basal and fluoride-stimulated activity was unchanged. Concomitantly, a 60% decrease in beta-adrenergic binding sites was found by saturation analysis. This decrease in binding was due to a decrease in the concentration of binding sites, with no change in receptor affinity.

Thyrotropin-Releasing Hormone (TRH)

Hinkle and Tashjian (47) have recently studied the effect of TRH on the TRH receptor in a membrane fraction from a clonal strain of rat pituitary cells. With chronic, but not acute, exposure to TRH there is a time- and concentration-dependent fall in the TRH receptor concentration, with no change in affinity. The dose response for this negative receptor regulation is in the same range as that for the binding of TRH to receptors and for biological activity in their system. This down regulatory effect is reversible in 96 hr. In addition, the ability of analogues to regulate receptor concentration corresponds to their biological activity and binding potency, suggesting that the site for down regulation is the same as for binding and other biological responses.

Prolactin

The prolactin receptor in rat liver appears to be under positive regulatory control by prolactin itself (41,48); Posner et al. (41) studied ^{125}I-human growth hormone (hGH) binding to a membrane fraction of rat liver as a measure of the lactogenic receptor and found binding in the male was less than in the female, and that estrogen treatment increased the prolactin receptor in intact males. Hypophysectomy decreased the basal prolactin receptor concentration in females and rendered males unresponsive to estrogen induction. They credit this effect of hypophysectomy to the absence of prolactin, in that implantation of a pituitary gland under the renal capsule prevented much of the fall of prolactin binding in hypophysectomized males again responsive to estrogen induction. Pituitary glands in an extrasellar site secrete prolactin, as evidenced by radioimmunoassay, but do not prevent the weight loss and the atrophy of thyroid, adrenals, and gonads otherwise seen with hypophysectomy. Costlow et al. (48) studied [^{125}I]ovine prolactin binding to female rat liver and found that hypophysectomy decreased specific prolactin binding. Furthermore, a single injection of 2 mg of ovine prolactin produced a 7-fold increase in the prolactin receptor, whereas estradiol, progesterone, and hydrocortisone were without effect. Prolactin thus appears to display positive receptor regulation; insufficient data are presented to determine whether this effect is due to a change in receptor concentration alone. Whether the induction of the prolactin receptor by estrogen is a direct effect on the membrane permissive with prolactin or mediated through an estrogen effect on prolactin secretion remains unclear, although the latter possibility is more likely (41). As noted previously, Posner et al. (41) also observed a rise in insulin binding with hypophysectomy, confirming the observations of Kahn et al. (32), and indicating that the effect of hypophysectomy on the prolactin binding is probably not due to nonspecific alterations in membrane metabolism.

Steroid Hormones

Receptor regulation has been demonstrated for the progesterone receptor in guinea pig uterus (49). The progesterone receptor concentration shows definite variation with the estrus cycle, with the receptor concentration rising during late diestrus and proestrus (when serum estrogen is high and progesterone low) and falling during estrus, as progesterone is secreted. Estrogen given to oophorectomized guinea pigs induces a rise in the progesterone receptor, with a maximal effect at about 24 hr, and a subsequent slow decay over the next 6 days. A prior injection of either cycloheximide or actinomycin D inhibits this estrogen effect. In contrast, when the cycloheximide administration is delayed so that protein synthesis is inhibited between 20.5 and 32 hr after the estrogen injection, the estrogen-induced rise in the progesterone receptor is not suppressed, and the rate of decay over the subsequent 6 days is not accelerated. These data indicate that estrogen induction of the progesterone receptor results from RNA transcription and protein synthesis occurring within the first 20 hr after injection. The persistent elevation of the receptor over the next 6 days is not dependent on continued rapid receptor synthesis. Progesterone given in physiological amounts 20.5 hr after the estrogen injection induces a rapid fall in the concentration of the progesterone receptor. As indicated above, the rate of synthesis of the receptor is low at this time, and indeed the down regulation of the progesterone receptor by progesterone proceeds at the same rate despite cycloheximide and actinomycin D administration. Consequently, the down regulation of the progesterone receptor by progesterone is not due to a turnoff of synthesis and therefore probably results from accelerated inactivation of the receptor. Whether this process is due to alteration of the binding site or to actual enzymatic degradation of the receptor molecule is unclear.

Recent studies of estrogen receptors in mammary tumors have highlighted the potential importance of a hormone receptor in determining tumor response to hormones and drugs. Breast cancer in women can be classified by the presence or absence of a cytosol estrogen receptor (50). The data available at present indicate that those patients with tumors that possess the estrogen receptor have a 60 to 75% chance for regression with endocrine ablative procedures, compared to a 5% probability of response with no estrogen receptor (50). Thus it appears that the presence of an estrogen receptor renders many of these tumors hormonally responsive. Further support for this conclusion comes from studies with nafoxidine, which competes with estrogen for binding sites and leads to regression of hormone responsive tumors (51). Tumor hormonal autonomy has been reported in spontaneous mammary tumors in mice where estrogen receptors are present in the cytosol, but nuclear binding sites are lacking (52). A similar mechanism might account for patients whose tumors possess estrogen receptors, yet do not appear to be hormonally dependent in that ablation does not cause regression.

There are no data regarding the mechanisms regulating the estrogen receptor in human breast carcinoma. However, there are data concerning control of the estrogen receptor in animal models. Leung et al. (52), using explanted mouse uterus and mammary tissue in short-term culture, have shown positive regulation of the estrogen receptor by prolactin with this prolactin stimulation of the estrogen receptor inhibited by progesterone. Costlow et al. (48) observed that hypophysectomy produced a fall in the concentration of a high-affinity estrogen binding protein in female rat liver. The concentration of this estrogen binding protein was largely restored within 18 hr following a single injection of prolactin. Sasaki et al. (54) have studied estrogen binding to cytosol protein in mammary tumors induced in mice by anthracene. These tumors are estrogen-responsive, stimulated by prolactin, and regress with progesterone and endocrine ablation. In short-term culture of tumor explants, prolactin increases the estrogen receptor. This stimulation is tissue specific, and is inhibited by high doses of progesterone. The implications of these preliminary studies are obvious.

Calcitonin

Parathyroidectomy, which, among other effects, decreases serum calcitonin, increases calcitonin binding to a membrane fraction of rat kidney (55). The increase in binding is apparently the result of an increase in receptor concentration, rather than a change in affinity. Preliminary data indicate that calcitonin administration prevents this effect of parathyroidectomy (55), suggesting that the phenomenon results from negative receptor regulation by calcitonin, rather than from the concomitant changes in parathyroid hormone or calcium metabolism.

Melanocyte-Stimulating Hormone (MSH)

Varga et al. (56), studying synchronized mouse melanoma cells in culture, have noted stimulation of tyrosinase activity by MSH only during the G_2 phase of the cell cycle, while cyclic AMP stimulated tyrosinase throughout the cell cycle. Since specific surface membrane receptors for [^{125}I]MSH were found only during G_2, it appears that the presence of the receptor accounts for cellular sensitivity to hormone.

Adrenocorticotropic Hormone (ACTH)

Hypersensitivity of isolated adrenal cortical cells to ACTH has been demonstrated following hypophysectomy (57); however, studies of receptor dynamics under these circumstances are not available.

The ACTH receptor on isolated fat cells does respond to certain manipulations. Phenoxazones (actinomycin D and actinocin) enhance ACTH-stimulated lipolysis and cyclic AMP production in isolated rat adipocytes. Lang et al. (58) have shown that this increased sensitivity of fat cells to ACTH induced by phenoxazones is related to an increase in the ACTH receptor concentration, with no change in receptor affinity. This increase is almost instantaneous and is seen equally with actinomycin D and actinocin, and since actinocin lacks the antibiotic properties of inhibiting RNA transcription, it appears that these lipid-soluble compounds induce an increase in the ACTH receptor by a direct effect on the membrane. Phenoxazones thus apparently activate preformed and otherwise inaccessible receptors, thus increasing the concentration of ACTH receptors and simultaneously increasing cellular sensitivity to hormone, again suggesting that hormone receptors are active modulators of hormone action.

Cholinergic Receptors

Following denervation of muscle, an increase in sensitivity to acetylcholine is associated with an increase in extrajunctional acetylcholine binding sites (59). There is some indication that exposure of denervated muscle to acetylcholine prevented this increase in binding sites (60); however, increased sensitivity to acetylcholine persists during acetylcholine exposure and the number of binding sites is not decreased after a brief wash. Therefore, denervation represents an example of how variations in hormone binding affect sensitivity to hormone; however, this variation in receptor concentration does not appear to be caused simply by regulation of a receptor population by its specific hormone. (See the chapter by Fambrough and Devreotes, *this volume*.)

CONCLUDING COMMENTS

The binding of a hormone to its surface membrane receptor appears to be a potential regulatory step in determining cellular sensitivity to hormone. The presence of decreased insulin binding in obesity, glucocorticoid excess, and the syndrome of insulin resistance and acanthosis nigricans, the absence of estrogen receptors in hormonally autonomous mammary tumors, and the association of a decreased stimulation of adenylate cyclase by beta-adrenergic agents and a decreased beta-adrenergic receptor concentration in catecholamine-exposed erythrocyte membranes are all examples of decreased binding of hormone mediating decreased cellular response to hormone. Increased hormone action appears to be partially accounted for by increased receptor concentration with increased sensitivity to insulin that follows hypophysectomy and adrenalectomy, with the phenoxazone-induced hypersensitivity of adipocytes to ACTH, and with the increased sensitivity of denervated muscle to acetylcholine. The intricate regulation of the estrous cycle depends not only upon carefully regulated secretion of releasing hormones, gonadotropins, and steroids, but also appears to depend on the variations in receptor concentrations in the uterus and probably also in the hypothalamus, pituitary, and ovary. There are some conditions such as fasting, aging, and growth hormone excess where hormone binding to its receptor does not appear to exert rate-limiting control over cellular sensitivity to the hormone.

Regulation of receptor concentration by the hormone that binds to that receptor appears now to be a well-established phenomenon. Negative receptor regulation where sustained elevations of its hormone cause down regulation of receptor concentrations pertains to insulin, growth hormone, TRH, progesterone, and catecholamines (Table 3). Other instances of negative receptor regulation are found where sustained decreases in hormone lead to an increase in receptor concentration, as with the insulin receptor following hypophysectomy or a period of fasting, and probably the calcitonin receptor following parathyroidectomy. Prolactin is the only example, at present, of positive receptor regulation by its own hormone, although there are several examples where a receptor can be induced or activated by another hormone or drug, such as the estrogen effect on the progesterone receptor, the prolactin effect on the estrogen receptor, and the phenoxazone effect on the ACTH receptor. Negative receptor regulation or repression by another drug or hormone also occurs as is evidenced by the progesterone effect on the prolactin stimulation of the estrogen receptor and possibly the glucocorticoid effect on the insulin receptor. A last example of what might have been considered negative receptor regulation is the blockade of receptors by antibodies in the syndrome of acanthosis nigricans and insulin resistance (37). In those instances where the question has been studied, sustained, but not acute, alterations of hormone concentration exert the regulatory effect. Also it appears that receptor regulation involves changes in receptor concentration rather than receptor affinity.

Without question, basal receptor concentration is maintained by a balance of synthesis and degradation. Information as to whether receptor regulation depends upon control of synthesis or of degradation is scant, and limited by the difficulty of interpreting a cycloheximide effect. Unfortunately, although cycloheximide directly inhibits protein synthesis and therefore receptor synthesis, it can also inhibit synthesis of components of a degrading system and thus attenuate degradation in those instances where the degrading mechanism is not present in adequate concentration in the basal state. It appears that the effect of estrogen on the progesterone receptor in guinea pig uterus is dependent largely upon stimulation of receptor synthesis, and negative regulation of the progesterone receptor by progesterone depends upon activation of preexisting mechanisms for receptor degradation, without requiring new protein synthesis. The recovery of the insulin and growth hormone receptors on cultured lymphocytes following down regulation is dependent upon *de novo* protein synthesis in that cycloheximide completely blocks recovery. The cycloheximide effect on negative receptor regulation by insulin and growth hormone in this system is of great interest because this agent partially blocks down regulation by insulin, but enhances down regulation by growth hormone. No explanation for this difference is available; one can only conclude that down regulation is a complex and variable process which can depend upon several mechanisms: (*i*) activation of preexisting degrading enzymes; (*ii*) activation of nonenzymic degrading mechanisms such as endocytosis or receptor shedding; (*iii*) turning-off of receptor synthesis; and (*iv*) *de novo* synthesis of components of the degrading system. Stimulation of receptor synthesis occurs at some point following the initiation of down regulation and must be accounted for in any quantitative assessment of receptor turnover during regulation. The exact nature of receptor turnover and regulation will be clarified only as new approaches and methodology become available.

ACKNOWLEDGMENTS

The author is indebted to Dr. C. Ronald Kahn, Dr. Jesse Roth, and Ms. Maxine Lesniak for their helpful comments and advice during the preparation of this manuscript, to Dr. M. I. Grossman for his review of the manuscript, and to Mrs. Ruth Abercrombie for excellent secretarial assistance.

REFERENCES

1. Roth, J. (1973): Peptide hormone binding to receptors: A review of direct studies *in vitro*. *Metabolism,* 8:1059–1073.
2. Kahn, C. R. (1975): Membrane receptors for polypeptide hormones. In: *Methods in Membrane Biology,* Vol. 3, edited by E. D. Korn. Plenum Press, New York.
3. Freychet, P., Roth, J., and Neville, D. M., Jr. (1971): Monoiodoinsulin: Demonstration of its biological activity and binding to fat cells and liver membranes. *Biochem. Biophys. Res. Commun.,* 43:400–408.
4. Freychet, P., Roth, J., Neville, D. M., Jr. (1971): Insulin receptors in the liver: Specific binding of ^{125}I-insulin to the plasma membrane and its relation to insulin bioactivity. *Proc. Natl. Acad. Sci. USA,* 68:1833–1837.
5. Freychet, P., Kahn, C. R., Roth, J., and Neville, D. M., Jr. (1972): Insulin interaction with liver plasma membranes: Independence of binding of the hormone and its degradation. *J. Biol. Chem.,* 247:3953–3961.
6. Kahn, C. R., Freychet, P., Neville, D. M., Jr., and Roth, J. (1974): Quantitative aspects of the insulin-receptor interaction in liver plasma membranes. *J. Biol. Chem.,* 249:2249–2257.
7. DeMetys, P., Roth, J., Neville, D. M., Jr., Gavin, J. R., III, and Lesniak, M. A. (1973): Insulin interactions with its receptors: Experimental evidence for negative cooperativity. *Biochem. Biophys. Res. Commun.,* 55:154–161.
8. Soll, A. H., Kahn, C. R., and Neville, D. M., Jr. (1975): The decreased insulin binding to liver plasma membranes in the obese hyperglycemic mouse (ob/ob): Demonstration of a decreased number of functionally normal receptors. *J. Biol. Chem.,* 250:4702–4707.
9. DeMeyts, P., and Roth, J. (1975): Cooperativity in ligand binding: A new graphic analysis. *Biochem. Biophys. Res. Commun.,* 66:1118–1126.
10. Stauffacher, W., Orci, L., Cameron, D. P., Burr, I. M., and Renold, A. E. (1971): Spontaneous hyperglycemia and/or obesity in laboratory rodents: An example of the possible usefulness of animal disease models with both genetic and environmental components. *Recent Prog. Horm. Res.,* 27:41–95.
11. Kahn, C. R., Neville, D. M., Jr., Gorden, P., Freychet, P., and Roth, J. (1972): Insulin receptor defect in insulin resistance: Studies in the obese-hyperglycemic mouse. *Biochem. Biophys. Res. Commun.,* 48:135–142.
12. Kahn, C. R., Neville, D. M., Jr., and Roth, J. (1973): Insulin-receptor interaction in the obese hyperglycemic mouse: A model of insulin resistance. *J. Biol. Chem.,* 248:244–250.
13. Freychet, P., Laudat, M. H., Laudat, P., Rosselin, G., Kahn, C. R., Gorden, P., and Roth, J. (1972): Impairment of insulin binding to fat cell plasma membrane in the obese hyperglycemic mouse. *Fed. Eur. Biochem. Soc. Lett.,* 25:339–342.
14. Soll, A. H., Goldfine, I. D., Roth, J., Kahn, C. R., and Neville, D. M., Jr. (1974): Thymic lymphocytes in obese (ob/ob) mice: A mirror of the insulin receptor defect in liver and fat. *J. Biol. Chem.,* 249:4127–4130.
15. Forgue, M-E., and Freychet, P. (1974): Insulin receptors in the heart muscle: Demonstration of specific binding sites and impairment of insulin binding to the plasma membrane of the obese hyperglycemic mouse. *Diabetes,* 24:715–723.
16. Soll, A. H., Kahn, C. R., Neville, D. M., Jr., and Roth, J. (1975): Insulin receptor deficiency in genetic and acquired obesity. *J. Clin. Invest.,* 56:769–780.
17. Chang, K. J., Huang, D., and Cuatrecasas, P. (1975): The defect in insulin receptors in obese-hyperglycemic mice: a probable accompaniment of more generalized alterations in membrane glycoproteins. *Biochem. Biophys. Res. Commun.,* 64:566–573.
18. Chang, K. J., Bennett, V., and Cuatrecasas, P. (1975): Membrane receptors as general markers for plasma membrane isolation procedures: The use of ^{125}I-labeled wheat germ agglutinin, insulin, and cholera toxin. *J. Biol. Chem.,* 250:488–500.
19. Baxter, D., Gates, R. J., and Lazarus, N. R. (1973): Insulin receptor of the New Zealand obese mouse (NZO): Changes following the implantation of islets of Langerhans. 8th Congress of the International Diabetes Federation. *Excerpta Medica,* 280:161.
20. Goldfine, I. D., Soll, A., Kahn, C. R., et al. (1973): The isolated thymocyte: A new cell for the study of insulin receptor concentrations. *Clin. Res.,* 21:492.

21. Livingston, J. N., Cuatrecasas, P., and Lockwood, D. (1972): Insulin insensitivity of large fat cells. *Science,* 177:626–628.
22. Mahler, R. J., and Szabo, O. (1971): Amelioration of insulin resistance in obese mice. *Am. J. Physiol.,* 221:980–983.
23. Loten, E. G., Freychet, P., et al. (1975): *Manuscript in preparation.*
24. Rabinowitz, D. (1970): Some endocrine and metabolic aspects of obesity. *Annu. Rev. Med.,* 21:241–258.
25. Salans, L. B., Bray, G. A., Cushman, S. W., Danforth, E., Jr., Glennon, J. A., Horton, E. S., and Sims, E. A. H. (1974): Glucose metabolism and the response to insulin by human adipose tissue in spontaneous and experimental obesity: Effects of dietary composition and adipose cell size. *J. Clin. Invest.,* 53:848–856.
26. Marinetti, G. V., Shatz, L., and Reilly, K. (1972): Hormone–membrane interactions. In: *Insulin Action,* edited by I. B. Fritz, pp. 207–276. Academic Press, New York.
27. Archer, J. A., Gorden, P., Gavin, J. R., III, Lesniak, M. A., and Roth, J. (1973): Insulin receptors in human circulating lymphocytes: Application to the study of insulin resistance in man. *J. Clin. Endocrinol.,* 36:627–633.
28. Archer, J. A., Gorden, P., and Roth, J. (1975): Defect in insulin binding to receptors in obese man: Amelioration with calorie restriction. *J. Clin. Invest.,* 55:166–174.
29. Schwartz, R. H., Bianco, A. R., Handwerger, B. S., and Kahn, C. R. (1975): Demonstration that monocytes rather than lymphocytes are the insulin-binding cells in preparations of human peripheral blood mononuclear leukocytes: Implications for studies of insulin-resistant states in man. *Proc. Natl. Acad. Sci. USA,* 72:474–478.
30. Bar, R. S., Gorden, P., Roth, J., Kahn, C. R., and DeMeyts, P. (1976): Regulation of the affinity and concentration of insulin receptors in man. *Clin. Res.,* 24:269A.
31. Amatruda, J. M., Livingston, J. N., and Lockwood, D. H. (1975): Insulin receptor: Role in the resistance of human obesity to insulin. *Science,* 188:264–266.
32. Kahn, C. R., Goldfine, I. D., Neville, D. M., Jr., Roth, J., Garrison, M., and Bates, R. W. (1973): Insulin receptor defect: A major factor in the insulin resistance of glucocorticoid excess. *Endocrinology,* 92 (Suppl.):240.
33. Goldfine, I. D., Kahn, C. R., Neville, D. M., Jr., Roth, J., Garrison, M. M., and Bates, R. W. (1973): Decreased binding of insulin to its receptors in rats with hormone-induced insulin resistance. *Biochem. Biophys. Res. Commun.,* 53:852–857.
34. Olefsky, J. M., Johnson, J., Liu, F., Jen, P., and Reaven, G. M. (1975): The effects of acute and chronic dexamethasone administration on insulin binding to isolated rat hepatocytes and adipocytes. *Metabolism,* 24:517–527.
35. Archer, J. A., Gorden, P., Kahn, C. R., Gavin, J. R., III, Neville, D. M., Jr., Martin, M. M., and Roth, J. (1973): Insulin receptor deficiency states in man: Two clinical forms. *J. Clin. Invest.,* 52:4a.
36. Kahn, C. R., Bar, R. S., Gorden, P., and Roth, J. (1975): The syndrome of extreme insulin resistance and acanthosis nigricans: A primary insulin receptor defect in man. *Clin Res.,* 23:324A.
37. Flier, J. S., Kahn, C. R., Roth, J., and Bar, R. S. (1975): Circulating antibodies that impair insulin receptor binding in patients with an unusual diabetic syndrome and extreme insulin resistance. *Science,* in press.
38. Patrick, J., Lindstrom, J., Culp, B., and McMillan, J. (1973): Studies on purified eel acetylcholine receptor and anti-acetylcholine receptor antibody. *Proc. Natl. Acad. Sci. USA,* 70:3334.
39. Alford, R. P., Martin, F. I. R., and Pearson, M. J. (1971): The significance and interpretation of mildly abnormal oral glucose tolerance. *Diabetologia,* 7:173–180.
40. Olefsky, J. M., and Reaven, G. M. (1974): Decreased insulin binding to lymphocytes from diabetic subjects. *J. Clin. Invest.,* 54:1323–1328.
41. Posner, B. I., Kelly, P. A., and Friesen, H. G. (1975): Prolactin receptors in rat liver: Possible induction by prolactin. *Science,* 188:57–59.
42. Abraham, R. R., and Beloff-Chain, A. (1971): Hormonal control of intermediary metabolism in obese hyperglycemic mice. I. The sensitivity and response to insulin in adipose tissue and muscle in vitro. *Diabetes,* 20:522–534.
43. Gavin, J. R., III, Roth, J., Neville, D. M., Jr., DeMeyts, P., and Buell, D. N. (1974): Insulin-dependent regulation of insulin receptor concentrations: A demonstration in cell culture. *Proc. Natl. Acad. Sci. USA,* 71:84–88.

44. Lesniak, M. A., Bianco, A. R., Roth, J., and Gavin, J. R., III (1974): Regulation by hormone of its receptors on cells: Studies with insulin and growth hormone. *Clin. Res.*, 22:343A.
45. Lesniak, M. A., and Roth, J. (1975): Regulation of receptor concentration by homologous hormone: effect of human growth hormone on its receptor in IM-9 lymphocytes (submitted for publication).
46. Mukherjee, C., Caron, M. G., and Lefkowitz, R. J. (1975): Catecholamine induced subsensitivity of adenylate cyclase associated with loss of beta-adrenergic receptor binding sites. *Proc. Natl. Acad. Sci. USA*, 72:1945–1949.
47. Hinkle, P. M., and Tashjian, A. H., Jr. (1975): Decrease in the number of receptors for thyrotropin releasing hormone (TRH) in GH_3 pituitary cells after prolonged incubation with TRH. *Endocrinology*, 96 (Suppl. 3):A–3.
48. Costlow, M. E., Buschow, R. A., Chamness, G. C., and McGuire, W. L. (1975): Autoregulation of prolactin receptors. *Endocrinology*, 96 (Suppl):A–58.
49. Milgrom, E., Luu Thi, M., and Baulieu, E. E. (1973): Control mechanism of steroid hormone receptors in the reproductive tract. In: Transactions of 6th Karolinska Symposia on Research Methods in Reproductive Endocrinology: Protein Synthesis in Reproductive Tissue, Geneva, May 21–23, 1973. *Acta Endocrinol.* [*Suppl.*] (*Kbh*), 180:380–403.
50. Jensen, E. V. (1974): Some newer endocrine aspects of breast cancer. *N. Engl. J. Med.*, 291:1252–1254.
51. Engelsman, E., Persign, J. P., Korsten, C. B., and Cleton, F. J. (1973): Oestrogen receptor in human breast cancer tissue and response to endocrine therapy. *Br. Med. J.*, 2:750–752.
52. Shyamala, G. (1972): Estradiol receptors in mouse mammary tumors: Absence of the transfer of bound estradiol from the cytoplasm to the nucleus. *Biochem. Biophys. Res. Commun.*, 46:1623–1630.
53. Leung, B. S., and Sasaki, G. H. (1973): Prolactin and progesterone effect on specific estradiol binding in uterine and mammary tissues *in vitro*. *Biochem. Biophys. Res. Commun.*, 55:1180–1187.
54. Sasaki, G. H., and Leung, B. S. (1975): On the mechanism of hormone action in 7,12 Dimethylbenz(a) Anthracene-induced mammary tumor. I. Prolactin and progesterone effects on estrogen receptor in vitro. *Cancer*, 35:645–651.
55. Sraer, J., Ardaillou, R., and Couette, S. (1974): Increased binding of calcitonin to renal receptors in parathyroidectomized rats. *Endocrinology*, 95:632–637.
56. Varga, J. M., Dipasquale, A., Pawelek, J., McGuire, J. S., and Lerner, A. B. (1974): Regulation of melanocyte stimulating hormone action at the receptor level: Discontinuous bindinf of hormone to synchronized mouse melanoma cells during the cell cycle. *Proc. Natl. Acad. Sci. USA*, 71:1590–1593.
57. Sayers, G., and Beall, R. J. (1973): Isolated adrenal cortex cells; hypersensitivity to adrenocorticotropic hormone after hypophysectomy. *Science*, 179:1330–1331.
58. Lang, U., Karlaganis, G., Vogel, R., and Schwyzer, R. (1974): Hormone receptor interactions. Adrenocorticotropic hormone binding site increase in isolated fat cells by phenoxazones. *Biochemistry*, 13:2626–2634.
59. Fambrough, D. M. (1974): Acetylcholine receptors: revised estimates of extrajunctional receptor density in denervated rat diaphragm. *J. Gen. Physiol.*, 64:468–472.
60. Miledi, R., and Potter, L. T. (1971): Acetylcholine receptors in muscle fibers. *Nature*, 233:599–603.

Biogenesis and Turnover of Membrane Macromolecules,
edited by John S. Cook. Raven Press, New York 1976.

Early Membrane Events in Lymphocyte Blastogenesis: The Action of Ouabain and of Protease Inhibitors

[1]J. G. Kaplan, [2]M. R. Quastel, and [1]Jacques Dornand

[1]*Department of Biology, University of Ottawa, Ottawa, Canada,* and [2]*Ben Gurion University and Soroka Medical Center, Beer Sheba, Israel*

We shall discuss several aspects of our work which bear on variations in the stability, quantity, and activity of certain functional membrane sites during lymphocyte transformation. We recently reviewed much of our early work on the role of cation transport during blastogenesis (1) and shall simply outline some major points which are more fully treated therein as an introduction to what is to follow.

At concentrations of approximately 10^{-7} M and above, the cardiotonic steroid ouabain arrests, after a short lag, virtually all the parameters of stimulation of human lymphocytes at all stages of transformation, regardless of how the cells were stimulated (2,3). The inhibition was found to be reversible, either by addition of excess K^+ to the medium or by washing. Stimulation by the mitogen phytohemagglutinin (PHA) caused a rapid increase in the influx of K^+, but no change in the ouabain-insensitive efflux (4). The effect did not require protein synthesis but was highly temperature-sensitive; the V_{max} of K^+ transport was changed by activation of the lymphocytes, not the K_m (5).

These data were consistent with the hypothesis that one of the essential early changes induced by stimulation of resting lymphocytes is an increase in the number of sites for the transport of monovalent cation at the cell surface, either by exposure of preformed but previously cryptic pump sites or by aggregation of inactive subunits to active oligomers in the membrane (1,6). The increase in the number of functional Na-K-ATPase sites would result in increasing the internal K^+ level above a threshold concentration required for almost all of the train of events that links the membrane stimulus to the synthesis of DNA and the mitosis that ultimately ensue.

EVIDENCE FOR PHYTOHEMAGGLUTININ-INDUCED INCREASE IN PUMP SITES

Our early work on activation of K^+ influx following stimulation by mitogen was extended by Averdunk (7), who demonstrated an increase in K^+ uptake within a minute of adding PHA to the medium. This author and Lauf have now shown (8) a significant activation of the ouabain-sensitive Na-K-ATPase of lymphocytes caused by PHA and concanavalin A (Con A).

Preliminary evidence in favor of our hypothesis came from studies of the binding of ^{42}K. After removal of unbound ^{42}K by washing with sucrose or saline solutions, two distinct K^+ cell compartments were demonstrated, one of which was internal. The other was identified as at the cell surface, since it was highly labile, easily removed by washing with KCl or RbCl (but not with NaCl) or by treatment with trypsin and very rapidly equilibrating; this compartment increased markedly after PHA treatment, as well as did the slowly equilibrating, presumably internal compartment (6).

A direct test of our hypothesis was undertaken by studying binding of [^{3}H]ouabain, on the assumption that an increase in number of pump sites would be reflected in an increase in the number of ouabain-binding sites. We first showed a concentration- and time-dependent binding of ouabain with saturation kinetics; the binding was significantly increased by PHA within a few hours of initiation of culture, the increase depending on the concentration of PHA over a range of 10–200 μg/ml culture (9). Inhibition of protein synthesis and of respiration had no effect on the kinetics of ouabain binding, on the amount bound, or on the PHA-induced increase; on the other hand, binding was extremely temperature-sensitive, with very little binding occurring at 5°C in presence or in absence of PHA (9). Subsequent unpublished work on the effect of temperature showed that very little binding of [^{3}H]ouabain occurred after a 1-hr pulse with or without PHA at temperatures of 20°C or lower. It is interesting that the Na-K-ATPase has a critical temperature lying between 18 and 20°C; below this temperature the phosphoenzyme is assumed to lose its ability to assume a K^+-sensitive conformation (10).

We have recently reported that PHA and its purified derivative, leukoagglutinin, within minutes of their being added to the culture, caused a significant increase in binding of [^{3}H]ouabain (11). The increased binding occurred during the same period as the enhanced transport of K^+, and was also insensitive to inhibition of RNA and protein synthesis; the magnitude of the increase (80%) was quantitatively similar to that of K^+ transport (85%), and there was an increase in V_{max} and no change in the K_m of binding. Further, the degree of inhibition of ^{42}K uptake at 1.4×10^{-7} M ouabain was $53 \pm 7\%$ (4) and at that concentration the saturation of the available ouabain sites was estimated to be $66 \pm 6\%$ (11); there is thus a parallel behavior between the sites which transport K^+ on the one hand, and those which bind ouabain on the other. The binding of ouabain could be strongly inhibited by the simultaneous presence of high K^+ as could its other effects on transformation; it was also readily reversed by adding nonradioactive ouabain to the medium (9).

Different lymphocyte preparations showed approximately the same number of binding sites per cell for [^{3}H]ouabain under conditions of saturation: 1.25×10^5 per cell in control cultures and 2.3×10^5 per cell following PHA treatment (11). The combined weight of this evidence favors the conclusion that ouabain binds to specific sites, that these sites are at the cell surface, that they are also specific for K^+ transport, and, finally, that they increase significantly in number following stimulation of the lymphocytes with PHA or leucoagglutinin. We cannot exclude the possibility that mitogens also increase the rate of cation transport by each pump site, as proposed by Averdunk and Lauf (8,11*a*); however, our data strongly indicate an increase in total number of pump sites occurring within minutes of addition of mitogen to the lymphocyte cultures.

ACTIVATION, INACTIVATION, AND REGENERATION OF THE STIMULATORY SITES RESPONSIBLE FOR THE MIXED LYMPHOCYTE REACTION

Lymphocytes from two genetically dissimilar donors will stimulate each other to proliferate, a phenomenon known as the mixed lymphocyte reaction (MLR) (11*b*). Responders and stimulators can be experimentally distinguished by pretreating one of the populations with mitomycin C, which prevents any subsequent DNA synthesis in the cells (11*c*). The treated cells retain their ability to stimulate in the MLR; all [^{3}H]thymidine incorporation in the mixed population can be ascribed to the untreated cells, or responders. This is called the one-way MLR.

Treatment of human lymphocytes for more than 40 hr in medium containing 2×10^{-7} M ouabain caused an irreversible loss of ability of the cells to show the MLR or to proliferate in response to antigens in the medium, but the cultures remained responsive to mitogens including $NaIO_4$ (9,12). At even higher concentrations of ouabain (8×10^{-7} M and above), the cultures would no longer respond to PHA unless they were rescued by including high concentrations of K^+ (20 mM) in the medium (12); we have recently confirmed this (Dornand and Kaplan, *unpublished*). The loss of the MLR and of antigen response could not be prevented by high K^+.

One of us and his colleagues then showed that the persistent effects of prolonged ouabain pretreatment on the MLR were of two opposite kinds: at high concentration ($>10^{-6}$ M), it caused an irreversible loss of the capacity to stimulate in the MLR without modifying significantly the capacity to respond to allogeneic cells by DNA synthesis and proliferation; at lower concentration (10^{-7} M) it caused the treated cells to become superstimulators, i.e., to cause an incorporation of thymidine into the DNA of the responding cells more than double that of the same cells when stimulated by controls (13). We recently confirmed these findings (13*a*). Table 1 shows an experiment (Christen and Sasportes, *personal communication*) which indicates that the increase in incorporation of thymidine into the DNA of

TABLE 1. *The effect of ouabain pretreatment on proliferation in the MLR*

Combination	Cell counts ($\times 10^{-6}$)	
ABm	9.0	
	After 2×10^{-6} M ouabain	After 10^{-7} M ouabain
ABmo	5.4	16.0
AoBm	8.9	9.1
AoBmo	4.9	14.3

The data shown represent cell counts after 6 days MLR. In all of the cell combinations, there were 8×10^6 cells present at day 0. Responding cells from individual A and stimulating cells from individual B [subsequently treated with mitomycin C; all techniques have been described (13)] were incubated either with or without ouabain at the concentration shown, washed, and then mixed in various combinations until harvested at day 6. These data are from an experiment done by our colleagues Christen and Sasportes (*personal communication*). m, mitomycin C treated; o, ouabain treated.

responders in mixed cultures in which the stimulators had been pretreated with 10^{-7} M ouabain was due to an increase in the number of cells proliferating under these conditions.

One such experiment, illustrated in Figs. 1–3, involved the treatment for 48 hr of the responding cells of individual A with 10^{-7} M ouabain (o) and the stimulating cells [treated with mitomycin C (m)] of individual B with either 10^{-7} M or 2×10^{-6} M ouabain. Figure 1 shows the effects of these pretreatments on incorporation of thymidine into DNA during MLR; the bottom curve, BBm, shows the extremely low level of basal, unstimulated DNA synthesis when autologous cells are mixed. The control, illustrated by the heavy curve, ABm, shows the normal MLR with peak incorporation at day 5 as is usual. The curve immediately beneath, Ao(10^{-7})Bm, shows the effect of treating the responding cells with 10^{-7} M ouabain; this combination behaves as do the controls. When the stimulators were pretreated with 2×10^{-6} M ouabain, the MLR was essentially abolished. Conversely, after pretreatment with 10^{-7} M (top curve) there was a lag of a day or two in reaching the peak of DNA synthesis, but the peak level of incorporation was twice that of the controls.

Figure 2 shows the effect on RNA synthesis of the same treatments on the same cells; the results are comparable to those observed in the case of DNA synthesis, except that in the case of the MLR involving stimulators pretreated with 10^{-7} M ouabain, the peak of incorporation of [^{3}H]uridine into the RNA of the responders, although delayed by 1 day over that of the controls, was not quantitatively significantly different. Figure 3 shows the

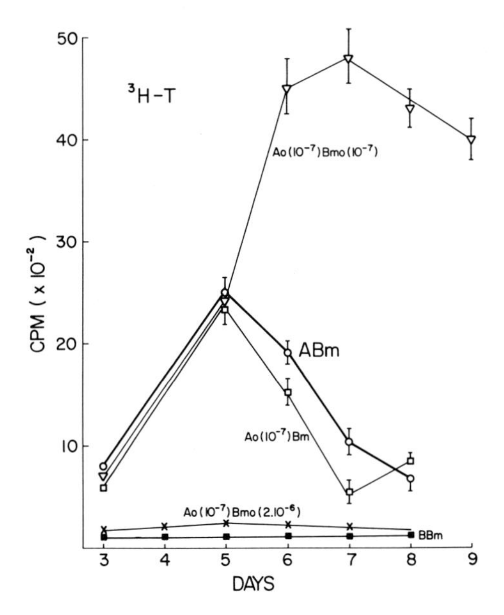

FIG. 1. Uptake of [^{3}H]thymidine into DNA of responding lymphocytes in the MLR. Responding cells of individual A and stimulating cells of individual B (Bm) were incubated for 48 hr either without or with ouabain at the two concentrations shown. The control MLR, ABm is illustrated by the heavy curve. Means and standard deviations of four replicate determinations shown here and in all other figures. This experiment, and all of the others illustrated, utilized human lymphocytes.

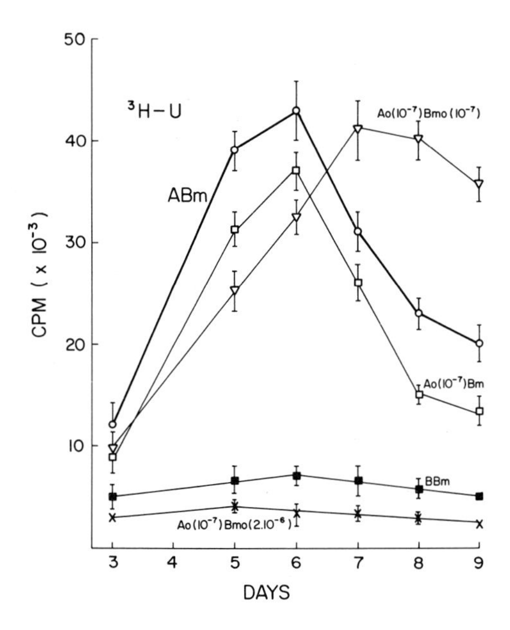

FIG. 2. Uptake of [^{3}H]uridine into the RNA of responding lymphocytes in the MLR. Conditions as in Fig. 1.

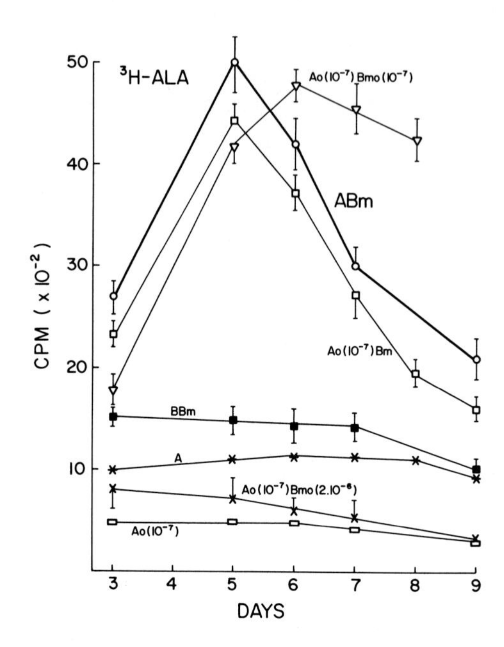

FIG. 3. Uptake of [^{3}H]alanine into the protein of responding lymphocytes in the MLR. Conditions as in Fig. 1.

effect on protein synthesis of the same treatments. Results are as in the case of RNA synthesis: pretreatment of the stimulators with 2×10^{-6} M abolished the MLR-induced protein synthesis, whereas 10^{-7} M ouabain pretreatment caused a lag of 1 day in attainment of peak incorporation of [^{3}H]alanine, but no change in peak levels of incorporation. Similar experiments have been presented and more fully discussed elsewhere (13*a*).

Pretreatment of lymphocytes with 10^{-7} M ouabain under these conditions causes several persistent effects: a 24-hr lag in RNA and protein synthesis in MLR when these cells are used as stimulators and a similar lag when they are exposed to stimulation by PHA or Con A or by antigens (13*a*). Yet cells in these treated cultures have the necessary membrane receptors and internal biochemical machinery to recognize allogeneic cells and to respond like the controls to stimulation in the MLR. This suggests that certain membrane receptors (those for mitogens and antigens) are inactivated during the 48-hr treatment with ouabain and regenerate when the cardiac glycoside is removed, while others (those for recognition of allogeneic cells) are not affected by such treatment; however, we did note one case in which DNA synthesis in responders pretreated with 2×10^{-6} M ouabain also showed a lag of 1 day following addition of allogeneic cells in the MLR (13).

The fact that cells treated with high concentrations of ouabain lose irreversibly the capacity to stimulate in MLR but, at least in high K^+, can be stimulated to proliferate by mitogens raises the following question: if cultures are given a tandem treatment with 2×10^{-6} M ouabain and, after washing, with PHA, do they recover their ability to stimulate in the MLR? When cells are given this tandem treatment and then mixed with allogeneic responding cells (ABm, pha, K^+, o), we in fact observed an incorporation of [^{3}H]thymidine into DNA like that of a normal MLR, ABm, as shown in Fig. 4. However, when we mixed the tandem-treated stimulators with autologous cells (BBm, pha, K^+, o) in what we thought would be a blank (i.e., like BBm; see Fig. 1), we observed a DNA synthesis which was just as great as in the allogeneic combination. We repeated this last combination without prior ouabain treatment (BBm, pha, K^+) and observed exactly the same phenomenon, as shown in Fig. 5. It is apparent that mitogen stimulation has caused the appearance at the cell surface of novel antigens, which are recognized as foreign by autologous cells, a remarkable phenomenon which has been observed in the case of cells pretreated with neuraminidase (14), $NaIO_4$ (15,16) and with protein mitogens (17,18). It is possible that during PHA-induced blastogenesis there was regeneration of the antigens which normally stimulate allogeneic cells, but this is technically difficult to demonstrate because of the appearance of the novel antigens which strongly stimulate autologous cells.

Our data permit us to distinguish four distinct types of action of ouabain on human lymphocytes: (*i*) the (usual) *transient reversible inhibition* of all parameters of stimulation, however achieved, caused by any concentration higher than 5×10^{-8} M during a treatment of 24 hr or less and reversed by washing or high K^+; (*ii*) the *persistent reversible inhibition* (24-hr lag) of DNA synthesis in cells pretreated with 10^{-7} M for 48 hr prior to stimulation

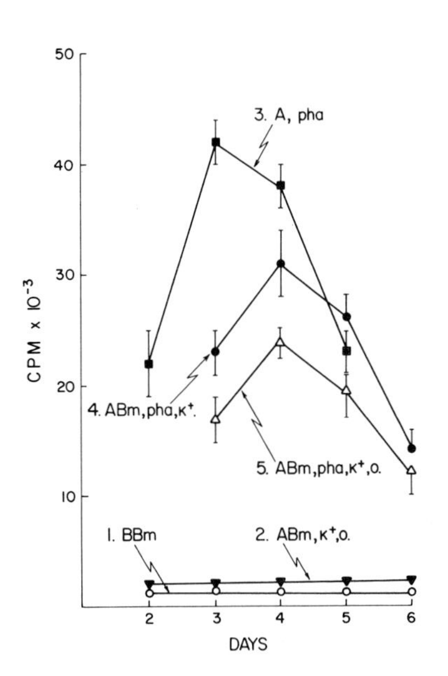

FIG. 4. Uptake of [^{3}H]thymidine into the DNA of responding allogeneic lymphocytes (A) after stimulation by cells (B) which were subjected to tandem treatment with ouabain (2×10^{-6} M for 48 hr) and PHA (in medium containing 20 mM K^+). (1) Autologous blank; (2) stimulating cells pretreated with ouabain, followed by 3 days in 20 mM K^+; (3) cells of individual A treated with PHA, [^{3}H]thymidine added 5 hr before harvesting; (4) one-way MLR, in which stimulators had been pretreated with PHA; (5) one-way MLR in which stimulators had been subjected to tandem treatment as described.

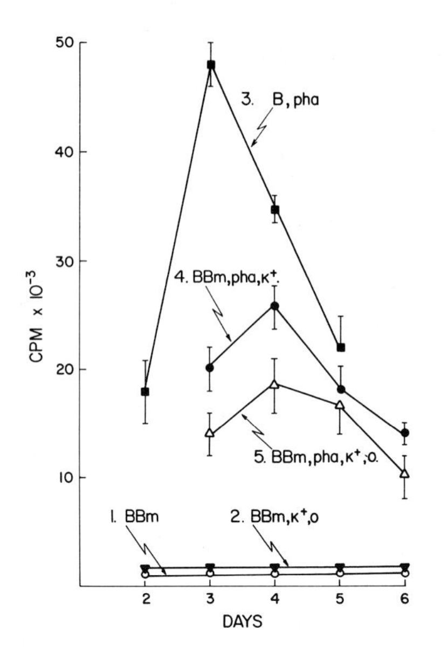

FIG. 5. Same as Fig. 4, except that following tandem treatment of cells of individual B, they were mixed with untreated autologous cells taken from same donor.

with mitogens or antigens, or of RNA and protein synthesis in a MLR in which such pretreated cultures were used as stimulators; (*iii*) the *irreversible inhibition* of DNA, RNA, and protein synthesis in the MLR following pretreatment for 48 hr of stimulating cells with 2×10^{-6} M and of antigen-induced proliferation in cells pretreated with 2×10^{-7} M and higher; and (*iv*) the *persistent enhancement* (superstimulation) of the synthesis of DNA of responding cells in the MLR following pretreatment of the stimulators with 10^{-7} M.

IMPLICATION OF A CELL-SURFACE PROTEOLYSIS AS AN ESSENTIAL EARLY STEP IN BLASTOGENESIS

It has been known for some time that treatment with trypsin can relieve cells in tissue culture from contact (density-dependent) inhibition (19). A considerable literature now exists which suggests that a proteolytic step is involved in cell activation and transformation (20–22). Quite recently, Vischer (23) and Kaplan and Bona (24) demonstrated that proteolytic treatment of mouse splenic lymphocytes with trypsin caused a very large blastogenic response, exclusively on the part of the nonthymus-derived (B) cells. The latter authors, who also showed that pronase could produce a mild proliferative response in mouse thymus and human peripheral blood lymphocytes, advanced the hypothesis that proteases acted as mitogens by short-circuiting an endogenous protease step at the cell surface assumed to be an essential event in all types of lymphocyte proliferation (24).

In order to test this hypothesis, we have examined the effect of a variety of naturally occurring and synthetic protease inhibitors on blastogenesis induced by mitogens and the MLR. Other authors have described an inhibition of blastogenesis caused by ϵ-aminocaproic acid (EACA) (25), *N*-tosyl-L-arginine methyl ester (TAME) and *N*-α-tosyl-L-lysyl-chloromethane (TLCK) (26), which we have confirmed for the first and last of these three compounds which we also found to inhibit the MLR (27,27*a*). Several authors have reported failure to demonstrate inhibition of mitogen-induced blastogenesis by soybean tryptic inhibitor (SBI) (23,26).

We have shown that the proteinase inhibitor phenylmethylsulfonylfluoride (PMSF) inhibits blast transformation induced by both mitogen and MLR, DNA synthesis being arrested virtually completely by concentrations between 5×10^{-5} and 10^{-4} M (27*a*). SBI is also a potent inhibitor of blastogenesis induced by the mitogens Con A, leucoagglutinin, and $NaIO_4$ (27,27*a*). Some of our data showing that SBI strongly inhibits blastogenesis induced by mitogens and MLR are shown in Fig. 6. We then demonstrated that crystalline SBI covalently crosslinked to Sepharose beads remained an excellent inhibitor of blastogenesis induced by Con A and by the MLR (Fig. 6); at comparable concentrations of SBI, inhibition of thymidine incorporation is as great with the matrix-bound SBI as with the free protein. One other observation is of interest: inhibition of blastogenesis, induced either by mitogen or by MLR, is greatest when the proteinase inhibitor (SBI or TLCK) is added at the onset of stimulation and diminishes progressively as blast transformation proceeds. TLCK produces a considerably smaller degree of inhibition of incorporation of thymidine into DNA when added 1 hr after onset of stimulation by mitogen than when added at time 0 and after 48 hr, it no longer inhibits at all (27*a*).

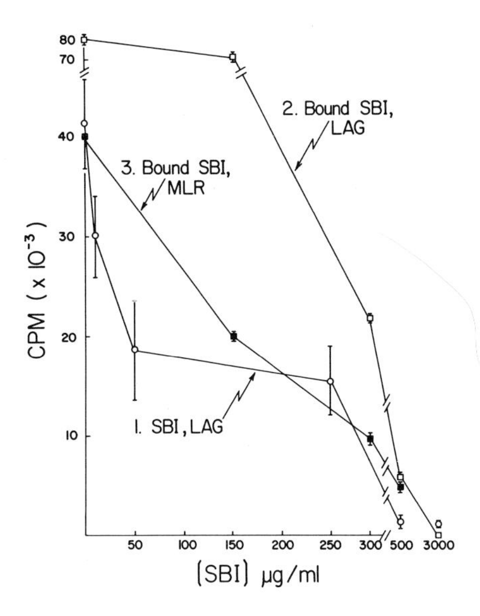

FIG. 6. Effect of concentration of matrix-bound and free crystalline soybean tryptic inhibitor (SBI) on incorporation of [^{3}H]thymidine into DNA of lymphocytes stimulated with leucoagglutinin (LAG) or in the MLR. (1) Stimulation by LAG, soluble SBI; (2) stimulation by LAG, inhibitor matrix-bound; (3) stimulation by MLR, inhibitor matrix-bound.

It is therefore evident that proteolytic cleavage of some critical molecule(s) at the cell surface appears to be one of the essential early events of lymphocyte blastogenesis, whether in MLR or induced by mitogens. Preliminary experiments have shown that the chymotrypsin-specific inhibitor TPCK causes significant inhibition of blastogenesis, as does the trypsin-specific TLCK; this suggests that more than one protease may be involved in the transformation.

BINDING, CAPPING, AND UNLOADING OF A B-CELL MITOGEN

Bona and his collaborators (27*b*) have studied the binding of labeled bacterial lipopolysaccharide (LPS), a B-cell mitogen or polyclonal antigen, by splenic lymphocytes of athymic homozygous nude mice and by thymic lymphocytes of mice heterozygous for this mutation. The thymocytes bind LPS appreciably with saturation kinetics, reaching a plateau at about 12 hr. Binding of the B-cell mitogen by the nonthymic lymphocytes (B cells) was different in that the rate and final level of binding were considerably greater. The most interesting difference was that the B cells, following a peak of binding at 12 hr, unloaded much of the label to the medium, such that the cell-bound radioactivity at 24 hr was 40–50% lower than that at 12 hr. If, following addition at 0 time of 10 μg/ml LPS, a second pulse is administered at 6 hr, a new peak is reached at 12 hr which is approximately the sum expected from the two pulses; however, by 24 hr, approximately half the label was unloaded into the medium. When the same experiment was done with the thymocytes, the two pulses were additive at 12 hr and radioactivity remained steady at that level until the experiment was discontinued at 48 hr. Unloading of label was never observed in the case of the T-cell suspensions, nor did it occur in B cells at 4°C.

After 1 hr at 37°C, autoradiographic studies with [^{3}H]LPS showed capping of the mitogen in the case of both T and B lymphocytes, suggesting the existence of receptors in both types of cells. Thereafter, the B cells quickly internalized the label, which was found after 6 hr to be localized in both cytoplasm and nucleus; in the case of the T cells, label was not internalized during the 48 hr duration of incubation.

It is of interest that the phenomenon of unloading did not require protein synthesis; if cells were pulsed at 5 hr following LPS with concentrations of cycloheximide sufficient to arrest protein synthesis, unloading of the labeled mitogen proceeded as in the controls. It seems to us possible that unloading is due to shedding of LPS–receptor complexes; whatever its explanation, it is striking that it occurs only in the case of the B cells which are stimulated by the mitogen. Internalization of the label occurs exclusively in B cells, and it seems possible that unloading is a consequence, direct or indirect, of the internalization of a portion of the membrane-bound LPS. In this connection, it is striking that while a variety of inhibitors, including sulfhydryl-specific reagents and azide, had no effect on the unloading phenomenon, the latter was completely abolished by both colchicine and vinblastine, which attack the microtubular apparatus. These two agents had no effect on the binding of LPS by the thymocytes in which the B-cell mitogen was neither internalized nor unloaded.

We hypothesize that the internalized LPS stimulates the B cells by direct action at the level of the chromatin, producing derepression of certain critical genes whose products are essential to the transformation of these cells.

We recently established (Truffa-Bachi, Kaplan, and Bona, *manuscript in preparation*) that LPS, unloaded from B cells, has undergone a metabolic modification—a considerable fraction of the mitogen (MW >2,000,000) has been transformed to a low-molecular-weight-form (MW <40,000) which is a potent B-cell mitogen, not only for splenic lymphocytes from wild-type mice, but also for those from strain C_3H/HeJ, which exhibit a very low response to the native LPS (27*b*). It appears that the metabolically transformed, low-molecular-weight product may be the active mitogen; those cells unable to effect this transformation (T cells and cells from the low LPS responder strain) are not stimulated by this mitogen.

SUMMARY AND CONCLUSIONS

1. Studies of the binding of ^{42}K and of [^{3}H]ouabain indicate that one of the effects of stimulation of human lymphocytes by the mitogen PHA is to increase the number of sites at the cell surface which are active as monovalent cation pumps; this increase does not require active concurrent protein synthesis, suggesting aggregation within the membrane of preformed Na-K-ATPase subunits or exposure of previously unavailable oligomeric pump sites. The data support our hypothesis that one of the necessary conditions of lymphocyte stimulation is an increase in the number of functional monovalent cation pump sites which in turn causes a rise in intracellular concentration of potassium. The exact intracellular function of K^+ at the molecular level is still a matter of conjecture. However, it has been shown to be the critical factor in differentiation of nerve and induction of notochord in the exciting work of the Barths (28) and in controlling macromolecular biosynthesis in bacterial and mammalian cells by Lubin (29); see also the review by Pestka (30).

2. Prolonged (48 hr) treatment of human lymphocytes with relatively high or low concentrations of ouabain causes, respectively, an irreversible loss of the capacity of the treated cells to stimulate in the MLR, or an increase in this capacity to double that of untreated controls. These superstimulators seem to act by inducing more responding cells to proliferate than do the control stimulators. Certain persistent effects are manifest after treatment with either high or low concentrations of ouabain: there is a 24-hr lag in the onset and in the peak of DNA, RNA, and protein synthesis in the

MLR and of DNA synthesis in the case of stimulation by mitogens and antigens. On the other hand, pretreatment did not affect capacity to recognize and respond to allogeneic-stimulating cells in MLR. Thus, prolonged treatment with ouabain seems to cause the reversible loss of certain membrane receptors (those for PHA and antigens after 10^{-7} M), irreversible loss of others (antigen receptors and stimulating sites for MLR after 2×10^{-6} M), and no change in yet others (receptor sites for response in MLR, after pretreatment at any concentration of ouabain). We distinguish four distinct types of action of ouabain on the transformation of human lymphocytes: (*i*) transient, reversible inhibition; (*ii*) persistent, reversible inhibition; (*iii*) irreversible inhibition; (*iv*) persistent enhancement (superstimulation). Mitogen-induced blastogenesis causes the appearance of novel membrane sites which are recognized as foreign by receptors on autologous cells. This occurs whether or not the stimulating cultures had been pretreated with high concentrations of ouabain and makes it technically difficult to establish whether blastogenesis was accompanied by regeneration of the normal alloantigens which are responsible for stimulation in the MLR.

3. The fact that a variety of synthetic protease inhibitors, including PMSF, strongly inhibits early stages of blastogenesis of human lymphocytes suggests that a proteolysis is an essential step in this process. The fact that SBI covalently bound to Sepharose beads inhibits both MLR and mitogen-induced transformation indicates that a proteolytic cleavage of some membrane-bound molecule is a critical event in the early events of blastogenesis. There is preliminary evidence that more than one protease may be involved in the process of lymphocyte transformation. These data support the hypothesis that mitogenic action of proteases (23,24) is due to a short-circuiting of some critical membrane proteolysis which must occur in all forms of lymphocyte blast transformation.

4. Both the B and T lymphocytes of the mouse bind the B-cell mitogen LPS, but the former bind at a faster rate and to a greater extent. In the case of the B cells, but not the T cells, peak binding of labeled LPS occurs at 12 hr, followed by a dramatic unloading of label to the medium; unloading does not occur at 4°C. Internalization of the label occurs only in the case of the B cells, and this process may directly or indirectly cause the unloading phenomenon. The existence of receptors for LPS in both B and T cells is indicated by the fact that capping of the mitogen occurs in both cases. It is suggested that unloading of labeled LPS is caused by a shedding of mitogen–receptor complexes.

ACKNOWLEDGMENT

All of the work done in the laboratory of the senior author has been supported from its inception by grants from the Medical Research Council and the National Research Council of Canada. Some of the work was done while

J.G.K. was at the Institut Pasteur, Paris; both he and J.D. were participants in the France–Canada Scientific Exchange Programme.

REFERENCES

1. Kaplan, J. G., and Quastel, M. R. (1975): Lymphocyte transformation and cation transport. In: *Immune Recognition,* edited by A. S. Rosenthal, pp. 391–403. Academic Press, New York.
2. Quastel, M. R., and Kaplan, J. G. (1968): Inhibition by ouabain of human lymphocyte transformation *in vitro. Nature,* 219:198–200.
3. Quastel, M. R., and Kaplan, J. G. (1970): Lymphocyte stimulation: the effect of ouabain on nucleic acid and protein synthesis. *Exp. Cell Res.,* 62:407–420.
4. Quastel, M. R., and Kaplan, J. G. (1970): Early stimulation of potassium uptake in lymphocytes treated with PHA. *Exp. Cell Res.,* 63:230–233.
5. Quastel, M. R., Dow, D. S., and Kaplan, J. G. (1970): Stimulation of K^{42} uptake into lymphocytes by phytohemagglutinin and the role of intracellular K^+ in lymphocyte transformation. In: *Proc. 5th Leukocyte Culture Conference,* edited by J. E. Harris, pp. 97–123. Academic Press, New York.
6. Quastel, M. R., Wright, P., and Kaplan, J. G. (1972): Potassium uptake and lymphocyte activation: Generality of the effect of ouabain and a model of events at the lymphocyte surface induced by phytohemagglutinin. In: *Proc. 6th Leucocyte Culture Conference,* edited by M. R. Schwarz, pp. 185–214. Academic Press, New York.
7. Averdunk, R. (1972): The effect of phytohemagglutinin and anti-lymphocyte serum on the transport of potassium ions, glucose and amino acids in human lymphocytes. *Hoppe-Seylers Z. Physiol. Chem.,* 353:79–87.
8. Averdunk, R., and Lauf, P. K. (1975): Effects of mitogens on sodium-potassium transport, ^3H-ouabain binding and adenosine triphosphatase activity in lymphocytes. *Exp. Cell Res.,* 93:331–342.
9. Wright, P., Quastel, M. R., and Kaplan, J. G. (1973): Potassium binding sites in lymphocyte transformation and the effect of their prolonged inactivation by ouabain. *Proc. 7th Leucocyte Culture Conference,* edited by F. Daguillard, pp. 87–104. Academic Press, New York.
10. Charnock, J. S., Cook, D. A., and Opit, L. J. (1971): Role of energized states of Na^+K-ATPase in the Na pump. *Nature,* 233:171–172.
11. Quastel, M. R., and Kaplan, J. G. (1975): Ouabain binding to intact lymphocytes: Action of phytohemagglutinin and leucoagglutinin. *Exp. Cell Res.,* 94:351–362.

11*a*. Lauf, P. K. (1975): Antigen–antibody reactions and cation transport in biomembranes: Immunophysiological aspects. *Biochim. Biophys. Acta,* 415:173–229.

11*b*. Bain, B., Vas, M., and Lowenstein, L. (1964): The Development of Large Immature Mononuclear Cells in mixed leucocyte culture, *Biochim. Biophys. Acta,* 23:108–116.

11*c*. Bach, F. H., and Voynow, N. K. (1966): One Way Stimulation in mixed leucocyte culture, *Biochim. Biophys. Acta,* 153:545, 547.

12. Wright, P., Quastel, M. R., and Kaplan, J. G. (1973): Differential sensitivity of antigen- and mitogen-stimulated human leucocytes to prolonged inhibition of potassium transport. *Exp. Cell Res.,* 79:87–94.
13. Christen, Y., Sasportes, M., Mawas, C., Dausset, J., and Kaplan, J. G. (1975): The mixed lymphocyte reaction: selective activation and inactivation of the stimulating cells. *Cell. Immunol.,* 19:137–142.

13a. Dornand, J., and Kaplan, J. G. (1976): Persistent effects of ouabain treatment on human lymphocytes: Synthesis of DNA, RNA and protein in stimulated and unstimulated cells. *Can. J. Biochem.,* 54:280–286.

14. Etheredge, E. E., Shons, A. R., and Najarian, J. S. (1972): Neuraminidase-induced autologous stimulation of human leucocyte cultures. In: *Proc. 6th Leucocyte Culture Conference,* edited by M. R. Schwarz, pp. 121–135. Academic Press, New York.
15. O'Brien, R. L., Parker, J. W., Paolilli, P., and Steiner, J. (1974): Periodate-induced lymphocyte transformation. IV. Mitogenic effect of $NaIO_4$ treated lymphocytes upon autologous lymphocytes. *J. Immunol.,* 112:1884–1890.

16. O'Brien, R. L., Parker, J. W., Paolilli, P., and Steiner, J. (1974): Remote stimulation of lymphocyte transformation by sodium periodate. In: *Lymphocyte Recognition and Effector Mechanisms,* edited by K. Lindahl-Kiessling and D. Osoba, pp. 19–24. Academic Press, New York.
17. Weksler, M. E. (1973): Lymphocyte transformation induced by autologous cells. *J. Exp. Med.,* 137:799–806.
18. Bluming, A. Z., Lynch, M. J., Kavanah, M., and Khiroya, R. (1975): Transformation antigens on stimulated lymphocytes. *J. Immunol.,* 114:117–121.
19. Burger, M. M. (1970): Proteolytic enzymes initiating cell division and escape from contact inhibition of growth. *Nature,* 227:170–171.
20. Reich, E. (1974): Secretion of enzymes by neoplastic cells and macrophages. In: *Proteinase Inhibitors,* edited by H. Fritz, H. Tschesche, L. J. Greene, and E. Truscheit, pp. 621–630. Springer-Verlag, New York and Heidelberg.
21. Goldberg, A. R., and Lazarowitz, S. G. (1974): Plasminogen activators of normal and transformed cells. In: *Proteinase Inhibitors,* edited by H. Fritz, H. Tschesche, L. J. Greene, and E. Truscheit, pp. 631–648. Springer-Verlag, New York and Heidelberg.
22. Kast, R. E. (1974): A theory of lymphocyte blast transformation and malignant change based upon proteolytic cleavage of a trigger peptide: The detendomer. *Oncology,* 29: 249–264.
23. Vischer, T. L. (1974): Stimulation of mouse B lymphocytes by trypsin. *J. Immunol.,* 113:58–62.
24. Kaplan, J. G., and Bona, C. (1974): Proteases as mitogens: the effect of trypsin and pronase on mouse and human lymphocytes. *Exp. Cell Res.,* 88:388–394.
25. Hirschhorn, R. Grossman, J., Troll, W., and Weissman, G. (1971): *J. Clin. Invest.,* 50: 1206–1217.
26. Darzynkiewicz, Z., and Arnason, B. G. W. (1974): Suppression of RNA synthesis in lymphocytes by inhibitors of proteolytic enzymes. *Exp. Cell Res.,* 85:95–104.
27. Moreau, P., and Kaplan, J. G. (1975): Antagonism by protease inhibitors of transformation of human lymphocytes. *Proc. Can. Fed. Biol. Soc.,* 18:87.
27*a*. Moreau, P., Dornand, J., and Kaplan, J. G. (1975): Inhibition of Lymphocyte Transformation: effect of soy bean trypsin inhibitor and synthetic anti-proteases. *Can. J. Biochem.,* 53:1337–1341.
27*b*. Watson, J., and Riblet, R. (1975): Genetic control of responses to bacterial lipopolysaccharides in mice: A gene that influences a membrane component involved in the activation of bone marrow-derived lymphocytes by lipopolysaccharides. *J. Immunol.,* 114: 1462–1468.
28. Barth, L. J. and Barth, L. G. (1974): Effect of the potassium ion on induction of notochord from gastrula ectoderm of *Rana pipiens.* Biol. Bull., 146:313–325.
29. Lubin, M. (1967): Intracellular potassium and macromolecular synthesis in mammalian cells. *Nature,* 213:451–453.
30. Pestka, S. (1970): Protein biosynthesis: mechanism, requirements, and potassium-dependency. In: *Membranes and Ion Transport,* edited by E. E. Bittar, Vol. 3, pp. 279–296. Wiley-Interscience, New York.

Biogenesis and Turnover of Membrane Macromolecules, edited by John S. Cook. Raven Press, New York 1976.

Nature of Membrane Sites Involved in Lymphocyte Activation

Abraham Novogrodsky

Department of Biophysics, The Weizmann Institute of Science, Rehovot, Israel

Lymphocytes are triggered to grow and divide upon interaction with a variety of agents. The explicit mechanism of the triggering process is largely unknown. However, it is generally considered that the triggering signal is confined to the cell membrane (1,2). It is important in understanding the mechanism of lymphocyte activation to ascertain whether different mitogens trigger the cells by affecting different membrane sites, or whether the triggering signal is localized at a single, unique site. Recently, a chemical approach has been used for the study of the mechanism of lymphocyte activation. Well-defined modifications of the cell membrane, induced chemically or enzymically were found to trigger the cells or alter its response to stimulation by other agents. This approach led to the identification of the membrane site(s) modified upon stimulation of the cells and also gave some insight on the nature of the triggering signal. These studies have strongly suggested that the lymphocyte activation site is unique.

LYMPHOCYTE TRANSFORMATION INDUCED BY GENERATION OF ALDEHYDE MOIETIES ON THE CELL SURFACE

Mild treatment of rat lymph node lymphocytes with sodium periodate ($NaIO_4$) induces extensive blastogenesis (3). Lymphocytes from human (4,5), mouse (6,7), rabbit (8), and guinea pig (9) are also stimulated by this agent. Treatment of lymphocytes with neuraminidase markedly reduces their response to $NaIO_4$ (6). Blastogenesis induced by $NaIO_4$ is decreased when the $NaIO_4$-treated lymphocytes are reacted with aldehyde-blocking agents such as KBH_4 or NH_2OH (6,7). These findings led to the conclusion that the $NaIO_4$ target site is a surface membrane sialyl residue that yields on oxidation an aldehyde moiety that is involved in the triggering process. This conclusion is further supported by the observation of Van Lenten and Ashwell (10) who showed that $NaIO_4$, under mild conditions, selectively oxidizes sialyl residues in sialoglycoproteins to yield a 7-carbon sialic acid analogue containing an aldehyde moiety. Liao et al. (11) have also shown that the main target of $NaIO_4$ oxidation of human erythrocytes is the sialic acid moieties of surface-membrane sialoglycoproteins.

The observation that $NaIO_4$ triggers lymphocytes by the formation of an aldehyde moiety on the cell surface prompted the search for an enzymic modification of the cells that will mimic the $NaIO_4$ effect. Sialic acid in glycoprotein always occupies a nonreducing terminal position and is glycosidically linked either to D-galactose or to *N*-acetylgalactosamine. Galactosyl residues exposed by the action of neuraminidase on plasma glycoprotein could be oxidized at the carbon-6 position to yield the 6-aldehydo analogues (12). It has been found that treatment of lymphocytes with galactose oxidase after incubation with neuraminidase induces extensive blastogenesis (13). Blastogenesis induced by galactose oxidase in neuraminidase-treated cells similar to that induced by $NaIO_4$ is also decreased upon reacting the cells with reagents that interact with aldehyde groups. The latter treatments do not affect transformation of the cells by concanavalin A (Con A). It has been suggested that galactosyl residues exposed by the action of neuraminidase on the cell membrane are oxidized by galactose oxidase, and that the aldehyde moiety thus formed is involved in the induction of blastogenesis (Fig. 1).

The function of the aldehyde moiety in the induction of blastogenesis is not known. It is possible that the aldehydes react with other functional groups on the cell membrane, and that the crosslinked structure thus formed

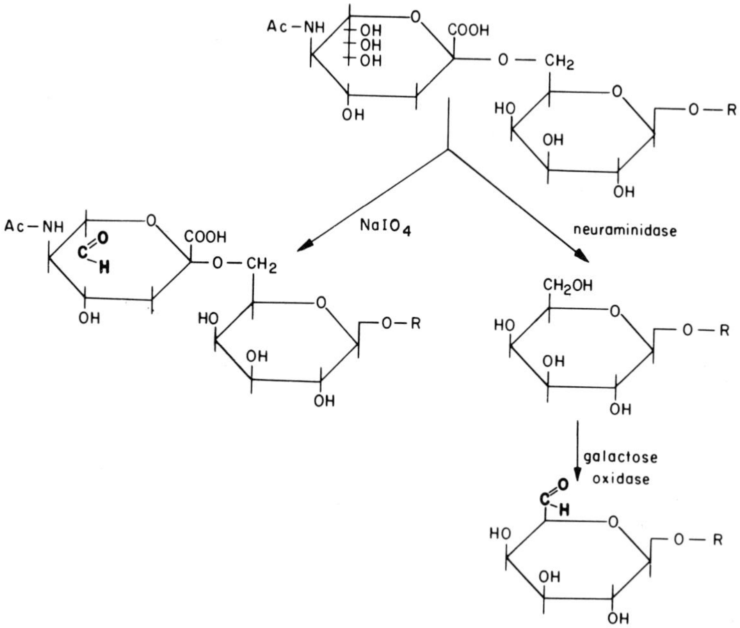

FIG. 1. Generation of aldehyde moieties on cell-surface sialyl and neuraminidase-exposed galactosyl residues.

might play a role in the triggering process. It has been suggested that cross-linkage and aggregation of specific membrane sites may be involved in the triggering of lymphocytes to undergo blastogenesis (14–16). The functional groups in the cell membrane that might react with the aldehyde group are: free amino groups (*N*-terminal amino groups or ε-amino groups of lysyl residues) to form Schiff bases; alcoholic groups (saccharide moieties of glycoproteins or glycolipids) to form hemiacetals (the latter could react with additional alcoholic groups to form acetals); thiol groups (cysteinyl residues) to form hemimercaptals (or mercaptals). The Schiff base is probably the most stable structure among those which were outlined above. The possibility that aldehydes react to form a linkage between the lymphocyte and the macrophage is discussed below.

MITOGENIC ACTION OF PHYTOMITOGENS THAT BIND TO GALACTOSYL SITES EXPOSED AFTER NEURAMINIDASE TREATMENT

Galactosyl residues exposed after neuraminidase treatment serve also as the target site for the mitogenic action of soybean agglutinin (SBA) and peanut agglutinin (PNA). SBA agglutinates mouse lymphocytes only at high concentrations and is not mitogenic to untreated cells. Incubation of lymphocytes with neuraminidase facilitates their agglutination by SBA and renders them responsive to stimulation by the lectin (17). Neuraminidase treatment increases also about threefold the amount of SBA bound to the lymphocytes. SBA-induced agglutination and transformation is inhibited specifically by *N*-acetyl-D-galactosamine and D-galactose.

PNA similar to SBA also stimulates lymphocytes only after neuraminidase treatment (18). It is of interest to note that PNA exhibits species-specificity in its mitogenic action. Although PNA binds similarly to neuraminidase-treated lymphocytes of man, rat, mouse, and guinea pig, it will stimulate only lymphocytes of man and rat. It thus differs from SBA that will stimulate neuraminidase-treated lymphocytes of all four species. PNA interacts specifically with galactosyl residues and is of more narrow specificity than SBA, which binds both galactosyl and *N*-acetylgalactosaminyl residues. Thus, we assume that PNA stimulates lymphocytes by interaction with galactosyl residues exposed on cell surface glycoproteins after neuraminidase treatment.

EFFECT OF β-GALACTOSIDASE ON THE RESPONSE OF NEURAMINIDASE-TREATED CELLS TO MITOGENS

Sequential treatment of lymphocytes with neuraminidase and β-galactosidase (purified from *Diplococcus pneumoniae* and kindly supplied by Dr. G. Ashwell, N.I.H.) markedly reduces the response of the cells to stimulation with galactose oxidase, soybean agglutinin, and peanut agglutinin, suggesting that β-galactosyl residues serve as the target site for these mitogens (18). These enzymic modifications of lymphocyte cell surface did not affect the response of the cells to stimulation by Con A (Fig. 2).

The above findings show that generation of aldehyde moieties on surface sialyl or neuraminidase-exposed galactosyl residues are sufficient to induce blastogenesis. Stimulation of the cells could also be obtained by interaction of soybean and peanut agglutinin with neuraminidase-exposed galactosyl sites. One may postulate that the chemical oxidizing agent, $NaIO_4$, the enzymic oxidizing agent, galactose oxidase, and the chemically inert lectins SBA and PNA trigger lymphocytes to undergo transformation by affecting the same glycoprotein that contains an oligosaccharide with the following sequence (see Fig. 3):

$$\text{Sialic acid} \rightarrow \text{galactose} \rightarrow$$

The same glycoprotein might also contain the receptor [$(Man)_3$-GlcNAc-GlcNAc] for PHA, Con A, and the mitogens from *Wisteria floribunda* and *Lens culinaris* (19). As a matter of fact, lymphocyte surface receptors, re-

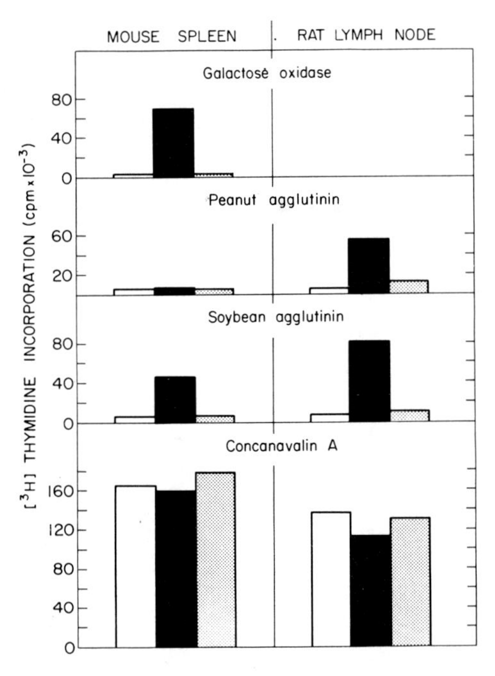

FIG. 2. Effect of β-galactosidase on the response of neuraminidase-treated lymphocytes to mitogens. *Open bars,* Untreated cells; *filled bars,* neuraminidase-treated cells; *dotted bars,* neuraminidase-, β-galactosidase-treated cells. Lectins were added at the following final concentrations: PNA (150 μg/ml); SBA (50 μg/ml); Con A (2 μg/ml). Stimulation was estimated by measuring [^{3}H]thymidine incorporation for 2 hr after incubation of the galactose oxidase-treated cells for 48 hr and the lectin-treated cells for 72 hr.

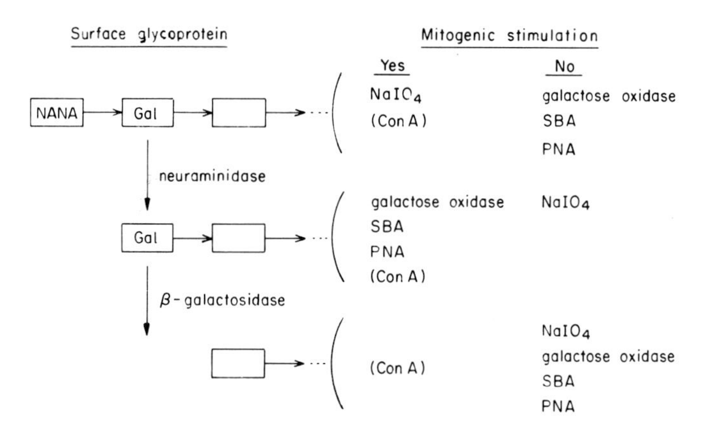

FIG. 3. Effect of enzymic modifications of lymphocytes on their response to mitogens.

sembling the carbohydrate moiety of glycoproteins such as fetuin, IgG, erythrocyte glycopeptide, and thyroglobulin, might serve as the target site for most of the mitogenic lectins.

A CHEMICAL APPROACH FOR THE MAPPING OF LYMPHOCYTE TRIGGERING SITES

A chemical approach has been used for mapping of the triggering sites on the lymphocyte membrane (20). The main goal was to elucidate whether lymphocytes can be triggered by interaction affecting different sites on the cell membrane, or whether the cells are triggered by interaction at a specific unique site.

Reagents were used to attach various ligands onto the cell via different functional groups on the membrane. It was expected that interaction of the antiligand with ligands conjugated to the lymphocyte triggering site(s) would stimulate the cells to undergo blastogenesis.

Reagents, listed in Table 1, group A, were used to attach various ligands via different functional groups onto the lymphocyte membrane protein. Under the coupling conditions used the cells were not damaged as shown by the trypan blue exclusion test. Furthermore, such modified cells still retained their ability to respond to concanavalin A. The 2,4-dinitrophenol (DNP) group was attached to the cell via the ϵ-amino group of lysyl, the sulfhydryl group of cysteinyl or the tyrosyl residues on cell surface protein. Attachment of DNP to the cell was demonstrated by agglutination of the modified cells with anti-DNP antibody and by lysis of the modified cells in the presence of antibody and complement. However, although such cells interacted with the antibody, it did not stimulate them. Such a result was obtained when the DNP group was introduced directly onto the membrane [2,4,6-trinitrobenzene sulfonic acid (TNBS) modification] or via a spacer (*N*-DNP-ϵ-

TABLE 1. *Antiligand-induced stimulation of ligand-conjugated cells*

Reagents	Functional group on cell	Ligand introduced	Antiligand	Stimulation
Group A				
Trinitrobenzene sulfonic acid	ϵ-NH_2	TNP	Anti-DNP Ig[b]	−
N-DNP-ϵ-aminocaproyl *N*-hydroxysuccinimide ester	ϵ-NH_2	TNP	Anti-DNP Ig	−
α-*N*-Bromoacetyl-ϵ-*N*-DNP-lysine[a]	ϵ-NH_2	DNP	Anti-DNP Ig	−
1,3-difluoro-2,4-dinitrobenzene	ϵ-NH_2	DNP	Anti-DNP Ig	−
2,4-dinitro-*N*,*N*-di(2-chloroethyl)aniline (*N*-DNP-nitrogen mustard)	ϵ-NH_2	DNP	Anti-DNP Ig	−
Diazotized *m*-nitroaniline	Tyrosine	MNP	Anti-DNP Ig[b]	−
Diazotized *p*-arsanilic acid	Tyrosine	Arsanilic acid	Anti-arsanilic acid Ig	−
Biotin *N*-hydroxysuccinimide ester	ϵ-NH_2	Biotin	Avidin	−
Group B				
Biotin hydrazide	Aldehyde ($NaIO_4$ generated)	Biotin	Avidin	+
DNP-hydrazine	Aldehyde ($NaIO_4$ generated)	DNP	Anti-DNP Ig	+

Reagents outlined in Group A were reacted with untreated mouse or rat spleen cells. Reagents outlined in Group B were reacted with $NaIO_4$-treated cells. Cells were then cultured with the appropriate antiligand and [^3H]thymidine incorporation within 2 hr was measured after 72 hr (Group A) and 48 hr (Group B).

[a] Also reacts with —SH groups

[b] Anti-DNP Ig cross-reacts with the 2,4,6-trinitrophenol group (TNP) and the *m*-nitrophenol group (MNP).

aminocaproyl *N*-hydroxysuccinimide ester modification). Furthermore, TNBS modifications were carried out over a wide range of reagent concentration and at different pH, but the modified cells were not stimulated by anti-DNP antibody.

Several chemical reagents were used to crosslink cell surface ϵ-amino groups. One such reagent [2,4-dinitro-*N*,*N*-di(2-chloroethyl)aniline(*N*-DNP-nitrogen mustard)] agglutinated the cells, but did not stimulate them. In addition, anti-DNP antibody also did not stimulate such cells.

Other ligands were also introduced via the cell surface protein. Diazotized *p*-arsanilic acid introduced the phenyl-arsonic group via tyrosyl residues and biotin *N*-hydroxysuccinimide ester introduced the biotin moiety via ϵ-amino groups. Such cells were agglutinated by antiarsanilic acid antibody and avidin, respectively, although neither antiligand induced stimulation of the modified cells.

Cell surface sugar residues have been shown to be important in the interaction of many mitogens with the cell surface (21) and may well play a key role in the activation of lymphocytes. Unfortunately, they are not readily available to simple chemical reactions that will not destroy the cell viability. We have utilized the aldehyde group, formed after mild periodate oxidation of cell-surface sialyl residues, to attach various ligands onto the cell membrane. Thus the biotin and DNP moieties were attached to periodate-treated cells by reaction with biotin hydrazide or DNP-hydrazine (Fig. 4). This method has also been used to label erythrocyte membranes for visualization with ferritin-avidin in the electron microscope (22).

It can be seen from Fig. 5 that cells treated sequentially with periodate, biotin hydrazide, and borohydride (biotin-conjugated cells) were stimulated with avidin. Avidin had very little effect, if any, on control cells. The avidin

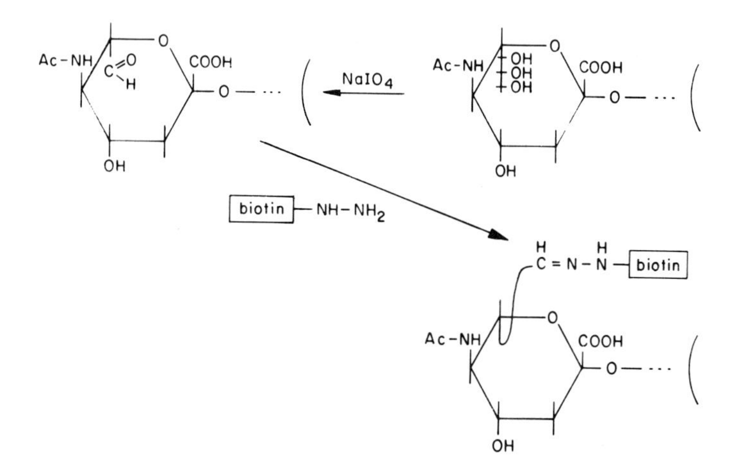

FIG. 4. Conjugation of biotin hydrazide to $NaIO_4$-treated cells.

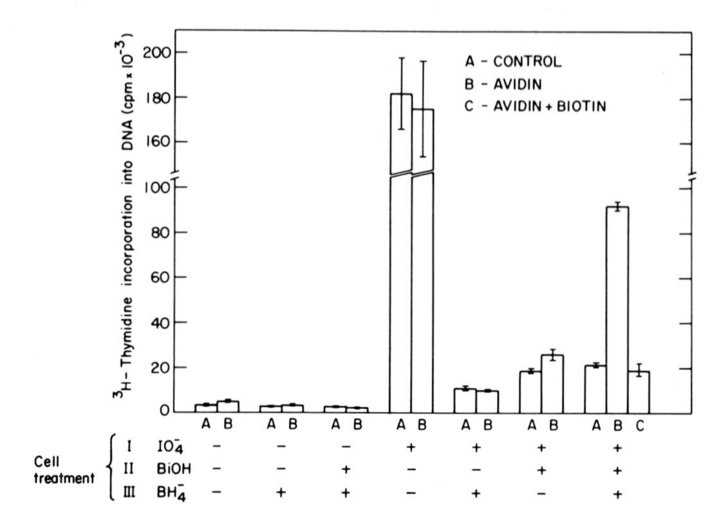

FIG. 5. Avidin-induced DNA synthesis in biotin conjugated rat spleen cells. Rat spleen cells were treated sequentially (where indicated) with $NaIO_4$, biotin hydrazide (BioH) and KBH_4 (BH_4^-). The cells were then washed, suspended in culture medium, and incubated in the presence of Control, A; avidin (5 μg/ml), B; or avidin 5 μg/ml premixed with biotin (10 μg/ml), C. [^{3}H]thymidine incorporation for 2 hr was measured after incubation of the cells for 48 hr.

stimulation was shown to be specific for the attached biotin on the cell, since avidin mixed with biotin did not stimulate the modified cells. Furthermore, binding studies showed that only cells sequentially treated with periodate, biotin hydrazide, and borohydride, or cells sequentially treated with periodate and biotin hydrazide only, specifically bound ^{125}I-labeled avidin. The extent of ^{125}I-avidin binding in both cases was 1–1.5 μg avidin per 10^6 cells. The maximal stimulation by avidin was obtained at an avidin concentration of 3 μg/ml (Fig. 6), and concentrations of avidin up to 40 μg/ml did not decrease this maximum. The avidin stimulation of biotin-conjugated cells was not due to reexposure of the periodate generated aldehyde groups, since such stimulation was still found in the presence of cysteine at concentrations that markedly inhibited the periodate response.

It is of interest to note that cells treated sequentially with periodate and biotin hydrazide respond poorly to avidin, whereas further treatment with borohydride strongly potentiated the avidin response. The explanation for this observation is not known, although two possibilities may be considered. (*i*) Borohydride reduces the unsaturated hydrazone derivative formed on reaction of biotin hydrazide and the aldehyde group, or (*ii*) borohydride reduces free aldehyde groups that did not react initially with biotin hydrazide. Avidin stimulation of cells treated sequentially with periodate and biotin hydrazide is also enhanced by further treatment of the cells with cysteine. In this case, cysteine acts only as an aldehyde reagent. Therefore,

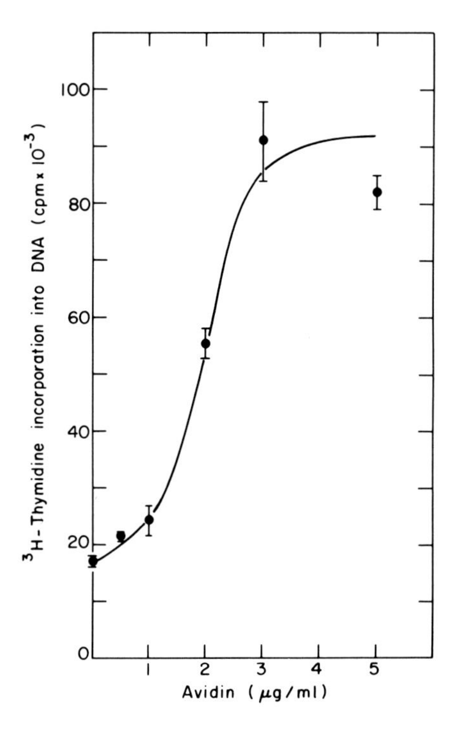

FIG. 6. Effect of avidin at different concentrations on the induction of DNA synthesis in biotin-conjugated cells.

it is suggested that borohydride potentiates the avidin response of biotin-conjugated cells by reacting with the free aldehyde groups on the cell.

The experimental results obtained with rat spleen cells outlined in Fig. 5 have also been verified in the mouse. However, biotin-conjugated mouse spleen cells had a low response to avidin as compared to rat cells. Preliminary results have shown that rat spleen cells, sequentially treated with periodate, DNP-hydrazine, and borohydride, could be stimulated by anti-DNP Ig.

The results outlined above have shown that only ligands conjugated to cells via the periodate-induced aldehyde group were transformed by the antiligand. To our knowledge this is the first demonstration of an insertion of a mitogenic site onto the lymphocyte membrane.

It has been suggested above that the chemical agent sodium periodate, galactose oxidase, and the lectins soybean agglutinin and peanut agglutinin induced mitogenesis by affecting the carbohydrate moiety of the same glycoprotein. It has also been suggested that other phytomitogens may bind to the cell surface via a similar receptor (19). Our mapping data strongly support this view and suggest that the lymphocyte activation site is unique.

CELL–CELL INTERACTION AND THE INDUCTION OF LYMPHOCYTE TRANSFORMATION

Previous studies have shown that macrophages are necessary for lymphocytes to undergo blastogenesis in response to specific antigens (23). Conflicting data have been obtained on the requirement of macrophages for PHA-induced blastogenesis (23,24).

Removal of phagocytic cells from human peripheral lymphocytes markedly reduced their blastogenic response to treatment with $NaIO_4$ or with neuraminidase and galactose oxidase (NAGO), and had less effect on PHA-induced stimulation (5,25). Transformation of lymphocytes by $NaIO_4$ or NAGO provides a suitable system for studying the separate effects of the mitogens on lymphocytes or macrophages alone in the induction of blastogenesis. The blastogenic response of $NaIO_4$ or NAGO-treated purified lymphocytes was markedly enhanced upon incubation on macrophage monolayers. In addition, we have found that untreated purified lymphocytes were stimulated upon incubation on $NaIO_4$ or NAGO-treated macrophage monolayers (Fig. 7). Trials to demonstrate a soluble factor which mediates the above-mentioned effects of the macrophages have been as yet unsuccessful. It is possible that a direct interaction between the lymphocyte and the macrophage is required for the induction of blastogenesis. This interaction might result from a reaction between aldehyde moieties on the lymphocyte (or macrophage) surface and functional groups on the macrophage (or the lymphocyte). The requirement for macrophage–lymphocyte interaction in guinea pig lymphocyte proliferation induced by generation of aldehydes on cell membranes has been studied recently (9).

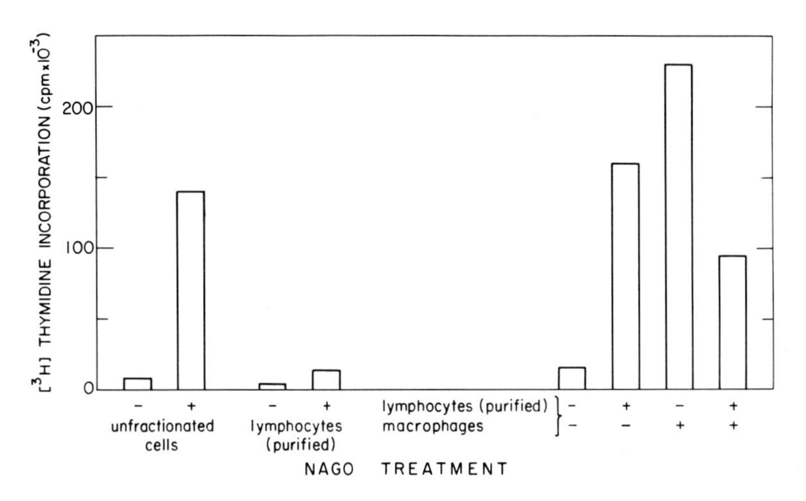

FIG. 7. Effect of macrophages on the induction of DNA synthesis in human lymphocytes by neuraminidase and galactose oxidase (NAGO).

The results were similar to those outlined above for human peripheral lymphocytes.

Cytotoxic effector cells can be formed in several different ways. Specific effector cells can be obtained from immunized animals or by the exposure of the lymphocyte to antigenically foreign target cells *in vitro*. Immunologically, nonspecific effector lymphocytes can be obtained by incubation of normal lymphocytes with different mitogens (26). Since the mitogenic agents $NaIO_4$ or NAGO can be easily removed after modification of the cells, it was possible to study the effect of modification of the effector or target cells alone on the development of cytotoxicity. It was found that treatment of mouse spleen cells with $NaIO_4$ or with NAGO rendered the cells cytotoxic to mastocytoma (P815) target cells. Treatment of target cells (P815 cells and turkey erythrocytes) with $NaIO_4$ or with NAGO rendered them susceptible to cytolysis by untreated mouse spleen cells. The cytotoxicity induced by $NaIO_4$ was reduced upon reacting the $NaIO_4$-treated effector or target cells with KBH_4 or NH_2OH. Thus, the formation of free surface aldehydes on either the effector or the target cell was sufficient to induce a cytotoxic effect (27). It is postulated that a linkage between the effector and the target cell initiates the cytotoxic effect.

The cytotoxic potential of the lymphocyte might be an inherent property of the normal, nonactivated cell or alternatively, might be induced shortly after the formation of the linkage between the effector and target cell. It is also possible that the mitogen-induced structural alterations in the target cell surface membrane are essential for the induction of cytotoxicity. Cytotoxicity induced by $NaIO_4$ or by NAGO is immunologically nonspecific and is independent of major antigenic differences between effector and target cells. Phagocytic cells are not involved in $NaIO_4$ or NAGO-induced cytotoxicity toward P815 target cells.

CONCLUSIONS

The observation that defined modifications of lymphocyte surface saccharides induce blastogenesis provides additional support for the fact that the triggering signal is confined exclusively to the cell membrane. The findings that the formation of aldehyde moieties on the cell surface is sufficient to activate the cells facilitate our understanding of the nature of the triggering signal.

The chemical oxidizing agent $NaIO_4$, the enzymic oxidizing agent galactose oxidase, the chemically inert lectins, SBA and PNA, and possibly many other phytomitogens might trigger lymphocytes by affecting the carbohydrate moiety of the same glycoprotein. Mapping of the lymphocyte triggering sites by studying the mitogenic effect of antiligands on ligand-conjugated cells also supported the conclusion that the triggering sites on the lymphocyte membrane are unique. Clustering of unique sites on the lymphocyte membrane induced by the mitogens might lead to crucial functional alterations resulting in cell activation.

Activation of lymphocytes by chemical or enzymic agents permitted the study of the separate effects of the mitogen on different cell types which are involved in blastogenesis (lymphocytes and macrophages) or lymphocyte-mediated cytotoxicity (effector and target cells). The conclusions derived from these studies might also be applicable to immunological systems in which lymphocytes are activated by specific antigens.

ACKNOWLEDGMENTS

This research was supported by grants from the United States–Israel Binational Science Foundation (No. 387) and the Israeli Commission for Basic Research. The skillful technical assistance of Mrs. Segula Halmann is greatly appreciated.

REFERENCES

1. Robbins, J. H. (1964): Tissue culture studies of the human lymphocyte. *Science*, 146:1648–1654.
2. Ling, N. R. (1968): *Lymphocyte Stimulation*. North-Holland, Amsterdam.
3. Novogrodsky, A., and Katchalski, E. (1971): Induction of lymphocyte transformation by periodate. *FEBS Lett.*, 12:297–300.
4. Parker, J. W., O'Brien, R. L., Lukes, R. J., and Steiner, J. (1972): Transformation of human lymphocyte by sodium periodate. *Lancet*, i:103–104.
5. Biniaminov, M., Ramot, B., and Novogrodsky, A. (1974): Effect of macrophages on periodate-induced transformation of normal and chronic lymphatic leukemia lymphocytes. *Clin. Exp. Immunol.*, 16:235–242.
6. Novogrodsky, A., and Katchalski, E. (1972): Membrane site modified on induction of the transformation of lymphocytes by periodate. *Proc. Natl. Acad. Sci. USA*, 69:3207–3210.
7. Zatz, M. M., Goldstein, A. L., Blumenfeld, O. O., and White, A. (1972): Regulation of normal and leukaemic lymphocyte transformation and recirculation by sodium periodate oxidation and sodium borohydride reduction. *Nature* [*New Biol.*], 240:252–255.
8. Ono, M., and Hozumi, M. (1973): Effect of cytochalasin B on lymphocyte stimulation induced by concanavalin A or periodate. *Biochem. Biophys. Res. Commun.*, 53:342–349.
9. Greineder, D. K., and Rosenthal, A. S. (1975): The requirement for macrophage–lymphocyte interaction in T lymphocyte proliferation induced by generation of aldehydes on cell membranes. *J. Immunol.*, 115:932–938.
10. Van Lenten, L., and Ashwell, G. (1971): Studies on the chemical and enzymic modifications of glycoproteins. A general method for the tritiation of sialic acid-containing glycoproteins. *J. Biol. Chem.*, 246:1889–1894.
11. Liao, T., Gallop, P. M., and Blumenfeld, O. O. (1973): Modification of sialyl residues of sialoglycoprotein(s) of human erythrocyte surface. *J. Biol. Chem.*, 248:8247–8253.
12. Morell, A. G., Van den Hamer, C. J. A., Scheinberg, I. H., and Ashwell, G. (1966): Physical and chemical studies on ceruloplasmin. IV. Preparation of radioactive, sialic acid-free ceruloplasmin labeled with tritium on terminal D-galactose residues. *J. Biol. Chem.*, 241:3745–3749.
13. Novogrodsky, A., and Katchalski, E. (1973a): Induction of lymphocyte transformation by sequential treatment with neuraminidase and galactose oxidase. *Proc. Natl. Acad. Sci. USA*, 70:1824–1827.
14. Fanger, M. W., Hart, D. A., Wells, J. V., and Nisonoff, A. (1970): Requirement for cross-linkage in the stimulation of transformation of rabbit peripheral lymphocytes by anti-globulin reagents. *J. Immunol.*, 105:1484–1492.

15. Greaves, M., and Janossy, G. (1972): Elicitation of selective T and B lymphocyte responses by cell surface binding ligands. *Transplant. Rev.*, 11:87–130.
16. Lotan, R., Lis, H., Rosenwasser, A., Novogrodsky, A., and Sharon, N. (1973): Enhancement of the biological activities of soybean agglutinin by cross-linking with glutaraldehyde. *Biochem. Biophys. Res. Commun.*, 55:1347–1355.
17. Novogrodsky, A., and Katchalski, E. (1973b): Transformation of neuraminidase-treated lymphocytes by soybean agglutinin. *Proc. Natl. Acad. Sci. USA*, 70:2515–2518.
18. Novogrodsky, A., Lotan, R., Ravid, A., and Sharon, N. (1975): Peanut agglutinin, a new mitogen that binds to galactosyl sites exposed after neuraminidase treatment. *J. Immunol.*, 115:1243–1248.
19. Toyoshima, S., Fukuda, M., and Osawa, T. (1972): Chemical nature of the receptor site for various phytomitogens. *Biochemistry*, 11:4000–4005.
20. Wynne, D., Wilchek, M., and Novogrodsky, A. (1976): A chemical approach for the localization of membrane sites involved in lymphocyte activation. *Biochem. Biophys. Res. Commun.*, 68:730–739.
21. Lis, H., and Sharon, N. (1973): The biochemistry of plant lectins (phytohemagglutinins). *Annu. Rev. Biochem.*, 42:541–574.
22. Heitzmann, H., and Richards, F. M. (1974): Use of the avidin-biotin complex for specific staining of biological membranes in electron microscopy. *Proc. Natl. Acad. Sci. USA*, 71:3537–3541.
23. Oppenheim, J. J., Leventhal, B. G., and Hersh, E. M. (1968): The transformation of column-purified lymphocytes with nonspecific and specific antigenic stimuli. *J. Immunol.*, 101:262–270.
24. Levis, W. R., and Robbins, J. H. (1970): Effect of glass-adherent cells on the blastogenic response of purified lymphocytes to phytohemagglutinin. *Exp. Cell. Res.*, 61:153–158.
25. Biniaminov, M., Ramot, B., Rosenthal, E., and Novogrodsky, A. (1975): Galactose oxidase-induced blastogenesis in human lymphocytes and the effect of macrophages on the reaction. *Clin. Exp. Immunol.*, 19:93–98.
26. Perlmann, P., and Holm, G. (1969): Cytotoxic effects of lymphoid cells *in vitro*. *Adv. Immunol.*, 11:117.
27. Novogrodsky, A. (1975): Induction of lymphocyte cytotoxicity by modification of the effector or target cells with periodate or with neuraminidase and galactose oxidase. *J. Immunol.*, 114:1089–1093.

Biogenesis and Turnover of Membrane Macromolecules,
edited by John S. Cook. Raven Press, New York 1976.

A Model for Fibroblast Growth Control

Thomas A. Cebula and Stephen Roth

Department of Biology and the McCollum-Pratt Institute, The Johns Hopkins University, Baltimore, Maryland 21218

The glycosyltransferases are a class of enzymes which catalyze the transfer of a sugar moiety from its nucleotide derivative to an appropriate sugar acceptor. The general reaction is summarized in Fig. 1, where the sugar donor may be any nucleotide sugar and the acceptor may be a simple sugar, oligosaccharide, glycoprotein, mucin, or glycolipid (1–6). The nomenclature of the enzymes is straightforward; for example, those enzymes that utilize uridine diphosphate galactose (UDP-galactose) as a sugar donor are galactosyltransferases. The enzymes are further classified by the nature of their acceptors. Thus, a galactosyltransferase that utilizes glycoprotein as an acceptor is designated a glycoprotein: galactosyltransferase.

Until recently, these enzymes were thought to be exclusively internally localized and to function in the biosyntheses of carbohydrates. In 1970, Roseman (3) suggested a surface (plasma membrane) localization for some glycosyltransferases. Since that time, numerous reports in the literature have correlated cell-surface glycosyltransferases with known physiological and developmental events.

Examples of such correlations include human platelet adhesion to collagen (7–12), intercellular adhesion in embryonic chick cells (13), morphogenesis in the chick embryo (14,15), differentiation along the rat intestinal villus (16), mating between gametes of *Chlamydomonas* (17), and concanavalin A binding to erythrocytes (18).

Additionally, there are data that have implicated both protein and sugar moieties in the growth control shown by cultured nonmalignant fibroblasts. If trypsin (19) or *N*-acetylneuraminidase (20) is added to confluent, contact-inhibited cells, these cells initiate mitosis and begin to divide. Likewise, the addition of some free sugars (21) or monovalent lectins (22) to some malignant cells make these cells more sensitive to growth control. Models for growth control have been proposed that employ protein moieties interacting with carbohydrate moieties (3,21,23). The purpose of this chapter is to review some of the data suggesting that the interaction of cell surface glycosyltransferases with their appropriate acceptors is a primary signal in growth control *in vitro* (24). We will draw heavily from the data obtained with the murine Balb/c 3T3 and Balb/c 3T12 cell strains and, for this reason, will briefly discuss the nature of these strains.

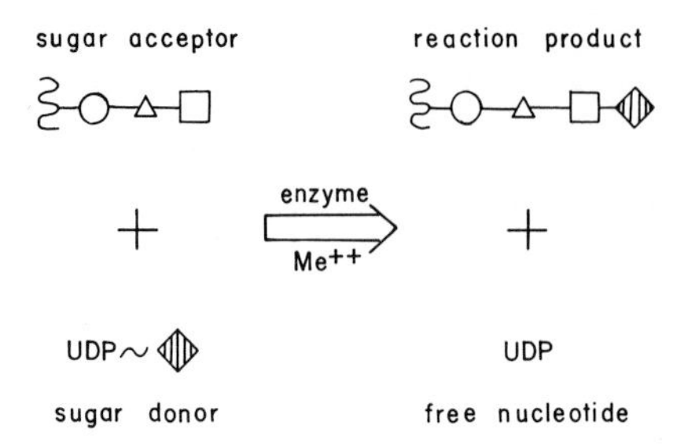

FIG. 1. Schematic of the glycosyltransferase reaction. In this representation, the monosaccharide units are depicted by circles, triangles, squares, and diamonds. The acceptor in this scheme is a trisaccharide, linked at its reducing end to a protein or lipid (wavy line). The "diamond"-transferase transfers the sugar moiety from the sugar donor (uridine diphosphate-"diamond") to the nonreducing terminus of a specific acceptor, in this case, a trisaccharide terminating with a square. Such catalysis yields free nucleotide and the reaction product (the tetrasaccharide terminating with a diamond).

CELL STRAINS

Mouse Balb/c 3T3 and Balb/c 3T12 cells were originally derived by Aaronson and Todaro (25,26). Briefly, the method employed was to prepare primary cultures from the disaggregated tissue minces of 14–17-day Balb/c mouse embryos. Their findings indicated that if the cells were passed every 3 days at an initial density of 3×10^5 cells/50-mm Petri dish, the resulting cell strain

1. Was sensitive to reduction in serum concentration such that it doubled five times more slowly in 1% than in 10% serum
2. Was anchorage-dependent; it would not grow in suspension
3. Grew to a saturation density of 5×10^4 cells/cm^2

These cells were designated Balb/c 3T3 cells. However, if primary cultures of the same mouse embryos were passed every 3 days at an initial density of 12×10^5 cells/50-mm petri dish, the cell strain that resulted from this protocol

1. Did not display any difference in growth rate when the serum levels were reduced from 10 to 1%
2. Could grow in suspension
3. Reached saturation densities of at least 25×10^4 cells/cm^2

Such cells were designated Balb/c 3T12. When 10^7 Balb/c 3T3 cells were injected subcutaneously into newborn or irradiated mice, no tumors were evident in the injected mice even after 11 months. Conversely, when 10^6 Balb/c 3T12 cells were injected, tumors formed in 8 out of 10 mice tested within 2 months (25,26).

In subsequent discussion, Balb/c 3T3 and Balb/c 3T12 cells will be referred to simply as 3T3 and 3T12 cells, respectively. These cells should not be considered identical to the cell lines derived from randomly bred Swiss mouse embryos by Todaro and Green (27).

CONTACT INHIBITION OF GROWTH AND MOTION

Over the last two decades, it has become apparent that nonmalignant cells (cells which do not form tumors when injected into animals) behave very differently from malignant cells when cultured *in vitro.* When two nonmalignant cells come into contact, rather than climb over or under each other, the cells stop moving at the point of contact. This behavior has been termed *contact inhibition of motion* (28,29). It was also demonstrated that nonmalignant cells, grown in monolayer culture, would stop dividing after a confluent monolayer had been reached. This process has been termed by Stoker (30) *contact inhibition of growth.* In malignant cells, both types of contact inhibition are less pronounced (30,31).

Since then, it has been shown that modifying normal culture conditions can affect the social behavior of cells. For example, by adding serum (32), serum factors (33), urea (34), or changing the pH of the medium (35), some nonmalignant cell types could be made to overlap and grow past their normal saturation densities.

Similarly, cells can sometimes be made to stop dividing. Thus, by the addition of monovalent plant lectins (22), free sugars (21), cyclic AMP or its derivatives (36), or phosphodiesterase inhibitors (37), some cell types have been shown to reduce cellular overlaps and cease division. The use of extrinsic factors in effecting contact inhibition or loss of contact inhibition has been previously discussed (23).

From these data, it is clear that many factors can affect the behavior of cells in culture. However, it is also very clear that when cultured under identical conditions, nonmalignant cells and malignant cells do behave differently, and these differences are probably due to their ability to send, receive, and respond to signals from neighboring cells.

LATERAL DIFFUSION ON THE CELL SURFACE

Some interesting differences between malignant and nonmalignant cells are observed when plant lectins are used to probe the cell surfaces of these cells. Although both malignant and nonmalignant cells seem to bind the same amounts of monovalent lectins (38,39), many malignant cells are more agglutinable by polyvalent lectins than are their nonmalignant counterparts (40–42). This agglutinability difference is not thought to be a quantitative or qualitative change in lectin binding sites, but rather a difference in the distribution of these sites (41,43,44). If lateral diffusion of surface components were less restricted on malignant cells (45,46) then, in the presence of lectin, more sites could cluster in specific regions of the cells. This, in turn, could allow cells to be held together more easily.

The agglutinability difference between malignant and nonmalignant cells is often not apparent during mitosis. Nonmalignant cells in mitosis are as agglutinable by lectins as malignant cells at all parts of the cell cycle (47,48). Many authors interpret these results to mean that the malignant cell surface displays a surface architecture throughout its cell cycle that can only be detected in nonmalignant cells during mitosis.

Assuming that the lectin studies reflect a difference between the organization of the plasma membranes in nonmalignant and malignant cells, the data are consistent with the hypothesis that malignant cells, and nonmalignant cells in mitosis, allow less restricted movement of surface moieties than do nonmalignant cells at other parts of the cell cycle.

GLYCOSYLTRANSFERASES AND CONTACT INHIBITION OF GROWTH

Because of the striking differences between Balb/c 3T3 and Balb/c 3T12 cells both *in vitro* and *in vivo,* the mouse fibroblastic system has become a paradigm for studies of both malignancy and cellular recognition. The model to be presented proposes that catalyzes by cell-surface glycosyltransferases may be the primary signal that leads to growth cessation.

A necessary requirement for surface glycosyltransferase involvement in growth control is that the enzymes are indeed located at the cell periphery. At the present time, most data on the localization of surface glycosyltransferases have come from work done on cultured malignant and nonmalignant cells. These systems include the Balb/c mouse fibroblast strains 3T3 and 3T12 (24,49,50), virally transformed 3T3 cells (51–53), nonmalignant and transformed hamster cells (54,55), the mouse lymphoma cell line L5178Y (56), the murine leukemic line L1210 (57,58), rat dermal fibroblasts (59), melanoma (60), and nonmalignant and virally transformed chick fibroblasts (61,62). These data have been critically reviewed (63) and may be summarized, briefly, by the following points.

When incubated with sugar nucleotides, whole cells were able to transfer the sugars from the sugar donors to endogenous acceptors (7–18,24,49–62). Since whole cells were able to transfer sugars to some, but not all, exogenous acceptors, and since control supernatants did not show any glycosyltransferase activities, these data imply that the results were not due to leaked internal enzymes or the result of acceptor uptake and internal glycosylation. Surface localization of glycosyltransferases was also suggested by the electron and light microscope autoradiographic studies that showed peripheral labeling patterns of cells (24,50,58).

These results are probably not due to transport of the sugar nucleotide

(52,63). However, it is imperative to distinguish surface glycosyltransferase activity from internal utilization of breakdown products of sugar nucleotides. That is, the activities observed could be the result of sugar nucleotide hydrolysis to the free sugar, uptake of the free sugar, and internal utilization. Yoggeeswaran et al. (54) have demonstrated glycosyltransferase activities of whole cells toward glycolipid acceptors derivatized to a glass surface. Since the acceptors were external to the cells, these results strongly infer that the cell periphery is the site of action for catalyses.

Kinetic experiments also rule unlikely that sugar nucleotide hydrolysis and internal utilization of sugars account for the observed activities. First, utilizing dual-labeled sugar and sugar nucleotide, Patt and Grimes (52) demonstrated that the sugar and the sugar nucleotide entered different pools prior to incorporation, and that the site of transferase action was distinct from the cytoplasmic sugar nucleotide pools. Second, whole cell glycosyltransferase activities were not inhibited under conditions where the breakdown of the sugar nucleotide was inhibited (15,64), or under conditions that totally blocked uptake and internal utilization of free sugars (15,52,64). Third, endogenous glycosyltransferase activities of whole cells were not inhibited by excesses of sugar nucleotides, free sugars, or sugar phosphates (63).

Finally, surface localization of glycosyltransferases has been shown in a number of systems through the isolation of plasma membranes (10–12,59, 65–68). These isolated membranes were able to catalyze the transfer of sugars from various sugar donors to endogenous and some exogenous acceptors (63).

CONTACT DEPENDENCE OF GALACTOSYLTRANSFERASE ACTIVITY

A number of differences between 3T3 and 3T12 cells became apparent when the cells were incubated with radioactive UDP-galactose. When sparse cultures of both cell strains were incubated with UDP-[^{3}H]galactose, washed, and examined autoradiographically, the cells from the 3T12 cultures were heavily labeled (Fig. 2), whereas 3T3 cells were not (Fig. 3).

However, when 3T3 cells were harvested from sparse cultures and assayed electrophoretically (13), they displayed a much greater endogenous galactosyltransferase activity than did 3T3 cells from confluent cultures. The 3- to 4-fold difference in endogenous transferase activity between sparse and confluent 3T3 cells seems due to availability of galactose acceptors since when an exogenous acceptor, *N*-acetylglucosamine, was utilized, there was no difference in the galactosyltransferase activities of these cells (24). It was also shown that 3T3 cells from sparse cultures transferred about two times as much galactose to endogenous acceptors as did 3T12 cells (50).

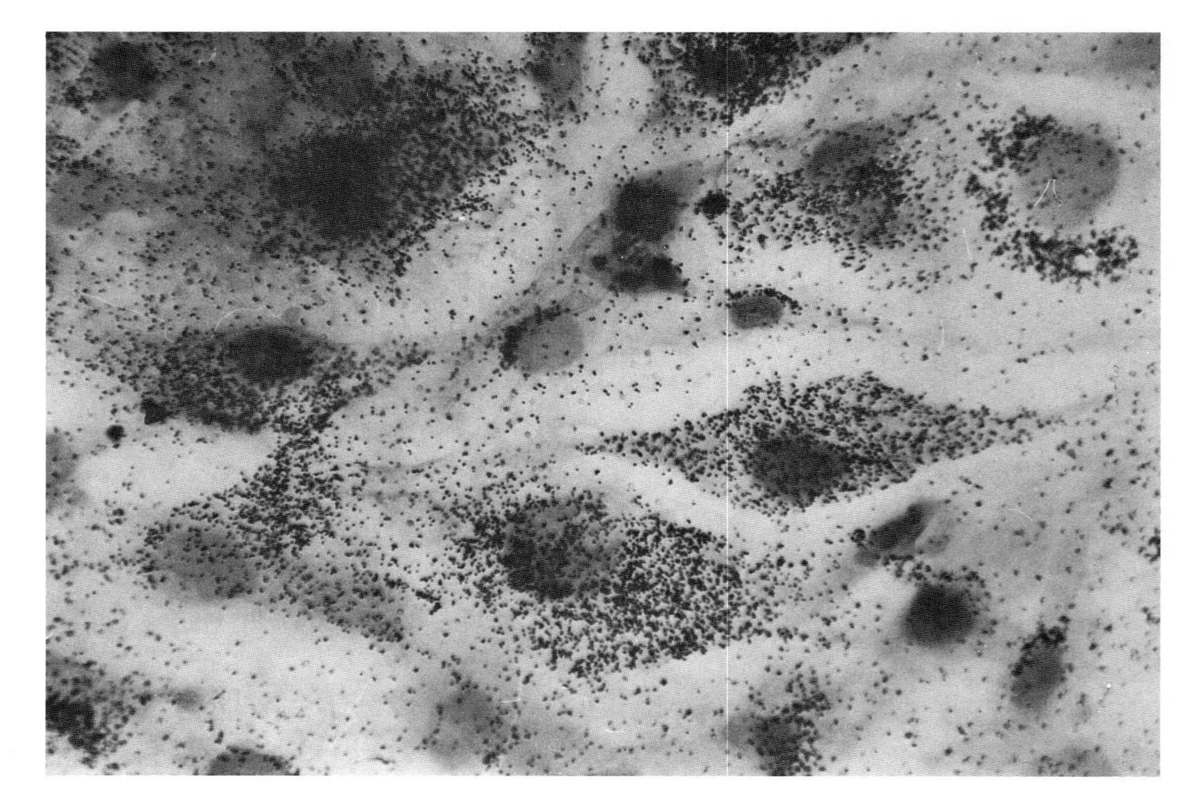

FIG. 2. Autoradiograph of a monolayer culture of 3T12 cells. A monolayer culture of 3T12 cells was incubated with UDP[^{3}H]galactose at 37°C for 3 hr, washed, and examined autoradiographically as previously described (24,50). ×1,188.

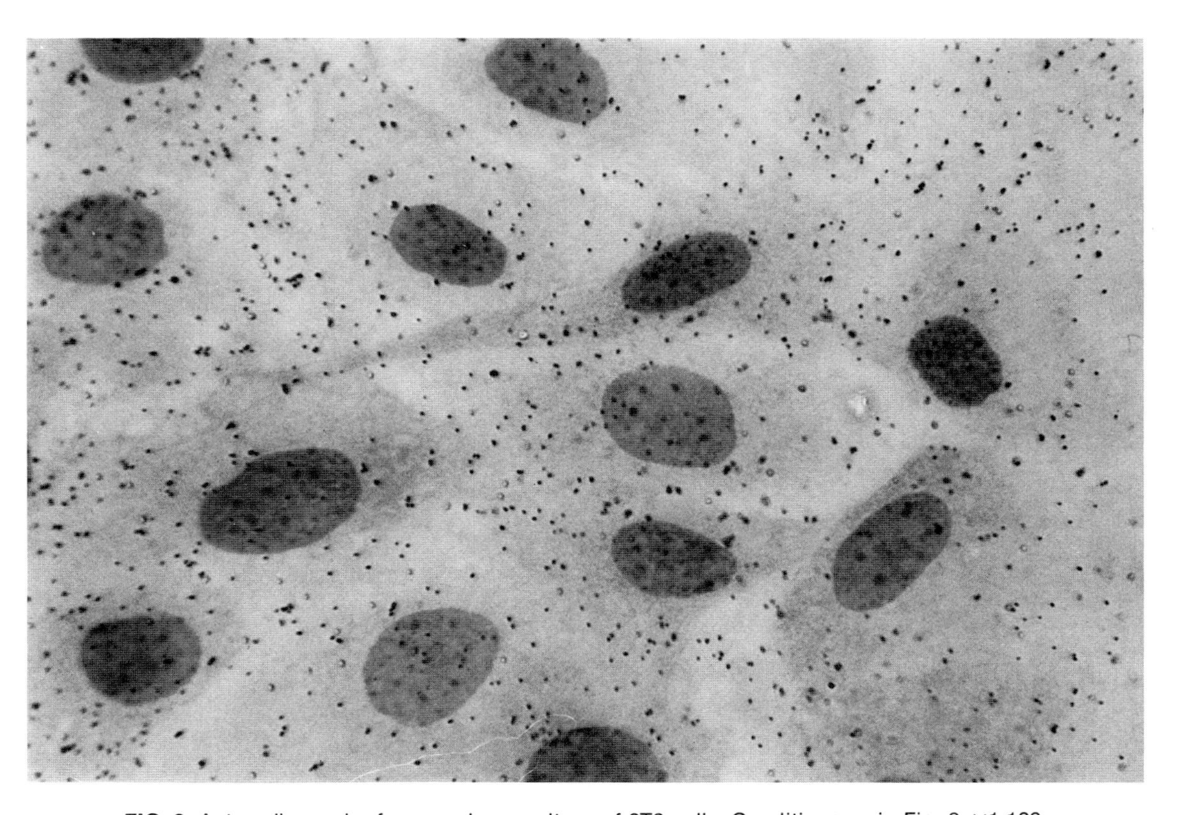

FIG. 3. Autoradiograph of a monolayer culture of 3T3 cells. Conditions as in Fig. 2. $\times$1,188.

The data are consistent with the hypothesis that 3T3 cells require cellular contact for glycosylation to occur, while 3T12 cells do not. That is, under optimal assay conditions, for 3T12 cells, the rate of reaction is dependent only on the availability of sugar nucleotide. For 3T3 cells, on the other hand, the rate of reaction is dependent both on the availability of sugar donor and the degree of cellular contact.

To test whether 3T3 cells required cellular contact for optimal endogenous transferase activity, an assay was developed to compare cell-surface endogenous transferase activity under conditions of varying degrees of cellular contact (24,49,50). This assay will be briefly discussed. Aliquots of identical cell suspensions were added to either a 3-ml conical centrifuge tube or a 2-ml flat-bottom shell vial to which a magnetic stirring bar was externally attached (Fig. 4). The cells in the centrifuge tube (stationary condition) formed a loose pellet after 5–10 min. The cells in the shell vial were spun at 200 rpm on an immersible magnetic stirrer (spinning condition). When spun at 200 rpm, these cells remained in suspension throughout the time course of the reaction. If contact were necessary for the galactosyltransferase endogenous reaction, then there should be a marked difference in endogenous transferase activities when measured by the spinning and stationary conditions.

When such assays were carried out, 3T3 cells were found to have 3 to 4 times the endogenous galactosyltransferase activity when assayed under stationary conditions than did the same cells assayed under spinning con-

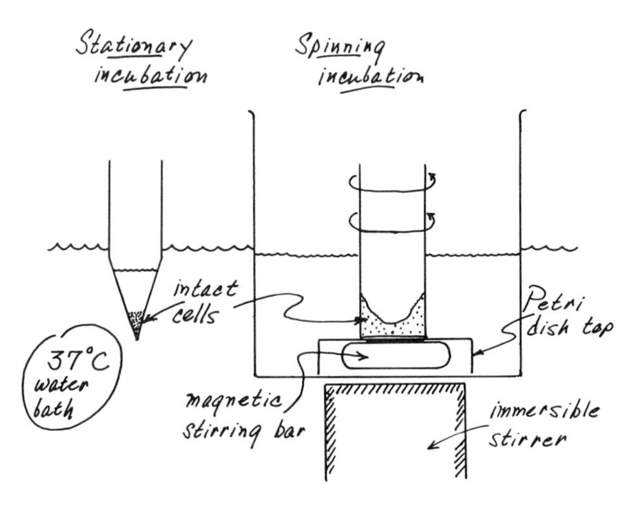

FIG. 4. Spinning and stationary assay for glycosyltransferase activity. Aliquots of cell suspensions are either placed in a conical 3-ml centrifuge tube (stationary condition) or a flat-bottom shell vial to which a magnetic stirring bar is externally attached. The cells in the shell vial are kept in suspension by spinning the shell vial at 200 rpm by means of an immersible magnetic stirrer (spinning condition). At various times, samples are removed and assayed electrophoretically (13) for product formation.

ditions. For 3T12 cells there was no difference in endogenous galactosyltransferase activity when assayed under spinning and stationary conditions (24,49,50).

Similarly, when colcemid-treated mitotic 3T3 cells were assayed for endogenous galactosyltransferase activity, there was no difference in enzymic rate when assayed under conditions of varying cell contact (50). Interphase 3T3 cells subjected to the same colcemid treatment displayed the endogenous galactosyltransferase activity difference that untreated 3T3 cells did. The results could not be attributed to differential cell viability or leakage of transferases or hydrolases (50).

The simplest interpretation of the results is that the surface components of mitotic 3T3 cells are more capable of lateral diffusion than are the surface components of 3T3 cells at other times in the cell cycle. Because of lateral diffusion, the surface galactosyltransferases are able to interact with their acceptors more readily, such that when sugar donor is made available, glycosylation can occur. Autoradiographs, using UDP[^{3}H]galactose, support this interpretation, since in a random 3T3 monolayer culture, about 5% of the cells are heavily labeled and have condensed nuclei characteristic of mitotic cells (50).

HYPOTHESIS

If, for nonmalignant interphase 3T3 cells, the lateral diffusion of surface components were restricted such as to prevent the interactions of galactosyltransferases with their appropriate acceptors on the same cell then, as indicated by the autoradiographs, endogenous galactosyltransferase activity should be low in sparse cultures. Under conditions that promote intercellular contact, however, galactosyltransferases of one cell could interact with appropriate acceptors of an apposed cell, allowing for catalysis (transglycosylation). Therefore, after cellular contact has occurred, available acceptors should decrease.

If on mitotic 3T3 cells and 3T12 cells at all times of the cell cycle, the enzymes and acceptors were located closer to each other, or if the surfaces of these cells allowed more interactions between the transferases and acceptors, then catalysis should not be contact-dependent. Accordingly, no difference in endogenous galactosyltransferase activity is detected for these cells when assayed under varying degrees of cell contact. Catalysis resulting from interactions between acceptors and enzymes on the same cell has been termed cis-glycosylation (24).

The model depicted in Fig. 5 states that nonmalignant cells, except during mitosis, can glycosylate their neighbors, but rarely themselves (transglycosylation). If such catalyses lead to a signal for growth cessation in nonmalignant cells (when the appropriate number of acceptors are glycosylated), then 3T12 cells and mitotic 3T3 cells, which have the potential for

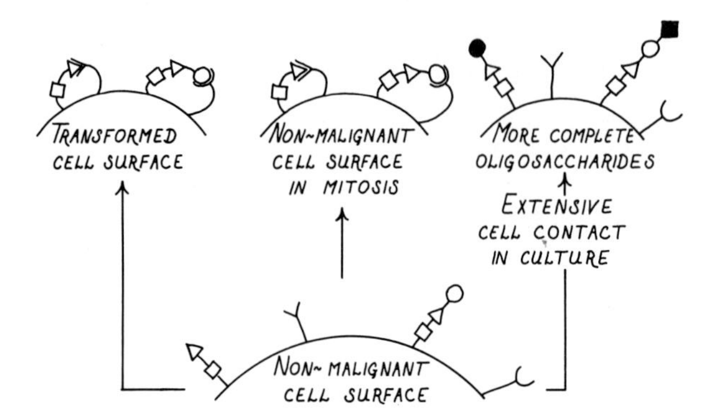

FIG. 5. Model for the spatial arrangement of surface glycosyltransferases and acceptors on nonmalignant and malignant cells. Nonmalignant cells have enzymes and acceptors which interact only rarely. However, when cells make contact, the acceptors of nonmalignant cells become more completed by means of trans-glycosylation (*top right*). Upon transformation, the surface membrane undergoes a change such that glycosyltransferases and their acceptors interact more readily. In the presence of exogenous sugar donor, cis-glycosylation occurs. It is proposed that the availability of sugar donor in some malignant cell types may account for the less complex acceptors observed on these cell types (*top left*). The membrane change occurring in transformed cells also occurs in normal mitotic cells (*top center*).

glycosylating themselves, might be expected to do so and cease growth when the necessary glycosylation occurred. Since for 3T12 cells this does not occur, it is necessary to postulate a second defect in 3T12 cells that interferes with glycosylation. For example, it is possible that 3T12 cells are defective in their ability to supply surface transferases with the appropriate sugar donors. A variant of this type would enjoy a great selective advantage and, under normal culture conditions, would continue to grow.

Consistent with this idea, Robbins and Macpherson (69) have demonstrated that nonmalignant hamster cells from confluent cultures have more than 10 times the amount of ceramide tetrahexoside than do nonmalignant cells from sparse cultures. Similarly, using baby hamster kidney cells (BHK), Hakomori (70) found that these cells at confluency contained more ceramide trihexoside than did growing BHK cells. Additionally, Hakomori showed that confluent, nonmalignant BHK cells contained more complex glycolipids than did their confluent, transformed counterparts (70).

In studies utilizing Balb/c 3T3 and Balb/c 3T12 cells, Nicolson and Lacorbierre (71) have shown that the number of *R. communis* binding sites (galactosides) were about 2.5 times greater in nonmalignant cells harvested from confluent cultures when compared with cells harvested from sparse cultures. Neither 3T12 cells nor virally transformed 3T3 cells displayed this contact-dependent increase in lectin binding sites (71).

Viewed in the framework of this model, 3T3 cells are cells that can both glycosylate and respond to glycosylation. Since 3T12 cells do not them-

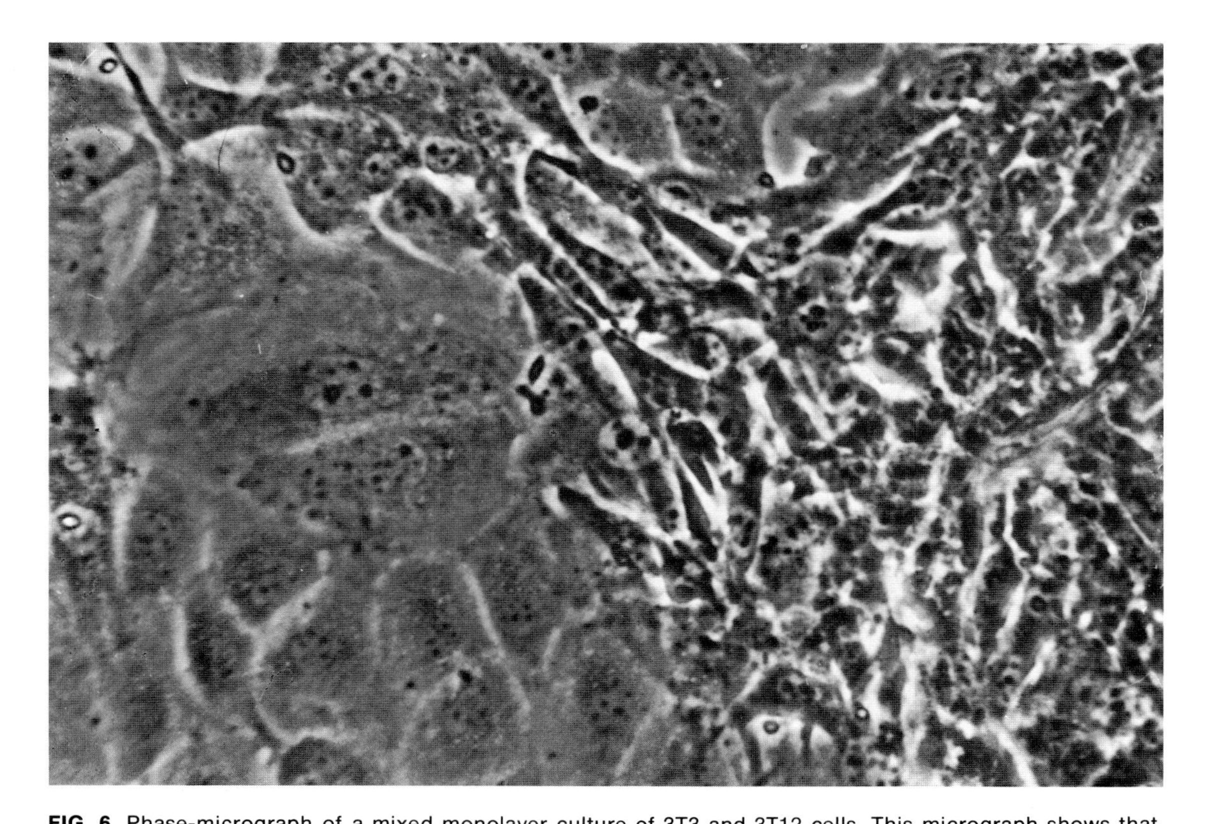

FIG. 6. Phase-micrograph of a mixed monolayer culture of 3T3 and 3T12 cells. This micrograph shows that, although 3T12 cells do not respect the boundaries of other 3T12 cells, they do respect the boundaries of 3T3 cells. $\times$1,188.

selves show contact inhibition, but are contact-inhibited by 3T3 cells (ref. 25 and Fig. 6), it is possible that this cell strain cannot glycosylate, but can respond to glycosylation.

PREDICTIONS OF THE MODEL

This simple model for growth cessation makes two predictions for 3T3 and 3T12 cells that can be experimentally tested. First, since the spinning and stationary assays showed that nonmalignant 3T3 cells displayed a stationary/spinning ratio of 3–4 and 3T12 cells a ratio of 1, it would be predicted that as saturation density for 3T3 cells increased toward that of 3T12 cells, the stationary/spinning ratio should decrease. To investigate this possibility, clones of 3T3 cells were selected for their ability to grow to higher saturation densities than did nonmalignant 3T3 cells. These clones showed an inverse relationship between saturation density and the stationary/spinning ratio; that is, as saturation density increased, the stationary/spinning ratio decreased toward 1, the value obtained for 3T12 cells (64).

Second, if 3T12 cells can receive but not send signals, it should be possible to alter the growth properties of these cells by incubating the cells with sugar nucleotides. Preliminary experiments indicate that forced glycosylation of 3T12 cells inhibited the growth rate of these cells (72). 3T12 cells were incubated in suspension with a mixture of UDP-galactose, UDP-*N*-acetylglucosamine, UDP-*N*-acetylgalactosamine, Mn^{2+}, and sodium azide for 20 to 120 min. After the incubation period, the cells were replated and the growth rate was monitored. Such incubations inhibited the growth rate of these cells over the next 24 hr, with the 2 hr incubation showing the greatest inhibition. Control incubations indicated that the inhibition was not due to Mn^{2+}, sodium azide, or a combination of these reagents (72).

SUMMARY

The data discussed in this chapter suggest a glycosyltransferase involvement in the process of growth control between some cultured fibroblasts. The model implies that, in nonmalignant 3T3 cells, trans-glycosylation signals the cells to stop dividing.

In the malignant 3T12 cells, there are at least two defects. The first defect leads to greater enzyme–acceptor interactions allowing for cis-glycosylation. If this were the only defect, these cells should immediately slow their growth rate. This slowed growth rate should strongly select for cells that can overcome this defect. One of many methods for accomplishing this would be an interference in the production or availability of sugar donors for surface glycosyltransferases. Variants of this type would obviously grow.

Although the secondary defect for 3T12 cells may be availability of sugar donors, it is unlikely that this is the case for SV_{40}3T3 cells, since these cells are neither inhibited by 3T3 nor SV_{40}3T3 cells (25,26). These cells seem unable to send or respond to the appropriate cessation signals.

A model of this type is consistent with at least a two-step mechanism leading to tumorigenesis. The results of the first step, *in vivo,* would be a group of cells showing aberrant histological and cell surface properties, but not necessarily uncontrolled growth. In fact, this type of cell could show a slower growth rate than surrounding tissue and could correspond to a benign tumor. The second defect could release the cells from self inhibition and cause the transition from a benign to a malignant tumor. This would be expected if the defect occurred at the level of membrane transduction so that glycosylation might take place but no longer be of any importance. Alternatively, the defect releasing the cells from self-inhibition could be in the pathway supplying the surface enzymes with sugar donors. This cell type would form a tumor that grows from within but not at its periphery since it would still be sensitive to growth control by the normal, surrounding tissue.

ACKNOWLEDGMENTS

This work is supported by grants from the National Institute of Child Health and Human Development and the American Cancer Society. T.A.C. is a postdoctoral fellow of the National Cancer Institute. This is contribution no. 858 from the McCollum-Pratt Institute.

REFERENCES

1. Roseman, S. (1959): Metabolism of connective tissue. *Annu. Rev. Biochem.,* 28:545–578.
2. Roseman, S. (1968): Biochemistry of glycoproteins and related substances. In: *Proceedings of the 4th International Conference on Cystic Fibrosis of the Pancreas (Mucoviscidosis),* pp. 244–269, edited by E. Rossi and E. Stoll. Karger, New York.
3. Roseman, S. (1970): The synthesis of complex carbohydrates by multiglycosyltransferase systems and their potential function in intercellular adhesion. *Chem. Phys. Lipids,* 5:270–297.
4. Roseman, S. (1973): The biosynthesis of cell-surface components and their potential role in intercellular adhesion. In: *The Neurosciences: III. Study Program,* pp. 795–804, edited by F. O. Schmitt. MIT Press, Cambridge, Mass.
5. Roseman, S. (1974): Complex carbohydrates and intercellular adhesion. In: *Biology and Chemistry of Eucaryotic Cell Surfaces, Miami Winter Symposia,* Vol. 7, pp. 317–354, edited by E. Y. C. Lee and E. E. Smith. Academic Press, New York.
6. Schacter, H., and Rodén, L. (1973): The biosynthesis of animal glycoproteins. In: *Metabolic Conjugation and Metabolic Hydrolysis,* Vol. 3, pp. 1–149, edited by W. H. Fishman. Academic Press, New York.
7. Bosmann, H. B. (1971): Platelet adhesiveness and aggregation: The collagen:glycosyl, polypeptide:*N*-acetylgalactosaminyl and glycoprotein:galactosyl transferases of human platelets. *Biochem. Biophys. Res. Commun.,* 43:1118–1124.
8. Jamieson, G. A., Urban, C. L., and Barber, A. J. (1971): Enzymatic basis for platelet: Collagen adhesion as the primary step in haemostasis. *Nature [New Biol.],* 234:5–7.
9. Jamieson, G. A. (1974): The adhesion and aggregation of blood platelets. In: *Biology and Chemistry of Eucaryotic Cell Surfaces, Miami Winter Symposia,* Vol. 7, pp. 67–79, edited by E. Y. C. Lee and E. E. Smith. Academic Press, New York.

10. Barber, A. J., and Jamieson, G. A. (1971): Platelet collagen adhesion characterization of collagen glucosyltransferase of plasma membranes of human blood. *Biochim. Biophys. Acta,* 252:533–545.
11. Barber, A. J., and Jamieson, G. A. (1971): Characterization of membrane-bound galactosyltransferase of human blood platelets. *Biochim. Biophys. Acta,* 252:546–552.
12. Bosmann, H. B. (1972): Platelet adhesiveness and aggregation. II. Surface sialic acid, glycoprotein: *N*-Acetylneuraminic acid transferase, and neuraminidase of human blood platelets. *Biochim. Biophys. Acta,* 279:456–474.
13. Roth, S., McGuire, E. J., and Roseman, S. (1971): Evidence for cell surface glycosyltransferases. *J. Cell Biol.,* 51:536–547.
14. Shur, B., and Roth, S. (1973): The localization and potential function of glycosyltransferases in chick embryos. *Am. Zool.,* 13:1129–1135.
15. Shur, B. (1975): *Manuscript in preparation.*
16. Weiser, M. M. (1973): Intestinal epithelial cell surface membrane glycoprotein synthesis. II. Glycosyltransferases and endogenous acceptors of the undifferentiated cell surface membrane. *J. Biol. Chem.,* 248:2542–2548.
17. McLean, R. J., and Bosmann, H. B. (1975): Cell–cell interactions: enhancement of glycosyl transferase ectoenzyme systems during *Chlamydomonas* gametic contact. *Proc. Natl. Acad. Sci. USA,* 72:310–313.
18. Podolosky, D. K., and Weiser, M. M. (1975): Role of cell membrane galactosyltransferase in concanavalin A agglutination of erythrocytes. *Biochem. J.,* 146:213–221.
19. Burger, M. M. (1970): Proteolytic enzymes initiating cell division and escape from contact inhibition of growth. *Nature,* 227:170–171.
20. Vaheri, A., Ruoslahti, E., and Nordling, S. (1972): Neuraminidase stimulates division and sugar uptake in density-inhibited cell cultures. *Nature* [*New Biol.*], 238:211–212.
21. Cox, R., and Gesner, B. (1965): Effect of simple sugars on the morphology and growth pattern of mammalian cell cultures. *Proc. Natl. Acad. Sci. USA,* 54:1571–1579.
22. Burger, M. M., and Noonan, K. D. (1970): Restoration of normal growth by covering of agglutinin sites on tumor cell surface. *Nature,* 228:512–515.
23. Roth, S. (1973): A molecular model for cell interactions. *Quart. Rev. Biol.,* 48:541–563.
24. Roth, S., and White, D. (1972): Intercellular contact and cell-surface galactosyltransferase activity. *Proc. Natl. Acad. Sci. USA,* 69:485–489.
25. Aaronson, S. A., and Todaro, G. J. (1968): Development of 3T3-like lines from Balb/c mouse embryo cultures: Transformation susceptibility to SV_{40}. *J. Cell. Physiol.,* 72:141–148.
26. Aaronson, S. A., and Todaro, G. J. (1968): Basis for the acquisition of malignant potential by mouse cells cultivated *in vitro. Science,* 162:1024–1026.
27. Todaro, G. J., and Green, H. (1963): Quantitative studies of the growth of mouse embryo cells in culture and their development into established lines. *J. Cell Biol.,* 17:299–313.
28. Abercrombie, M., and Heaysman, J. E. M. (1953): Observations on the social behaviour of cells in tissue culture. I. Speed of movement of chick heart fibroblasts in relation to their mutual contacts. *Exp. Cell Res.,* 5:111–131.
29. Abercrombie, M., and Heaysman, J. E. M. (1954): Observations on the social behaviour of cells in tissue culture. II. "Monolayering" of fibroblasts. *Exp. Cell Res.,* 6:293–306.
30. Stoker, M. (1964): Regulation of growth and orientation in hamster cells transformed by polyoma virus. *Virology,* 24:165–174.
31. Abercrombie, M., Heaysman, J. E. M., and Karthauser, H. M. (1957): Social behaviour of cells in tissue culture. III. Mutual influence of sarcoma cells and fibroblasts. *Exp. Cell Res.,* 13:276–291.
32. Todaro, G. J., Lazer, G. K., and Green, H. (1965): The initiation of cell division in a contact-inhibited mammalian cell line. *J. Cell Comp. Physiol.,* 66:325–333.
33. Paul, D., Lipton, A., and Klinger, I. (1971): Serum factor requirements of normal and simian virus 40-transformed 3T3 mouse fibroblasts. *Proc. Natl. Acad. Sci. USA,* 68:645–648.
34. Weston, J. A., and Hendricks, K. L. (1972): Reversible transformation by urea of contact-inhibited fibroblasts. *Proc. Natl. Acad. Sci. USA,* 69:3727–3731.
35. Ceccarini, C., and Eagle, H. (1971): pH as a determinant of cellular growth and contact inhibition. *Proc. Natl. Acad. Sci. USA,* 68:229–233.

36. Johnson, G. S., Friedman, R. M., and Pastan, I. (1971): Restoration of several morphological characteristics of normal fibroblasts in sarcoma cells treated with adenosine 3′,5′-cyclic monophosphate and its derivatives. *Proc. Natl. Acad. Sci. USA,* 68:425–429.
37. Bürk, R. R. (1968): Reduced adenyl cyclase activity in a polyoma virus transformed cell line. *Nature,* 219:1272–1275.
38. Cline, M. J., and Livingston, D. C. (1971): Binding of ^{3}H-concanavalin A by normal and transformed cells. *Nature* [*New Biol.*], 232:155–156.
39. Ozanne, B., and Sambrook, J. (1971): Binding of radioactively labeled concanavalin A and wheat germ agglutinin to normal and virus-transformed cells. *Nature* [*New Biol.*], 232: 156–160.
40. Aub, J. C., Tieslau, C., and Lankester, A. (1963): Reactions of normal and tumour cell surfaces to enzymes. I. Wheat germ lipase and associated mucopolysaccharides. *Proc. Natl. Acad. Sci. USA,* 50:613–619.
41. Inbar, M., and Sachs, L. (1969): Structural difference in sites on the surface membrane of normal and transformed cells. *Nature,* 223:710–712.
42. Burger, M. M. (1969): A difference in the architecture of the surface membrane of normal and virally transformed cells. *Proc. Natl. Acad. Sci. USA,* 62:994–1001.
43. Nicolson, G. L. (1971): Difference in topology of normal and tumor cell membranes shown by different surface distributions of ferritin-conjugated Concanavalin A. *Nature* [*New Biol.*], 233:244–246.
44. Rosenblith, J. Z., Ukena, T. E., Yin, H. H., Berlin, R. D., and Karnovsky, M. J. (1973): A comparative evaluation of the distribution of concanavalin A binding sites on the surfaces of normal, virally-transformed, and protease-treated fibroblasts. *Proc. Natl. Acad. Sci. USA,* 70:1625–1629.
45. Frye, L. D., and Edidin, M. (1970): The rapid intermixing of cell surface antigens after formation of mouse-human heterokaryons. *J. Cell Sci.,* 7:319–335.
46. Edidin, M., and Weiss, A. (1974): Restriction of antigen mobility in the plasma membranes of some cultured fibroblasts. In: *Control of Proliferation in Animal Cells. Cold Spring Harbor Conferences on Cell Proliferation,* Vol. 1, pp. 213–219, edited by B. Clarkson and R. Baserga. Cold Spring Harbor Laboratory.
47. Fox, T., Sheppard, J., and Burger, M. (1971): Cyclic membrane changes in animal cells: Transformed cells permanently display a surface architecture detected in normal cells only during mitosis. *Proc. Natl. Acad. Sci. USA,* 68:244–247.
48. Shoham, J., and Sachs, L. (1974): Differences in lectin agglutinability of normal and transformed cells in interphase and mitosis. In: *Control of Proliferation in Animal Cells. Cold Spring Harbor Conferences on Cell Proliferation,* Vol. 1, pp. 297–304, edited by B. Clarkson and R. Baserga. Cold Spring Harbor Laboratory.
49. Roth, S., Patteson, A., and White, D. (1974): Surface glycosyltransferases on cultured mouse fibroblasts. *J. Supramol. Struct.,* 2:1–6.
50. Webb, G. C., and Roth, S. (1974): Cell contact dependence of surface galactosyltransferase activity as a function of the cell cycle. *J. Cell Biol.,* 63:796–805.
51. Bosmann, H. B. (1972): Cell surface glycosyltransferases and acceptors in normal and RNA- and DNA-virus transformed fibroblasts. *Biochem. Biophys. Res. Commun.,* 48:523–529.
52. Patt, L. M., and Grimes, W. J. (1974): Cell surface glycolipid and glycoprotein glycosyltransferases of normal and transformed cells. *J. Biol. Chem.,* 249:4157–4165.
53. Patt, L. M., Van Nest, G. A., and Grimes, W. J. (1975): A comparison of glycosyltransferase activities and malignant properties in normal and transformed cells derived from Balb/c mice. *Cancer Res.,* 35:438–441.
54. Yogeeswaran, G., Laine, R. A., and Hakomori, S. (1974): Mechanism of cell contact-dependent glycolipid synthesis: further studies with glycolipid-glass complex. *Biochem. Biophys. Res. Commun.,* 59:591–599.
55. Sasaki, T., and Robbins, P. W. (1974): A comparative study of sialic acid incorporation into endogenous acceptors by normal and polyoma virus transformed hamster cells. In: *Biology and Chemistry of Eucaryotic Cell Surfaces, Miami Winter Symposia,* Vol. 7, pp. 125–157, edited by E. Y. C. Lee and E. E. Smith. Academic Press, New York.
56. Bosmann, H. B. (1974): Cell plasma membrane external surface glycosyltransferases: activity in the cell mitotic cycle. *Biochim. Biophys. Acta,* 339:438–441.

57. Bernacki, R. J. (1974): Plasma membrane ectoglycosyltransferase activity of L1210 murine leukemic cells. *J. Cell Physiol.*, 83:457–466.
58. Porter, C. W., and Bernacki, R. J. (1975): Ultrastructural evidence for ectoglycosyltransferase systems. *Nature*, 256:648–650.
59. Lloyd, C. W., and Cook, G. M. W. (1974): On the mechanism of the increased aggregation by neuraminidase of 16 c malignant rat dermal fibroblasts *in vitro*. *J. Cell Sci.*, 15:575–590.
60. Bosmann, H. B., Bieber, G. F., Brown, A. B., Case, K. R., Gersten, D. M., Kimmerer, T. W., and Lione, A. (1974): Biochemical parameters correlated with tumor cell implantation. *Nature*, 246:487–489.
61. Morgan, H. R., and Bosmann, H. B. (1974): Alterations of surface properties of chick embryo fibroblasts infected with four Rous sarcoma viruses producing distinctive cell transformation. *Proc. Soc. Exp. Biol. Med.*, 146:1146–1149.
62. Spataro, A. C., Morgan, H. R., and Bosmann, H. B. (1975): Cell surface changes accompanying viral transformation: *N*-Acetylneuraminic acid ectotransferase system activity. *Proc. Soc. Exp. Biol. Med.*, 149:486–490.
63. Shur, B., and Roth, S. (1975): Cell surface glycosyltransferases. *Biochim. Biophys. Acta.*, 415:473–512.
64. Webb, G. C., and Roth, S. (1975): *Manuscript in preparation.*
65. Bosmann, H. B. (1972): Sialyltransferase activity in normal and RNA- and DNA-virus transformed cells utilizing desialyzed, trypsinized cell plasma membrane external surface glycoproteins as exogenous acceptors. *Biochem. Biophys. Res. Commun.*, 49:1256–1262.
66. Aronson, N., Tan, L. Y., and Peters, B. P. (1973): Galactosyl transferase—the liver plasma membrane binding site for asialoglycoproteins. *Biochem. Biophys. Res. Commun.*, 53:112–118.
67. Warley, A., and Cook, G. M. W. (1973): The isolation and characterization of plasma membranes from normal and leukaemic cells of mice. *Biochim. Biophys. Acta*, 323:55–68.
68. Cebula, T. A., and Roth, S. (1975): *Manuscript in preparation.*
69. Robbins, P., and Macpherson, T. (1971): Glycolipid synthesis in a cultured hamster cell line. *Nature*, 229:569–570.
70. Hakomori, S. (1970): Cell density-dependent changes of glyco-lipid concentration in fibroblasts, and loss of this response in virus transformed cells. *Proc. Natl. Acad. Sci. USA*, 67:1741–1747.
71. Nicolson, G. L., and Lacorbierre, M. (1973): Cell-contact dependent increase in membrane D-galactopyranosyl-like residues on normal, but not virus- or spontaneously transformed, murine fibroblasts. *Proc. Natl. Acad. Sci. USA*, 70:1672–1676.
72. Roelke, M., and Roth, S. (1975): *Manuscript in preparation.*

Biogenesis and Turnover of Membrane Macromolecules,
edited by John S. Cook. Raven Press, New York 1976.

Cell-Surface Structure and Function in Rous Sarcoma Virus-Transformed Cells

Michael J. Weber,* Trent Buckman,*,† Arthur H. Hale,*
Tom M. Yau,** Terrance M. Brady,* Denise D. LaRossa*,‡

** Department of Microbiology, University of Illinois, Urbana, Illinois 61801, and ** Department of Radiology, Case Western Reserve University, Cleveland, Ohio 44106*

We have for some time been interested in the role that tumor virus-induced cell-surface alterations play in malignant transformation, and our examination of these virus-induced changes has proceeded on two fronts. First, we have studied the cell-surface changes themselves, in an attempt to understand the relationship between alterations in cell-surface structure and function. Second, we have used these cell-surface alterations as phenotypic markers of the transformed state, in an analysis of the mechanism by which viral activity modifies the cell. In the first case we use viral gene expression to probe the cell surface, while in the second case we use cell surface alterations as a probe of virus gene expression. Clearly, the two approaches are tightly linked; however, they do generate different sorts of experiments. In our analysis of cell-surface structure–function relationships, we have explored the possible role of virus-induced alterations in membrane bilayer chemistry and structure in the control of hexose transport. As part of our investigation of the mechanism by which viral activity modifies the cell surface, we have examined the role of virus-induced proteases in the alteration of the cell surface. In this chapter, we would like to describe some of our recent work along each of these lines. In all of the research described, we have used chicken embryo fibroblasts transformed by Rous sarcoma virus (RSV).

TRANSPORT CHANGES ASSOCIATED WITH GROWTH CONTROL AND MALIGNANT TRANSFORMATION

Our work started with an investigation of the alterations in membrane transport associated with growth control and malignant transformation. We chose this starting point primarily because the majority of workers studying transformed cell surfaces were taking a biochemical approach,

† *Present address:* Frederick Cancer Research Center, Frederick, Maryland 21701.
‡ *Present address:* Kodak Research Park, Eastman Kodak Company, Rochester, New York 14650.

looking for alterations in surface molecules, whereas the physiological or functional approach was relatively neglected. We found that, although uptake rates for many molecules varied with changes in the growth rate of cells, only hexose transport was specifically changed by the transformation process per se. This is shown in Table 1. It is clear that rates of uptake of hexoses, nucleosides, the amino acid analogue α-aminoisobutyric acid, phosphate, and potassium are from 2 to 6 times higher in normal, exponentially growing cells than in nongrowing, density-inhibited cells. This can be considered a "balanced growth" phenomenon: as cells slow their growth, the rate at which they take up molecules from their environment is slowed as well. The extent of the uptake change is presumably some function of the extent of growth inhibition and the rates of metabolism and efflux of the transported species. Of particular interest is the fact that density-inhibited cells transport potassium at a reduced rate, as do resting lymphocytes (Kaplan et al., *this volume*). In addition, we have found that density-inhibited fibroblasts have a somewhat reduced potassium content (Weber and LaRossa, *in preparation*).

If we compare the transport properties of normal exponentially growing cells to those of transformed cells growing at the same rate, we see that only transport of the glucose analogues 2-deoxyglucose and 3-*O*-methylglucose is changed: the transformed cells take up these hexoses at a rate which is 3–5 times faster than that obtained with the normal, growing cells. It should be emphasized that the normal, growing cells and the transformed cells have nearly identical rates of cellular multiplication (2) and nearly identical pool sizes of ATP (3,4), as well as the same volume and RNA and protein content (1). The excess glucose taken up during growth of the transformed cells is released into the growth medium as lactic acid (5). The increased rate of hexose transport is thus a specific marker of the transformed phenotype, independent of differences in growth. Transport differences between growing and density-inhibited cells we call "growth state-contin-

TABLE 1. *Relative rates for nutrient uptake in density-inhibited, normal, growing cells and transformed cells*

	Density-inhibited	Normal-growing	Transformed
2-Deoxyglucose	1.0	6.8	23.9
3-*O*-Methylglucose	1.0	6.5	32.8
Uridine	1.0	4.0	3.4
Adenosine	1.0	2.1	n.d.[a]
Thymidine	1.0	9.0	8.9
α-Aminoisobutyric acid	1.0	4.5	4.5
Phosphate	1.0	1.9	1.8
Potassium	1.0	1.8	1.6

Uptake into acid-soluble material was determined during a linear uptake period (generally 15 min) essentially as described (1).

[a] n.d. = not determined.

gent" changes, whereas the difference in hexose transport rate between normal growing and transformed cells we term "transformation-specific." It is important to distinguish between these two classes of transport change, because (although both are interesting regulatory phenomena) "growth state-contingent" differences seen between normal and transformed cultures more than likely are secondary consequences of the loss of growth control, whereas the "transformation-specific" changes may, in fact, underlie the altered growth behavior of the transformed cells.

The increased rate of hexose transport is not only a specific marker of transformation, it is also dependent on the continuous expression of the viral transforming gene (Fig. 1). To determine this, we took advantage of a temperature-conditional mutant of Rous sarcoma virus (RSV-T5) (6). This mutant is temperature-sensitive in a function required to maintain the transformed phenotype: cells infected with RSV-T5 are normal when held at 41°C, but become transformed when shifted to 36°C. Conversely, RSV-T5 infected cells held at 36°C are transformed, but revert to normal when shifted to 41°C. Since high titers of infectious virus are produced at both temperatures, RSV-T5 is temperature-conditional only in a gene product [termed "onc" or "sarc" (7)] which is required for oncogenesis. Cells infected with RSV-T5 and held at 41°C, which is the restrictive temperature, display a level of 2-deoxyglucose transport which is characteristic of normal cells (Fig. 1). However, when shifted to the permissive temperature of 36°C, these cells rapidly begin to increase their rate of hexose trans-

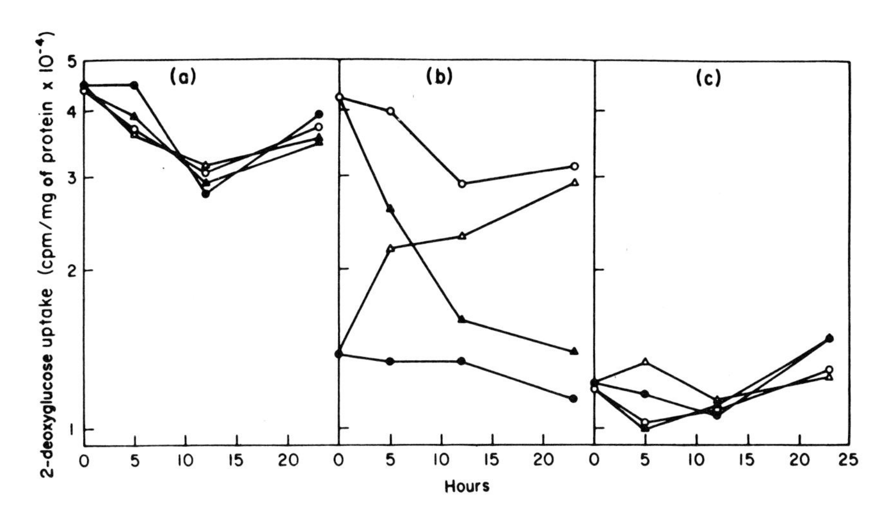

FIG. 1. Effect of temperature shifts on 2-deoxyglucose uptake. **a:** Cells infected with wild-type RSV. **b:** Cells infected with RSV-T5. **c:** Uninfected cells. (○) Cultures maintained at 36°C; (▲) cultures shifted from 36 to 41°C; (●) cultures maintained at 41°C; (△) cultures shifted from 41 to 36°C. Uptake was for 15 min at 37°C, with 0.5 μCi/ml [^{3}H]2-deoxyglucose. Uptake was linear for this time period. (Reprinted with permission from *Proc. Natl. Acad. Sci. USA.*)

port until, by 12–24 hr after the shift, the transport rate is at the fully transformed level. Conversely, cells held at 36°C have the transformed level of hexose transport, but rapidly reduce their hexose transport rate when shifted to 41°C. Thus, the change in the rate of hexose transport occurs rapidly in response to viral oncogenic information, and is one of the earliest manifestations of the transformed phenotype. Other experiments have shown that the increased transport rate does not require DNA synthesis, and may not require RNA synthesis, although it is dependent on new protein synthesis (8). We also have found that the increased transport rate can be induced at any stage in the cell cycle (9).

Properties of the Hexose Transport System

We next turned to a characterization of the hexose transport activity. We found, first, that uptake of 3-*O*-methylglucose displayed the same variations in activity that were seen when 2-deoxyglucose or glucose uptake was measured (Table 1). Since 3-*O*-methylglucose is transported, but not phosphorylated or further metabolized by these cells, this shows that it is, in fact, the transport activity which is changing (as opposed to some later step in sugar metabolism), and that the change in transport rate is sufficient to account for all of the increased uptake and metabolism of glucose (1,5). We found that the uptake system was saturable, that efflux of 3-*O*-methylglucose was increased to the same extent as influx, and that the analogue was not accumulated against a concentration gradient. These are all characteristics of a facilitated diffusion system (10). Thus, hexose transport in these cells is via facilitated diffusion, and it is the activation of a facilitated diffusion system which is responsible for the increased rate of hexose uptake in transformed cells (1).

We also found that there was a significant unsaturable component to the uptake of hexoses by these cells. The unsaturable component probably includes simple diffusion into the cell, trapping in the extracellular space, binding to the cell surface and uptake by other carriers which have a low affinity for the sugar. The high unsaturable component made it very difficult to obtain an accurate K_m and V_{max} for the uptake, especially in the case of the density-inhibited cells which have such a low transport capacity. But even though we could not determine an accurate absolute K_m and V_{max} we found that if unsaturable uptake was taken into account, the K_m for transport of hexoses was approximately the same for density-inhibited, growing, and transformed cells (approximately 1 mM for the sugar analogues used in our studies) and that only the V_{max} varied significantly and consistently. Similarly, the K_i for the phlorizin inhibition of hexose transport was the same for all three cell states. Thus, based on the available data, there is no need to invoke the synthesis of a new, virus-specific transport system in the transformed cells. We feel it is most likely that variations in the level of hexose transport occur via either additonal *de novo* synthesis or modification of the existing system (1).

ALTERATIONS IN MEMBRANE FATTY ACIDS AND FLUIDITY

We then asked what chemical and structural alterations in the membranes of the transformed cells could be responsible for the altered transport activity. One way in which the activity of existing transport systems could be modified is by altering the chemistry and structure of the membrane bilayer. A number of transport systems and membrane-associated activities are affected by the state of the membrane lipids, including the Na-K-ATPase in animal cells and sugar transport in bacteria (11–14). It had previously been shown that cells transformed by Rous sarcoma virus show little or no change in cholesterol or phospholipid composition and turnover (15–17). We therefore examined the acyl group composition of the lipids from these cells to determine whether they are altered by malignant transformation, whether these alterations are associated with changes in bilayer structure, and whether the changes in membrane chemistry and structure could account for some of the changes in membrane activity. The results indicate a somewhat decreased unsaturation of some of the fatty acids in the transformed cells and a corresponding small decrease in the flexibility ("fluidity") of the bilayer fatty acids (measured using spin-labeled probes). However, studies on the kinetics with which the chemical changes appear during transformation suggest that the changed fatty acid composition probably does not account for the altered biological properties of the transformed cells.

Fatty Acid Composition of Normal and Transformed Cells

Table 2 shows the fatty acid composition of the lipids extracted from normal, growing cells and transformed cells. It is clear that there is a substantial drop in the arachidonate (20 : 4) content of the transformed cells relative to the normal, growing cells, and a roughly equivalent increase in the percent oleate (18 : 1). Similar results have been reported for mouse and human cells transformed by SV40 (19,20). This trend toward a higher ratio of 18 : 1/20 : 4 held true for all of the separated phospholipids (18) and for isolated plasma membranes as well (*Biochemistry, accepted for publication*). Thus, although the ratio of saturated to unsaturated fatty acids remained approximately constant in the normal growing and transformed cells, the *degree* of unsaturation decreased in the transformed cells, as shown by the decreased Unsaturation Index.

TABLE 2. *Acyl group composition*[a] *of total lipids from normal, growing cells and transformed cells*

Acyl groups	Normal, exponentially growing cells	Rous sarcoma virus-transformed cells
16 : 0	20.9	16.8
18 : 0	20.6	24.3
18 : 1	23.8	30.7
18 : 2	12.6	10.4
20 : 3	3.2	3.6
20 : 4	14.5	10.1
22 : 4	1.8	1.8
22 : 5	1.2	1.0
22 : 6	1.6	1.4
Total saturated fatty acids	41.5	41.1
Ratio 18 : 1/20 : 4	1.6	3.0
Unsaturation index[b]	1.28	1.14
Hexose transport activation energy	10.6 kcal	14.2 kcal

[a] Each value is the average of two determinations, expressed as a weight percentage of the total. Error was less than 5%. Extraction and analysis as described (18).

[b] Double bonds per acyl group.

Structural Correlates of the Fatty Acid Changes

To determine whether the changes in acyl group composition alter the structure of the lipid bilayer, we measured the paramagnetic resonance spectra of a nitroxide-labeled stearic acid analogue, which was incorporated into the cells. This stearic acid analogue (Fig. 2) has a paramagnetic nitroxide at the 16 position ($m = 1$; $n = 14$) of the polymethylene chain and has been shown to be most sensitive to the environment in the center of the bilayer (21,22). When ESR spectra of the label were recorded for normal growing and transformed cells, small differences in motional freedom were observed which are summarized in Tables 3 and 4. Although there was substantial variability from experiment to experiment in the absolute values obtained [which seems not to be uncommon in biological systems (23)], the differences between cell types were qualitatively reproducible in repli-

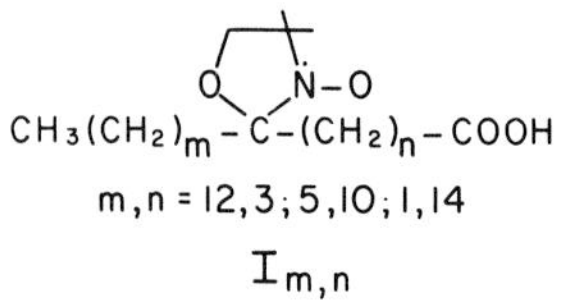

O N–O

$CH_3(CH_2)_m-C-(CH_2)_n-COOH$

m,n = 12,3; 5,10; 1,14

$I_{m,n}$

FIG. 2. Structure of nitroxide-labeled stearic acid analogues, where *m* and *n* are the number of CH_2 units.

TABLE 3. *Spin-label analysis of normal and transformed cells*

A. Typical[a] spectral data and τ_c values[b] for $I_{1,14}$

Cell type	W_0 (gauss)	h_0/h_1	h_0/h_{-1}	τ_c (sec × 10^{-10})
Normal, growing	3.0	1.21	2.78	14.9
Transformed	3.0	1.23	2.99	16.2

B. Typical[a] spectral data and τ_c values[b] for $I_{5,10}$

Cell type	W_0	h_0/h_1	h_0/h_{-1}	τ_c (sec × 10^{-10})
Normal, growing	3.4	1.68	5.49	36.0
Transformed	3.9	1.66	5.25	39.7

C. Typical[a] spectral data and order parameter[c] (S) values for $I_{12,3}$

Cell type	T'_1	T'_{11}	a/a'	S
Normal, growing	9.9	27.3	0.90	0.63
Transformed	10.0	28.1	0.88	0.64

[a] Data taken from single experiments, in which cells in each of the physiological states were simultaneously labeled with the stearic acid analog.
[b] Rotational correlation times were calculated as described by Sinensky (24).
S = Order parameter, calculated according to Hubbell and McConnell (22).

TABLE 4. *Differences between cell types in τ_c and S values for incorporated spin labels*

Label	Value
	$\tau_c^T - \tau_c^N$
$I_{1,14}$	1.7 ± 0.5 (12)
$I_{5,10}$	2.9 ± 1.4 (4)
	$S^T - S^N$
$I_{12,3}$	0.02 ± 0.01 (3)

$\tau_c^T, S^T, \tau_c^N, S^N$, = τ_c or S for transformed or normal growing cells, respectively.
τ_c = sec × 10^{-10}
± = SE
Values in parentheses are the number of independent experiments used in calculating the mean and the standard error.

cate experiments performed over a period of more than a year (Table 4). Label mobilities are given as τ_c (correlation time). The greater the value for τ_c the more restricted is the motion of the probe, and thus, we presume, the less flexible or "fluid" is the lipid bilayer in the region of the probe (21,22,24).

These data clearly show a decreased mobility of the label in the transformed cells, compared to the normal, growing cells, consistent with the idea that a decreased arachidonate content would allow better packing of the lipid acyl groups and give rise to a decrease in their motional freedom. Spectra from nitroxide-labeled stearic acid analogues in which the nitroxide group was closer to the polar end showed differences which were much smaller, and only marginally significant, suggesting that the changes in membrane composition primarily perturb the interior of the bilayer (Tables 3,4). Similar results were obtained with plasma membranes isolated from labeled cells (*Biochemistry, accepted for publication*).

Arrhenius Plots for Hexose Transport

Maintenance of a "fluid" lipid phase is important for many membrane-associated enzymic activities, including sugar transport in bacteria (11,12), and Na-K-dependent ATPase in animal cells (13,14). At the temperature of transition from a more fluid to a more gel-like membrane state these reactions display a marked increase in the apparent Arrhenius activation energy. The temperature at which the phase transition occurs is dependent in part on the fatty acid composition of the membrane, occurring at a low temperature when the fatty acids are more unsaturated, and at a higher temperature when the fatty acids are saturated. We thus expected that if hexose transport carriers were sensitive to the alterations in fatty acid composition, the temperature dependence of the transport reaction would be altered, the apparent Arrhenius activation energy for transport being higher in the cells with the less "fluid" bilayer. We therefore measured uptake of [^{3}H]2-deoxyglucose as previously described (1) at temperatures from 43°C to 0°C, and used the data to construct Arrhenius plots as shown in Fig. 3. The lines were drawn by the least-squares procedure, using a Wang computer. One can see that the slope of the Arrhenius plot is steeper for the transformed cultures than for the normal cultures. There were no obvious or reproducible inflections in the temperature dependence, suggesting that the phase transition did not occur sharply. (The slight break in the transformed cell curve at 8°C was not reproducible.) This behavior is typical for complex cholesterol-containing membranes (14,25–27). A similar change in the temperature dependence of enzyme activity with changed fatty acid composition has also been reported for the 1-acylglycerol-3-phosphate acyl transferase in *E. coli* (28). The apparent Arrhenius activation energy for hexose uptake, calculated from the slopes of the lines in Fig. 3,

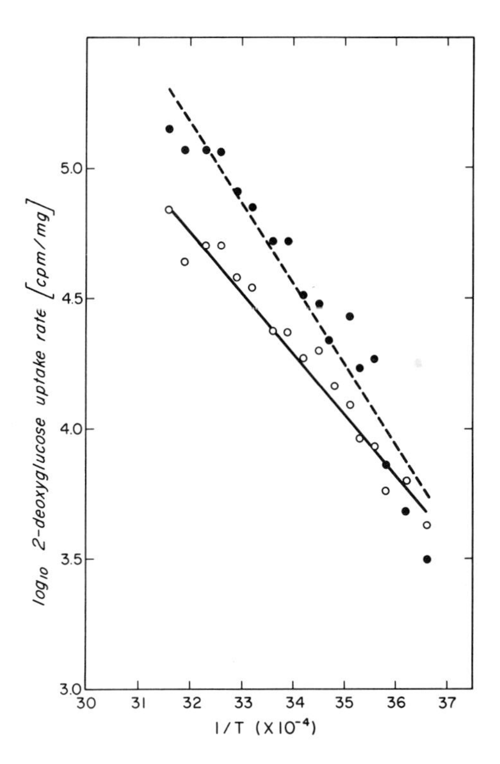

FIG. 3. Arrhenius plot of 2-deoxyglucose uptake in normal (○) and transformed (●) cultures. Break in slope at 35.5 (8°C) was not reproducible.

is shown in Table 2. (For technical reasons, the uptake measurements could not be performed at saturating substrate concentrations, so these values cannot be considered true activation energies.) It can be seen that the temperature dependence of the uptake reaction is inversely proportional to the Unsaturation Index of the membrane lipids, and is higher in the transformed cells. Bose and Zlotnick (29) have obtained similar results for cells transformed by murine sarcoma virus. These findings are consistent with the notion that the hexose transport carriers are sensitive to the alterations in membrane lipid acyl groups.

Kinetics of Appearance of Fatty Acid Changes

To examine the time course of appearance of the tumor virus-induced fatty acid alterations, we used the temperature-conditional mutant of Rous sarcoma virus, RSV-T5 (6). In Fig. 4 the time course of appearance of the fatty acid alterations are correlated with the changes in the rate of hexose transport which accompany transformation. Chick cells infected with RSV-T5 and held at 36°C had a transformed phenotype (including a trans-

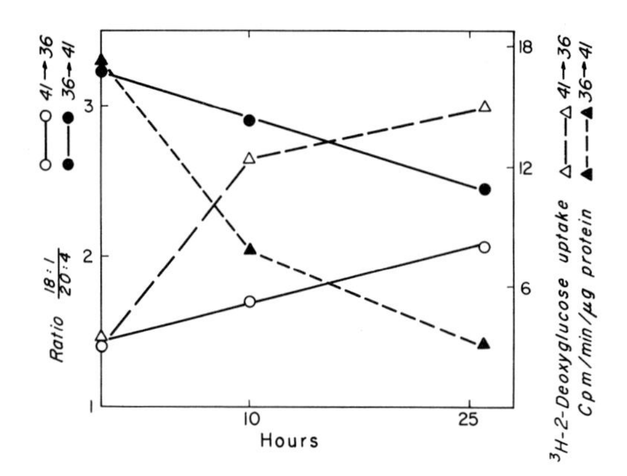

FIG. 4. Kinetics of change in 2-deoxyglucose uptake rate (triangles) and ratio of arachidonate to oleate (18 : 1/20 : 4) (circles) in cells infected with RSV-T5 following a shift in temperature from 36 to 41°C or from 41 to 36°C.

formed level of hexose transport) and had a fatty acid composition typical of transformed cells, whereas the cultures held at 41°C had a normal phenotype and a normal-type fatty acid composition. At time zero, cultures were shifted to the opposite temperature. At 10 and 26 hr after the shift, cultures were taken for fatty acid analysis and for measurement of their capacity to take up 2-deoxyglucose. It can be seen (Fig. 4) that, whereas the switch in the capacity to take up 2-deoxyglucose was complete within 26 hr [as shown previously (2,8)], the changes in fatty acid composition were less than half complete. The change in fatty acid composition occurred no faster in individual phospholipids (*Biochemistry, accepted for publication*). Thus, the functional alteration in the cell membrane which is characteristic of cells transformed by RNA tumor viruses precedes the completion of the gross compositional alterations.

Summary and Discussion of Fatty Acid and Fluidity Changes

The results presented here demonstrate that chick cells transformed by Rous sarcoma virus have less arachidonate and more oleate in their cellular phospholipids than do their normal, untransformed counterparts. This is true even when comparison is made between transformed cells and normal cells which are multiplying at the same rate. Moreover, cells infected with the RSV-T5 temperature-conditional mutant of Rous sarcoma virus and held at the permissive temperature have a fatty acid composition characteristic of transformed cells, but, when held at the restrictive temperature, they display a "normal" fatty acid composition, even though they produce virus at both temperatures. Thus, the change in fatty acid composition seems to be a specific property of the transformed state, not contingent on changes in growth rate or on viral infection. Since mouse 3T3 cells (20) and human WI-38 cells (19) transformed by SV40 virus show similar changes in acyl group composition, this alteration in membrane composition may be a universal characteristic of virally transformed cells. Hepatoma cells have also been reported to have a higher ratio of 18 : 1/20 : 4 than normal liver (30).

Associated with the increased ratio of 18 : 1/20 : 4 in the transformed cells was a decreased flexibility of membrane lipid acyl chains, as measured by the paramagnetic resonance spectra of incorporated nitroxide-labeled stearic acid analogues. We interpret the spectral differences observed between the different cell types as indicating that the interior of the lipid bilayer of the transformed cells is slightly less "fluid" than that of normal cell membranes. These findings can be readily understood in terms of the lipid fatty acid composition of these cells, since the higher arachidonate content of the normal cells would be expected to have a "fluidizing" effect. The cholesterol content and phospholipid composition of membranes from these cells does not change upon transformation by Rous sarcoma virus (15–17), and thus cannot be responsible for the observed change in membrane lipid "fluidity."

The functional significance of the small alterations in membrane chemistry and structure we have detected is by no means clear. Transformed cells are more readily agglutinated by plant lectins than are normal cells, a phenomenon which seems to be related to increased mobility of lectin binding proteins in the plane of the membrane bilayer (31). It has been suggested that increased mobility of lectin binding sites might be caused by increased "fluidity" of the membrane bilayer, but the results presented here demonstrate that changes in motion of incorporated spin-labeled fatty acids which occur upon transformation are very small and are in a direction opposite to that expected: transformed cells, which presumably have the highest binding-site mobility also have the lowest fatty acid "fluidity." These results are inconsistent with simple notions that gross changes in the "fluidity" of the lipid bilayer are responsible for transformation-specific changes in the mobility of lectin binding sites, consistent with the findings of Horwitz et al. (20).

Transformed cells also display a higher hexose transport rate than do normal cells. Thus, the rate at which these cells transport hexoses is directly proportional to the ratio of 18 : 1/20 : 4 and is inversely related to the measured "fluidity" of the lipid bilayer. Although transformation-specific changes in the apparent Arrhenius activation energy of hexose uptake were detected, consistent with the notion that the hexose transport carriers in fully transformed cells were sensitive to the chemical and structural changes in the bilayer, we doubt whether the large changes in transport rate could be brought about by these small changes in membrane microviscosity and composition. Moreover, the fact that the changes in fatty acid composition occur more slowly than the change in hexose transport rate strongly suggests that the compositional changes are not causing the functional changes. In fact, it seems possible to us that the increased transport of glucose could be responsible for the decreased unsaturation of the fatty acids by causing an increased flux through the glycolytic pathway (5), and thus perhaps an altered availability of substrates and cofactors for fatty acid biosynthesis.

A number of other groups have also investigated changes in membrane bilayer composition and structure, with varying results, depending on the cell type, the type of label and the labeling procedure (32–36). The fact that these measured variations in the average "fluidity" of membrane lipids do not correlate with malignant transformation suggests that gross changes in the "fluidity" of membrane lipids cannot be a primary cause of the transformed phenotype. This is consistent with our finding that the changes in acyl group composition are relatively slow to develop. Thus, although we began this line of investigation anticipating that we would be able to understand the molecular basis of transformation-specific alterations in membrane transport, our data are most consistent with the notion that the small differences in membrane composition and structure we have detected do not underlie the altered transport properties.

CELL-SURFACE ALTERATIONS AS MARKERS OF VIRUS GENE EXPRESSION: ROLE OF PROTEASES IN TRANSFORMATION

We will now briefly describe work of another type, in which transformation-specific changes in cell-surface properties are used as phenotypic markers of transformation, in an analysis of the mechanism by which tumor virus gene expression modifies the cell. Currently, it is fashionable to suspect that modification of the cell surface by virus-induced proteases plays an important role in the genesis of the transformed state. Three types of evidence support this suspicion. First, transformed cells have higher levels of proteolytic or other hydrolytic enzymes than do normal cells (37,38). This has been shown most clearly for the case of a cellular protease which converts the serum zymogen plasminogen to the active protease plasmin (39–42). Second, inhibitors of proteolytic enzymes are capable of reversing at least some manifestations of the transformed state (43–45) (but see also refs. 46 and 47). Finally, treatment of normal cells with proteases can cause them to behave transiently like transformed cells with respect to certain parameters (48,49). The work we have done has been primarily of the latter two types. We find that proteolysis of the cell surface (in particular by plasmin) is neither necessary nor sufficient for induction of the transformation-specific increase in hexose uptake. However, plasmin activity does seem to be necessary for transformation-specific changes in cellular adhesiveness and morphology.

Can the Transformed Phenotype Be Mimicked by Treatment of Normal Cells with Proteolytic or Hydrolytic Enzymes?

It is well known that chick cells can be released from density-dependent inhibition of growth by treatment with proteolytic or other hydrolytic enzymes (48,49), and that accompanying this growth stimulation is an increase in the rate of hexose transport (49). However, it has not been clear in these cases whether the hexose transport rate was increased up to the level seen in transformed cells, or whether the stimulation was only to the level characteristic of normal exponentially growing cells. Therefore, we have performed a quantitative comparison of the 2-deoxyglucose transport rate of normal and transformed cells and of normal cells treated with a selection of proteolytic or other hydrolytic enzymes. It can be seen (Fig. 5) that trypsin, plasmin, neuraminidase, and hyaluronidase are not capable of stimulating density-inhibited cells to the hexose transport level obtained with transformed cells. Treatment of growing cells with these enzymes was

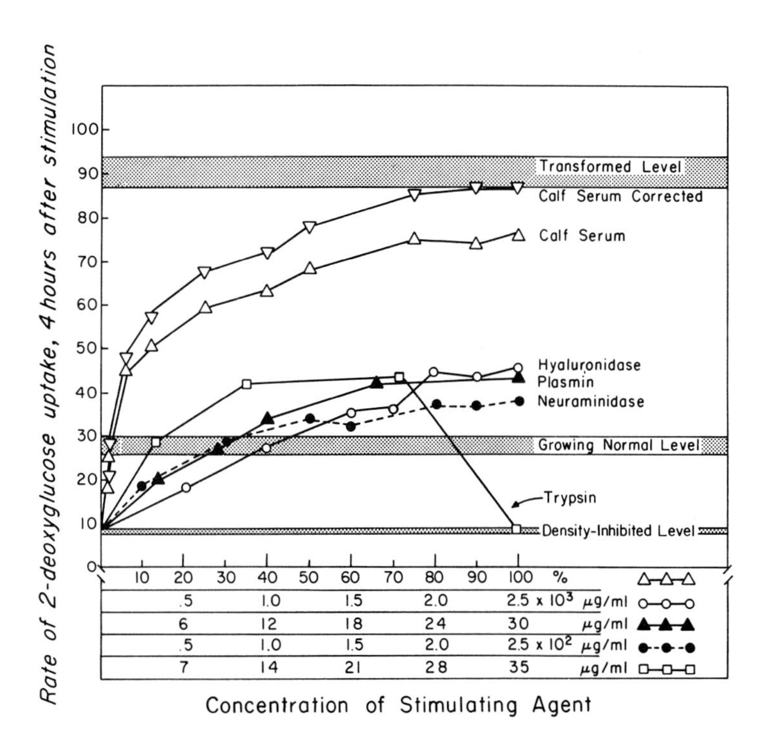

FIG. 5. Stimulation of 2-deoxyglucose uptake rate in density-inhibited cultures by hydrolytic enzymes or serum. Density-inhibited cultures were treated with the stimulating agent for 4 hr, which was the time giving maximum stimulation in all cases. Then 2-deoxyglucose uptake rate was measured, as described (45). Uptake rate = cpm/μg cell protein-15 min. (Reprinted with permission from *Cell.*)

no more effective (50). However, high concentrations of serum could stimulate the hexose transport capacity to levels nearly equal to that displayed by the transformed cells (Fig. 5). If allowance is made for the binding of serum proteins to the cells in calculating the counts per minute per milligram, then the stimulating ability of serum appears even better (calf-serum corrected). Thus, treatment of normal cells with these enzymes does not convert them into phenocopies of transformed cells with respect to hexose transport. If hydrolysis of the cell surface plays a role in controlling the rate of hexose transport, it may be by altering the sensitivity of cells to regulatory factors in the serum.

Effects of TLCK on the Transformed Phenotype

TLCK (tosyl-lysyl-chloromethyl ketone), a site-specific titrant for trypsin, is among the protease inhibitors reported to inhibit selectively the growth of transformed cells (44). We therefore have investigated in detail its effects on the transformation-specific changes in morphology, adhesiveness, and hexose transport. It can be seen (Figs. 6 and 7) that TLCK at 50 μg/ml was highly effective at restoring cellular morphology to normal. The rounded morphology of the transformed cells became flattened and elongated when treated with TLCK, and the complex cell surface with numerous microvilli which characterizes Rous-transformed cells became simple, smooth, and relatively free of microvilli (Fig. 7). Transformed cells treated with TLCK frequently could not be distinguished from normal cells when the observations were made single-blind.

TLCK also caused the cells to adhere more tightly to the culture dish (Fig. 8). Cells transformed by the temperature-conditional mutant of Rous sarcoma virus, RSV-T5, rapidly increased their adhesiveness when shifted to the restrictive temperature, demonstrating the close association between the decreased adhesiveness and the expression of viral oncogenic information. Addition of TLCK caused a similar, although slower, change in adhesiveness. The data demonstrate not only that the percentage of adherent cells in the TLCK-treated culture increases, but also that the absolute number of detachable cells declines, while the absolute number of adherent cells increases reciprocally. Thus, the protease inhibitor cannot be acting solely as a selective agent in this case, but is most likely converting transformed cells into phenotypically normal cells.

Finally, TLCK caused transformed cells to lower their rate of hexose transport down to the normal level (Fig. 9). The inhibitor did not lower the transport rate below that characteristic of normal exponentially growing cells, and was without effect when added to normal growing cells (45).

TLCK is somewhat growth-inhibitory to our cells at the concentrations used in these experiments, so it was important to determine whether the

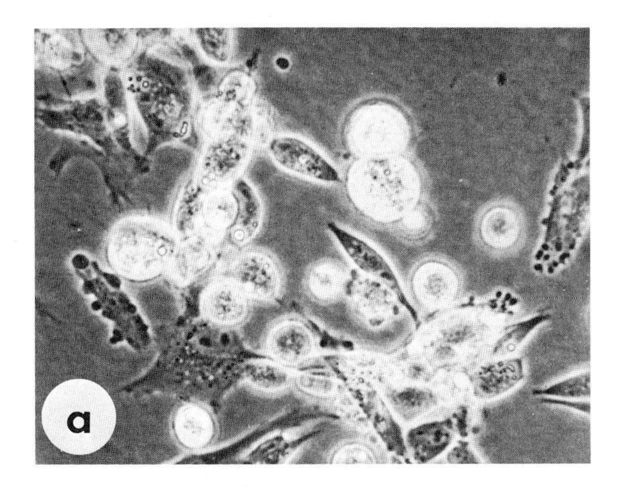

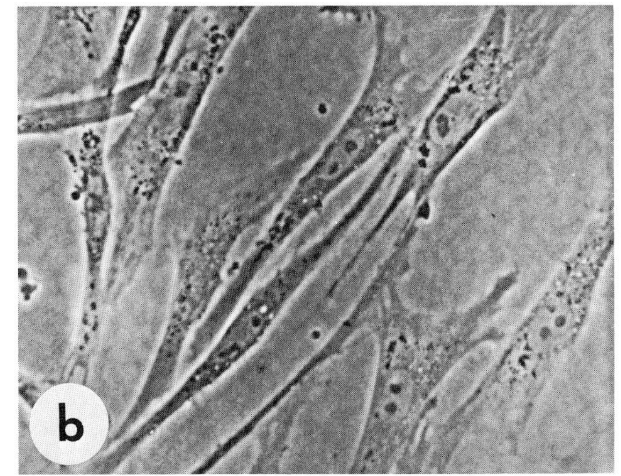

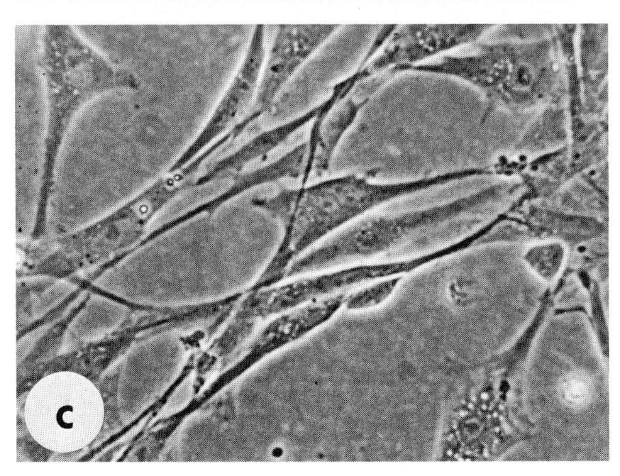

FIG. 6. Phase micrographs. **a:** Cells transformed by RSV-T5. **b:** Normal cells. **c:** RSV-T5-transformed cells treated 40 hr with 50 μg/ml TLCK. All cells were at 36°C × 344. (Reprinted with permission from *Cell*.)

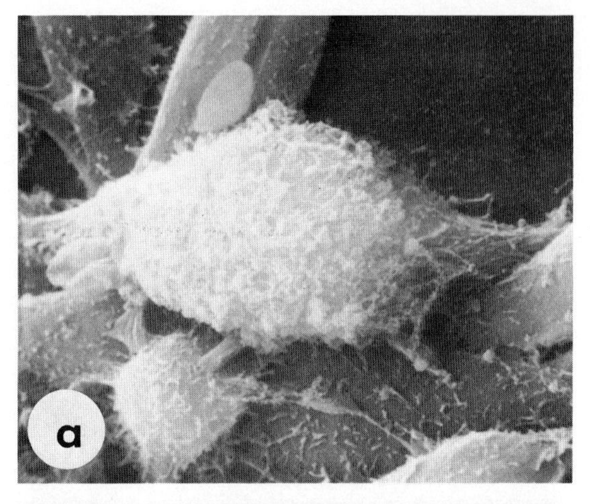

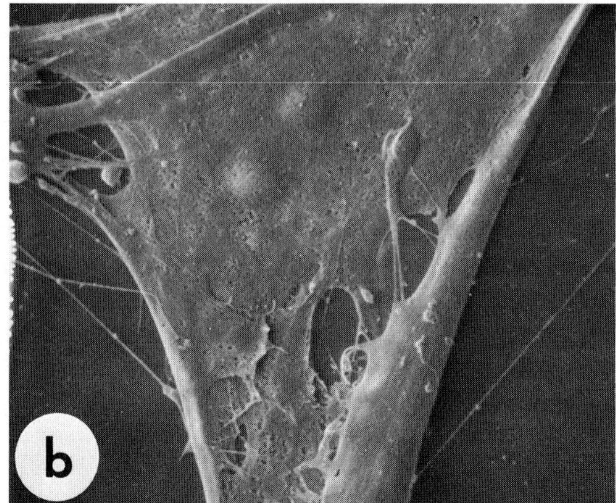

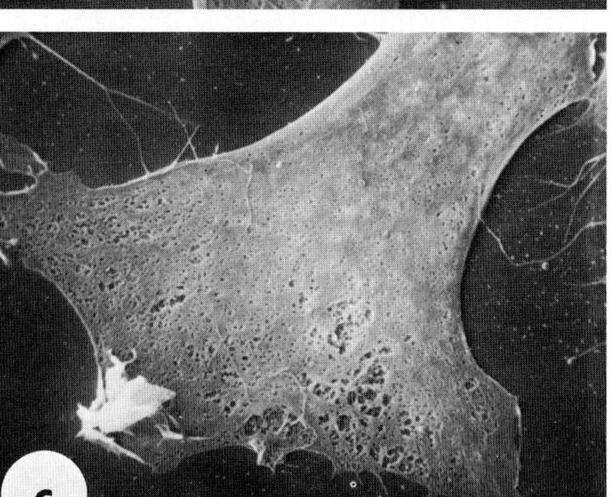

FIG. 7. Scanning electron micrographs. **a:** Cells transformed by RSV-T5. **b:** Normal cells. **c:** RSV-T5-transformed cells treated 40 hr with 50 μg/ml TLCK. All cells were at 36°C. ×3520. (Reprinted with permission from *Cell.*)

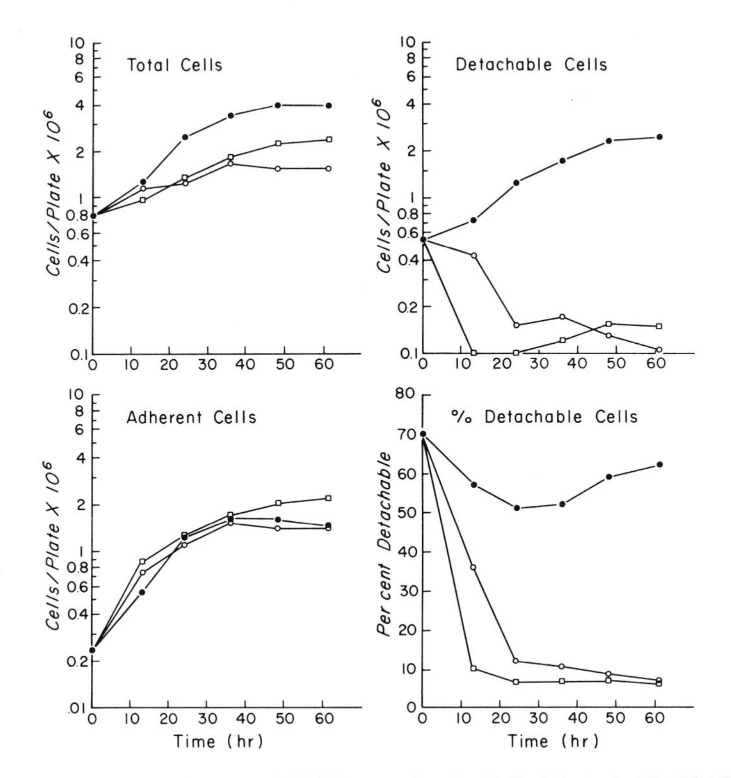

FIG. 8. Growth and adhesiveness of TLCK-treated cells. Cells infected with RSV-T5 were plated at 1.5×10^5 cells/35-mm dish at 36.5°C and allowed to remain in culture 48 hr. At this time, the cultures were changed to fresh medium either with or without 50 μg/ml TLCK, to start the experiment. One set of cultures without TLCK was shifted to 41°C to "switch off" viral oncogenic information and allow the cells to revert to a normal phenotype. Every 12 hr fresh TLCK at 25 μg/ml was added to the TLCK-treated cultures. Adhesion was measured by determining the number of cells which could be dislodged from the dish by a stream of medium, as described (45). (●) Untreated control; (○) TLCK-treated; (□) shifted to 41°C. (Reprinted with permission from *Cell*.)

effects of TLCK were due to growth inhibition. We found that growth inhibition by cycloheximide did not cause similar changes in cellular morphology and adhesiveness, although at high concentrations it did cause a decay in the hexose transport capacity of the cell, presumably due to turnover of the transport system. However, at concentrations which caused about 50% decrease in growth rate (the same degree of growth inhibition caused by 50 μg/ml of TLCK), cycloheximide had very little effect on the hexose transport capacity of the cells. Thus, we feel that simple growth inhibition does not account for the effects of TLCK on cellular morphology and adhesiveness, and may not account for all of the effects of TLCK on hexose transport (45).

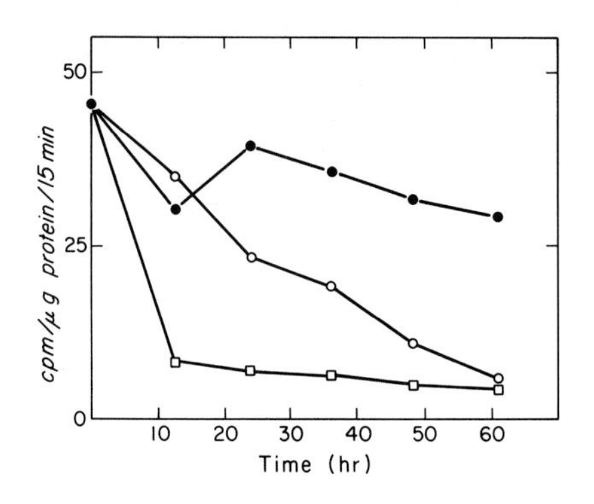

FIG. 9. Inhibition of 2-deoxyglucose uptake by TLCK. (●) Untreated 36°C control; (○) TLCK-treated 36°C; (□) shifted to 41°C. (Reprinted with permission from *Cell*.)

Effects of Other Protease Inhibitors

The use of protease inhibitors to demonstrate the involvement of proteases in the genesis of the transformed phenotype could be made more persuasive if an effective protease inhibitor could be found which did not inhibit the growth of the cells. To this end a wide variety of inhibitors and artificial peptides (20 in all) were screened for their ability to cause a flattening of the cellular morphology (45). Only the protease inhibitors specific for trypsin were found to be effective, and in addition, a synthetic dipeptide, L-lysyl-L-valine, was found to cause some flattening.

Thus the ability of these inhibitors to reverse the transformed morphology was correlated with their specificity, and was unrelated to the mechanism by which they inhibit: natural and synthetic, alkylating and nonalkylating trypsin inhibitors were found in the effective group. The effective inhibitors were then tested for their effects on growth, adhesion, and hexose transport, with the results shown in Table 5. The data demonstrate that all of the inhibitors effective at causing flattening were also capable of causing a significant increase in cellular adhesiveness, and that two of the trypsin inhibitors (ovomucoid and soy bean trypsin inhibitor) did not greatly inhibit cell growth. However, except for TLCK, none of them restored the rate of hexose transport to normal, including NPGB, which inhibited cell growth almost as much as TLCK.

Inhibition of Transformation-Associated Fibrinolysis

The best-documented example of an increased proteolytic activity associated with transformation is the finding of enhanced levels of a plasminogen activator in transformed cells and in culture fluids taken from the cells (39–42). We therefore examined the ability of some of the effective protease inhibitors to inhibit transformation-associated fibrinolysis. Fibrinolysis in this system can be stopped by inhibiting either plasmin, or the plasminogen activator produced by the transformed cells. We found that NPGB or soy

TABLE 5. *Effect of protease inhibitors on adhesion and 2-deoxyglucose transport by RSV-T5-transformed cells*

	Cells/plate × 10^6			2-Deoxyglucose transport (cpm/μg 15 min)
	Detachable	Total	% Detachable	
Untreated control	1.5	2.1	71	34.5
Soybean trypsin inhibitor (2.5 mg/ml)	0.3	1.9	16	30.9
Ovomucoid (50 mg/ml)	0.4	1.7	23	39.7
L-lysyl-L-valine (10 mg/ml)	0.4	1.2	33	27.1
Shifted to 41°C	0.2	1.8	11	4.5
Untreated control	1.7	3.4	50	32.7
NPGB[a] (50 μg/ml)	0.1	1.8	6	33.9
TLCK[b] (50 μg/ml, plus 25 μg/ml readded at 24 hr)	0.1	1.3	8	8.0
Shifted to 41°C	0.1	2.2	5	3.9

Cells were treated with the inhibitors for 45 hr, and then incubated with 0.5 μCi/ml [^{3}H]2-deoxyglucose in phosphate-buffered saline for 15 min to measure transport rate, or tested for adhesiveness, as described in the legend to Fig. 8, to determine the percent detachable cells/plate.

[a] Nitrophenyl-*p*-guanidinobenzoate.

[b] Tosyl-lysyl-chloromethyl ketone.

bean trypsin inhibitor at concentrations which caused increased adhesiveness and a morphological flattening completely inhibited transformation-associated fibrinolysis. TLCK at 50 μg/ml, on the other hand, was without effect on the fibrinolytic system, in agreement with the results of Unkeless et al. (39).

Effects of Plasminogen-Free Serum

To examine the specific function of plasmin in the genesis of the transformed phenotype, cells infected with the temperature-conditional mutant RSV-T5 were plated at the restrictive temperature (41°C) in medium containing plasminogen-free serum or regular serum. The plasminogen was removed from the serum by two or three passes over a lysine-Sepharose column (52). After 2 days in culture, the cells were shifted to the permissive temperature (36°C) for 14 hr, and their morphology, adhesiveness, and hexose transport were measured. It can be seen that the cells held in the plasminogen-free serum did not become morphologically transformed when shifted to the permissive temperature (Fig. 10), nor did they lose their adhesiveness (Table 6) the way sister cultures kept in plasminogen-containing medium did. When purified plasminogen equivalent to 5–10% serum was added back to the plasminogen-free medium at the time of the temperature

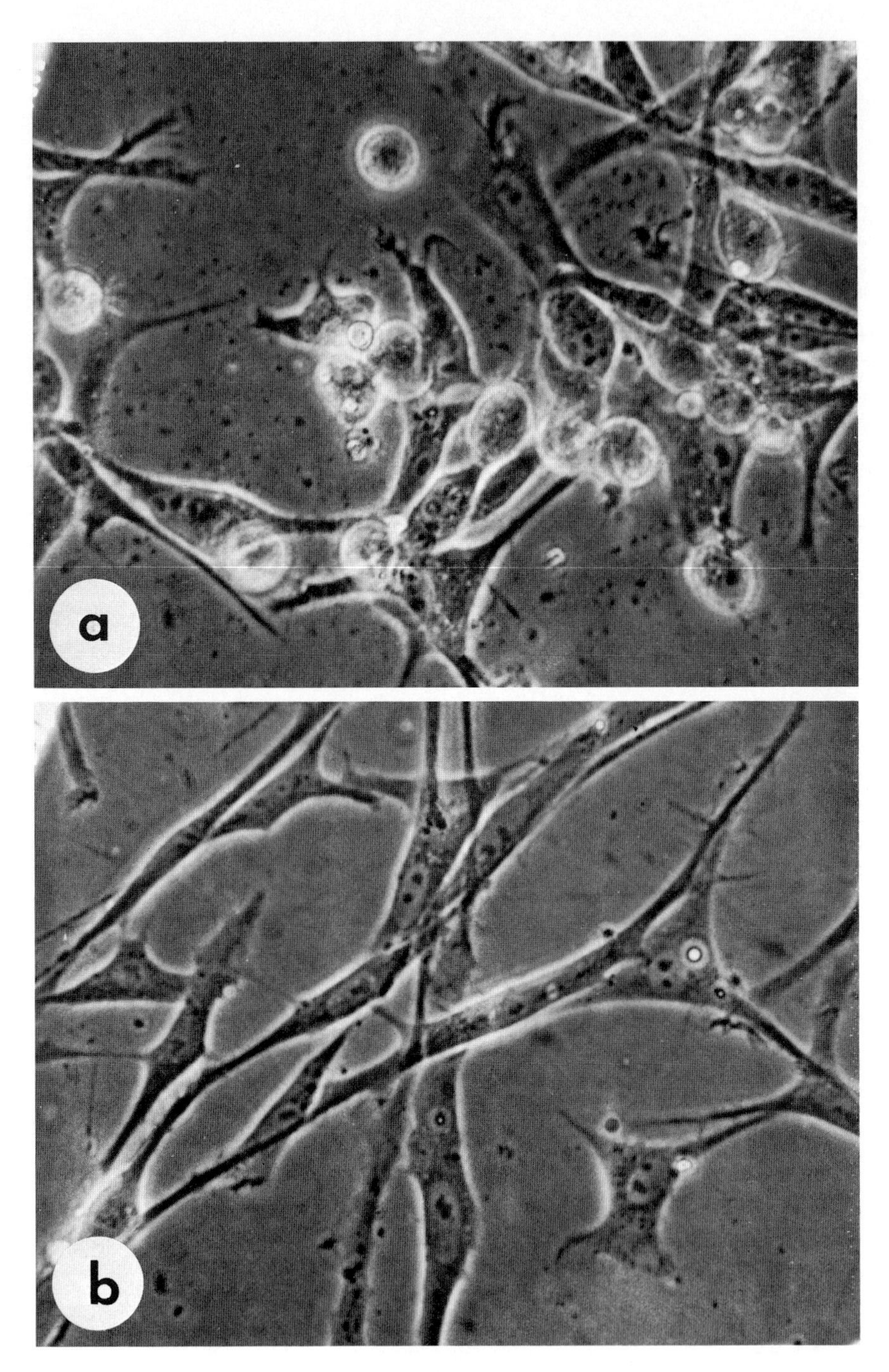

FIG. 10. Phase micrographs of cells infected with RSV-T5. **a:** Shifted to 36°C in normal serum. **b:** Shifted to 36°C in Plasminogen-free serum.

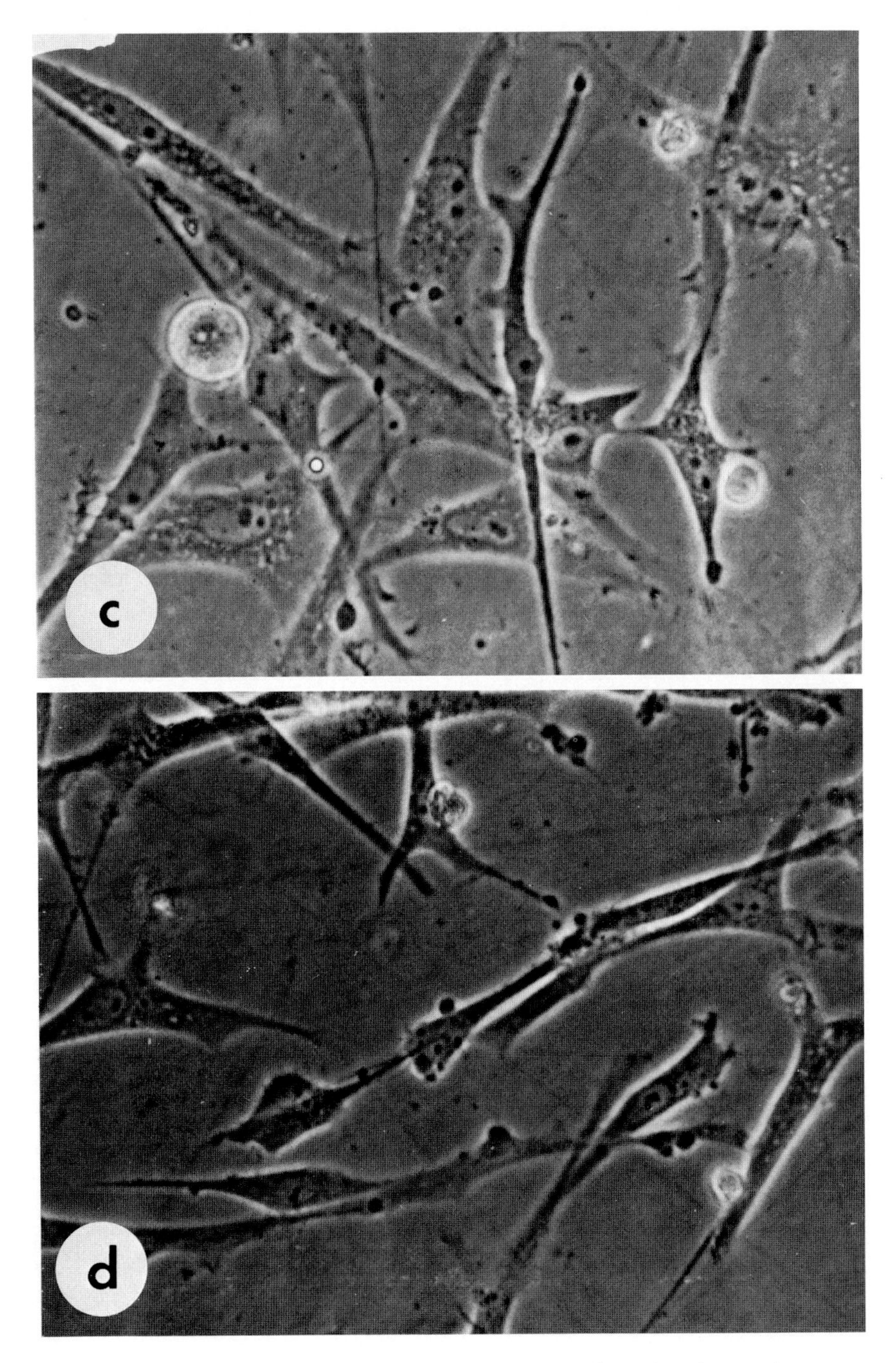

FIG. 10. c: Held at 41°C in normal serum. **d:** Held at 41°C in plasminogen-free serum. (Reprinted with permission from *Cell*.)

TABLE 6. *Expression of the transformed phenotype in plasminogen-free serum*

	% Detachable cells		Rate of 2-deoxy-glucose transport (cpm/μg protein-15 min)	
	41°C	36°C	41°C	36°C
Control	11	66	6.6	95
Plasminogen-free serum	10	21	4.7	127

RSV-T5 infected cells, held at 41°C, were trypsinized and plated at 41°C in growth medium containing 10% tryptose phosphate broth and either 5% plasminogen-free chicken serum or 5% normal chicken serum. After 48 hr half the plates were shifted to 36°C. After an additional 14 hr, 2-deoxyglucose transport and adhesiveness were assayed as described (45).

shift, the morphology of the infected cells became completely transformed, and the adhesiveness dropped so that at 14 hr, 85% of the cells were detachable. The plasminogen added to the serum was purified by the method of Deutsch and Mertz (52) and shows a single (although diffuse) band on gel electrophoresis.

Although the appearance of transformation-specific changes in morphology and adhesiveness seemed to be dependent on the presence of plasminogen, the activation of hexose transport occurred normally in cells infected with RSV-T5 and shifted to the permissive temperature in plasminogen-free serum (Table 6).

Summary and Discussion of Role of Proteases in Transformation

The work presented here demonstrates that proteolytic modification of the cell surface—in particular by plasmin—is neither necessary nor sufficient for the transformation-specific increase in hexose transport but is probably necessary for the transformation-specific changes in adhesiveness and morphology. Treatment of normal cells with plasmin or other proteolytic enzymes did not stimulate their hexose transport rate all the way up to the level which characterizes transformed cells. Addition of the trypsin inhibitors NPGB or soy bean trypsin inhibitor at concentrations which blocked transformation-associated fibrinolytic activity did not lower their hexose transport rate, although adhesiveness and morphology did return to normal. And growth of cells in plasminogen-free medium blocked the appearance of the transformation-specific alterations in adhesiveness and morphology, although the hexose transport rate did rise to the transformed level.

The fact that the adhesive and morphological properties of the cells did not vary coordinately with changes in hexose transport indicates that the transport change has a different proximal cause than these other manifestations of the transformed phenotype. This notion is supported by some of our other data (51) which demonstrates that treatment of transformed cells with dibutyryl cyclic AMP restores hexose transport to normal, but has only a partial effect on adhesiveness and morphology.

The results obtained with TLCK are complex and must be interpreted cautiously. It seems clear that TLCK, like the other effective protease inhibitors, causes transformed cells to return to normal with respect to adhesive and morphological properties. These effects of TLCK cannot be attributed to its growth-inhibitory properties, since growth inhibition by cycloheximide does not have similar effects. In fact, toxic doses of TLCK (in the range of 250–500 μg/ml) cause rounding of the cells and detachment from the dish, effects opposite to those reported here. TLCK was found not to inhibit fibrinolysis activated by the factor released from Rous-transformed chick cells, in agreement with the results of Unkeless et al. (39). This raises the intriguing possibility that there is yet another protease involved in the genesis of the transformed phenotype, separate from, and perhaps prior to, plasmin and plasminogen activator. This notion gains support from our finding that TLCK also restores the transformed level of hexose transport to normal. (However, as indicated in the text, the evidence that this effect is due to a specific inhibition of proteases is not conclusive.) In addition, we have found in preliminary experiments that, even though the activity of plasminogen activator released from transformed cells is insensitive to TLCK, the appearance of the activator in cell extracts is partially sensitive to TLCK, consistent with the results of Goldberg (42), who found that production of a proteolytic activity by transformed cells was sensitive to TLCK. Final determination of the role of TLCK in the reversal of the transformed phenotype, and elucidation of the interaction of proteolytic activities in malignant transformation, will require more detailed biochemical analysis of the earliest steps in viral oncogenesis.

ACKNOWLEDGMENTS

We thank S. Yau and B. Larraine for excellent technical assistance. Supported by USPHS Grant CA-12467 and American Cancer Society Grant NP-149. MJW is a recipient of a NIH Research Career Development Award.

REFERENCES

1. Weber, M. J. (1973): Hexose transport in normal and in Rous sarcoma virus-transformed cells. *J. Biol. Chem.*, 248:2978–2983.
2. Martin, G. S., Venuta, S., Weber, M. J., and Rubin, H. (1971): Temperature dependent alterations in sugar transport in cells infected by a temperature-sensitive mutant of Rous sarcoma virus. *Proc. Natl. Acad. Sci. USA*, 68:2739–2741.

3. Colby, C., and Edlin, A. (1970): Nucleotide pool levels in growing, inhibited and transformed chick fibroblast cells. *Biochemistry,* 9:917–918.
4. Weber, M. J., and Edlin, G. (1971): Phosphate transport, nucleotide pools, and ribonucleic acid synthesis in growing and in density-inhibited 3T3 cells. *J. Biol. Chem.,* 246: 1828–1833.
5. Bissell, M. J., Hatie, C., and Rubin, H. (1972): Patterns of glucose metabolism in normal and virus-transformed chick cells in tissue culture. *J. Natl. Cancer Inst.,* 49:555–565.
6. Martin, G. S. (1970): Rous sarcoma virus: a function required for the maintenance of the transformed state. *Nature,* 227:1021–1023.
7. Baltimore, D. M. (1974): Tumor viruses: 1974. *Cold Spring Harbor Symp. Quant. Biol.,* 34:1187–1200.
8. Kawai, S., and Hanafusa, H. (1971): The effects of reciprocal changes in temperature on the transformed state of cells infected with a Rous sarcoma virus mutant. *Virology,* 46:470–479.
9. Hale, A. H., Winkelhake, J. L., and Weber, M. J. (1975): Cell surface changes and Rous sarcoma virus gene expression in synchronized cells. *J. Cell Biol.,* 64:398–407.
10. Roseman, S. (1969): The transport of carbohydrates by a bacterial phosphotransferase system. *J. Gen. Physiol.,* 54:138–184.
11. Wilson, G., Rose, S., and Fox, C. F. (1970): The effect of membrane lipid unsaturation on glycoside transport. *Biochem. Biophys. Res. Commun.,* 38:617–623.
12. Overath, P., and Traüble, H. (1973): Phase transitions in cells, membranes, and lipids of *Escherichia coli.* Detection by fluorescent probes, light scattering and dilatometry. *Biochemistry,* 12:2625–2634.
13. Grisham, C. M., and Barnett, R. F. (1973): The role of lipid-phase transitions in the regulation of the (sodium + potassium) adenosine triphosphatase. *Biochemistry,* 12:2635–2637.
14. Kimelberg, H. K., and Papahadjopoulos, D. (1972): Phospholipid requirements for (Na^+ + K^+)-ATPase activity: head group specificity and fatty acid fluidity. *Biochim. Biophys. Acta,* 282:277–292.
15. Quigley, J. P., Rifkin, D. B., and Reich, E. (1971): Phospholipid composition of Rous sarcoma virus, host cell membranes and other enveloped RNA viruses. *Virology,* 46:106–116.
16. Quigley, J. P., Rifkin, D. B., and Reich, E. (1972): Lipid studies of Rous sarcoma virus and host cell membranes. *Virology,* 50:550–557.
17. Perdue, J. F., Kletzien, R., and Miller, K. (1971): The isolation and characterization of plasma membrane from cultured cells. I. The chemical composition of membrane isolated from uninfected and oncogenic RNA virus-converted chick embryo fibroblasts. *Biochim. Biophys. Acta,* 249:419–434.
18. Yau, T. M., and Weber, M. J. (1972): Changes in acyl group composition of phospholipids from chicken embryonic fibroblasts after transformation by Rous sarcoma virus. *Biochem. Biophys. Res. Commun.,* 49:114–120.
19. Howard, V., and Kritchevsky, D. (1969): The lipids of normal diploid (WI-38) and SV40-transformed human cells. *Int. J. Cancer,* 4:393–402.
20. Horwitz, A. F., Hatten, M. E., and Burger, M. M. (1974): Membrane fatty acid replacements and their effect on growth and lectin-induced agglutinability. *Proc. Natl. Acad. Sci. USA,* 71:3115–3119.
21. Mehlhorn, R. J., and Keith, A. D. (1972): Spin labeling of biological membranes. In: *Membrane Molecular Biology,* pp. 192–227, edited by C. F. Fox and A. C. Keith. Sinauer Assoc., Stamford, Conn.
22. Hubbell, W. L., and McConnell, H. M. (1971): Molecular motion in spin-labeled phospholipids and membranes. *J. Am. Chem. Soc.,* 93:314–326.
23. Butterfield, D. A., Chesnut, D. B., Roses, A. D., and Appel, S. H. (1974): Electron spin resonance studies of erythrocytes from patients with myotonic muscular dystrophy. *Proc. Natl. Acad. Sci. USA,* 71:909–913.
24. Stone, T. J., Buckman, T., Nordio, P. L., and McConnell, H. M. (1965): Spin-labeled biomolecules. *Proc. Natl. Acad. Sci. USA,* 54:1010–1017.
25. Marsh, D., and Smith, I. C. P. (1973): An interacting spin label study of the fluidizing and condensing effects of cholesterol on lecithin bilayers. *Biochim. Biophys. Acta,* 298:133–144.

26. Ladbrooke, B. D., Jenkinson, T. J., Kamat, V. B., and Chapman, D. (1968): Physical studies of myelin. I. Thermal analysis. *Biochim. Biophys. Acta,* 164:101–109.
27. Ladbrooke, B. D., Williams, R. M., and Chapman, D. (1968): Studies on lecithin-cholesterol-water interactions by differential scanning calorimetry and X-ray diffraction. *Biochim. Biophys. Acta,* 150:333–340.
28. Mavis, R. D., and Vagelos, P. R. (1972): The effect of phospholipid fatty acid composition on membranous enzymes in *Escherichia coli. J. Biol. Chem.,* 247:652–659.
29. Bose, S. K., and Zlotnick, B. J. (1973): Growth- and density-dependent inhibition of deoxyglucose transport in Balb 3T3 cells and its absence in cells transformed by Murine sarcoma virus. *Proc. Natl. Acad. Sci. USA,* 70:2374–2378.
30. Veerkamp, J. H., Mulder, I., and Van Deenen, L. L. M. (1962): Comparison of the fatty acid composition of lipids from different animal tissues including some tumours. *Biochim. Biophys. Acta,* 57:299–309.
31. Robbins, J. C., and Nicolson, G. (1975): Surfaces of normal and transformed cells. In: *Cancer: A Comprehensive Treatise,* Vol. III: *Biology of Tumors,* edited by F. F. Becker. Plenum Press, New York.
32. Gaffney, B. J., Branton, P. E., Wickus, G. G., and Hirschberg, C. B. (1974): Fluid lipid regions in normal and Rous sarcoma virus transformed chick embryo fibroblasts. In: *Viral Transformation and Endogenous Viruses,* edited by A. S. Kaplan. Academic Press, New York.
33. Gaffney, B. J. (1975): Fatty acid chain flexibility in the membranes of normal and transformed fibroblasts. *Proc. Natl. Acad. Sci. USA,* 72:664–668.
34. Barnett, R. E., Furcht, L. T., and Scott, R. E. (1974): Differences in membrane fluidity and structure in contact-inhibited and transformed cells. *Proc. Natl. Acad. Sci. USA,* 71:1992–1994.
35. Barnett, R. E., Furcht, L. T., and Scott, R. E. (1975): Correction: Differences in membrane fluidity and structure in contact-inhibited and transformed cells. *Proc. Natl. Acad. Sci. USA,* 72:1217.
36. Shinitzky, M., and Inbar, M. (1974): Difference in microviscosity induced by different cholesterol levels in the surface membrane lipid layer of normal lymphocytes and malignant lymphoma cells. *J. Mol. Biol.,* 85:603–615.
37. Schnebli, H. P. (1972): A protease-like activity associated with malignant cells. *Schweiz. Med. Wochenschr.,* 102:1194–1197.
38. Bosmann, H. B. (1972): Elevated glycosidases and proteolytic enzymes in cells transformed by RNA tumor virus. *Biochim. Biophys. Acta,* 264:339–343.
39. Unkeless, J. C., Tobia, A., Ossowski, L., Quigley, J. P., Rifkin, D. B., and Reich, E. (1973): An enzymatic function associated with transformation of fibroblasts by oncogenic viruses. I. Chick embryo fibroblast cultures transformed by avian RNA tumor viruses. *J. Exp. Med.,* 137:85–111.
40. Ossowski, L., Unkeless, J. C., Tobia, A., Quigley, J. P., Rifkin, D. B., and Reich, E. (1973): An enzymatic function associated with transformation of fibroblasts by oncogenic viruses. II. Mammalian fibroblast cultures transformed by DNA and RNA tumor viruses. *J. Exp. Med.,* 137:112–126.
41. Ossowski, L., Quigley, J. P., Kellerman, G. B., and Reich, E. (1973): Fibrinolysis associated with oncogenic transformation. *J. Exp. Med.,* 138:1056–1064.
42. Goldberg, A. R. (1974): Increased protease levels in transformed cells: a casein overlay assay for the detection of plasminogen activator production. *Cell,* 2:95–102.
43. Goetz, I., Weinstein, C., and Roberts, E. (1972): Effects of protease inhibitors on growth of hamster tumor cells in culture. *Cancer Res.,* 32:2469–2474.
44. Schnebli, H. P. (1974): Growth inhibition of tumor cells by protease inhibitors: Consideration of the mechanisms involved. In: *Control of Proliferation in Animal Cells. I,* pp. 327–337, edited by B. Clarkson and R. Baserga. Cold Spring Harbor Press, New York.
45. Weber, M. J. (1975): Inhibition of protease activity in cultures of Rous sarcoma virus-transformed cells: Effect on the transformed phenotype. *Cell,* 5:253–261.
46. Chou, I.-N., Black, P. H., and Roblin, R. (1974): Effects of protease inhibitors on growth of 3T3 and SV3T3 cells. In: *Control of Proliferation in Animal Cells. I,* pp. 339–350, edited by B. Clarkson and R. Baserga. Cold Spring Harbor Press, New York.
47. McIlhinney, A., and Hogan, B. (1974): Effect of inhibitors of proteolytic enzymes on the

growth of normal and polyoma transformed BHK cells. *Biochem. Biophys. Res. Commun.*, 60:348–354.
48. Burger, M. M. (1970): Proteolytic enzymes initiating cell division and escape from contact inhibition of growth. *Nature*, 227:170–171.
49. Sefton, B. M., and Rubin, H. (1970): Release from density-dependent growth inhibition by proteolytic enzymes. *Nature*, 227:843–845.
50. Hale, A. H., and Weber, M. J. (1975): Hydrolase and serum treatment of normal chick embryo cells: Effects on hexose transport. *Cell*, 5:245–252.
51. Weber, M. J. (1974): Reversal of the transformed phenotype by dibutyryl cyclic AMP and a protease inhibitor. In: *Mechanisms of Virus Disease*, pp. 327–345, edited by W. S. Robinson and C. F. Fox. Benjamin, Menlo Park, New York.
52. Deutsch, D. G., and Mertz, E. T. (1970): Plasminogen: Purification from human plasma by affinity chromatography. *Science*, 170:1095–1097.

Subject Index

7967469
3 1378 00796 7469